Water Operator Certification Study Guide

Sixth Edition

A Guide to Preparing for Water Treatment and Distribution Operator Certification Exams

By John Giorgi
Reviewed by the Association of Boards of Certification

American Water Works Association

ABC
Association of Boards of Certification

Water Operator Certification Study Guide, 6th edition

Copyright © 1979, 1983, 1989, 1993, 2003, 2012, American Water Works Association

All rights reserved. No part of this publication may be reproduced or transmitted in any form or by any means, electronic or mechanical, including photocopy, recording, or any information or retrieval system, including the Internet, except in the form of brief excerpts or quotations for review purposes, without the written permission of the publisher.

Disclaimer
Although this study guide has been extensively reviewed for accuracy, there may be an occasion to dispute an answer, either factually or in the interpretation of the question. Both AWWA and ABC have made every effort to correct or eliminate any questions that may be confusing or ambiguous. If you do find a question that you feel is confusing or incorrect, please contact the AWWA Publishing Group at books@awwa.org.

Additionally, it is important to understand the purpose of this study guide. It does not guarantee certification. It is intended to provide an operator, or aspiring operator, with an understanding of the types of questions he or she will be presented with on a certification exam and the areas of knowledge that will be covered. AWWA highly recommends that readers study and understand the reference material from which the questions are drawn, as well as the additional reference list at the end of the book.

AWWA Publications Manager: Gay Porter De Nileon
Production: Cenveo Publishing Services
Cover Art: Cheryl Armstrong

ISBN: 1-58321-852-5

Library of Congress Cataloging-in-Publication Data

Giorgi, John.
  Water operator certification study guide : a guide to preparing for water treatment and distribution operator certification exams / by John Giorgi. -- 6th ed.
     p. cm.
   Rev. ed. of: Operator certification study guide. c2003. 5th ed.
   ISBN 978-1-58321-852-5
   1. Water treatment plants -- Employees -- Certification. 2. Water -- Purification -- Examinations -- Study guides. 3. Sewage disposal plants -- Employees -- Certification. 4. Sewage -- Purification -- Examinations -- Study guides. 5. Water -- Distribution -- Examinations -- Study guides. I. Giorgi, John. Operator certification study guide. II. Title.

TD430.G45 2012
628.1'62076—dc23
                              2011033972

6666 W. Quincy Avenue
Denver, CO 80235-3098
303.794.7711
www.awwa.org

Contents

About AWWA & ABC, v

Preface, vii

ABC Operator Certification Education Requirements, ix

Acknowledgments, xi

Water Treatment Operator Requirements, 1

Monitor, Evaluate, & Adjust Treatment Processes, 3
Laboratory Analyses, 23
Comply with Drinking Water Regulations, 29
Operate and Maintain Equipment, 35
Perform Security, Safety, and Administrative Procedures, 41
Evaluate Characteristics of Source Water, 47
Additional Water Treatment Operator Practice Questions, 51
Math for Water Treatment Operators Levels I & II, 139
Math for Water Treatment Operators Levels III & IV, 147

Answers to Water Treatment Operator Questions, 155

Monitor, Evaluate, & Adjust Treatment Processes, 155
Laboratory Analyses, 165
Comply with Drinking Water Regulations, 167
Operate and Maintain Equipment, 169
Perform Security, Safety, and Administrative Procedures, 171
Evaluate Characteristics of Source Water, 172
Answers to Additional Water Treatment Operator Practice Questions, 175
Answers to Math Questions for Water Treatment Operators Levels I & II, 229
Answers to Math Questions for Water Treatment Operators Levels III & IV, 237

Water Distribution Operator Requirements, 247

Distribution System Information/Components, 249
Monitor, Evaluate, and Adjust Disinfection, 255
Laboratory Analysis, 261
Install Equipment, 267
Operate Equipment, 273
Perform Maintenance, 279
Perform Security, Safety, and Administrative Procedures, 285
Additional Distribution Operator Practice Questions, 291

Math for Water Distribution Operators Levels I & II, 323
Math for Water Distribution Operators Levels III & IV, 331

Answers to Water Distribution Operator Questions, 339

System Information/Components, 339
Monitor, Evaluate, and Adjust Disinfection, 341
Laboratory Analysis, 343
Install Equipment, 345
Operate Equipment, 347
Perform Maintenance, 349
Perform Security, Safety, and Administrative Procedures, 351
Answers to Additional Distribution Operator Practice Questions, 353
Answers to Math Questions for Water Distribution Operators
   Levels I & II, 371
Answers to Math Questions for Water Distribution Operators
   Levels III & IV, 381

Appendix A: Formulas for Water Treatment, Distribution, and Laboratory Exams, 393

Appendix B: Sample CT Tables for *Giardia* Inactivation, 399

Additional Resources, 401

About AWWA & ABC

American Water Works Association

The American Water Works Association is an international nonprofit educational association dedicated to safe water. Founded in 1881 as a forum for water professionals to share information and learn from each other for the common good, AWWA is the authoritative resource for knowledge, information, and advocacy for improving the quality and supply of water in North America and beyond.

AWWA brings together people from the water community with a variety of backgrounds, skills, perspectives, and experience. Our more than 50,000 members represent every segment of the profession, including small, medium, and large utilities serving rural communities and large cities. Members include engineers, scientists, managers, operators, and communicators, as well as manufacturers, distributors, and consultants, providing expertise and technology to customers. AWWA is the water profession's information source for everything from the design, construction, and operation of water facilities to the treatment, distribution, management, and conservation of water supplies to the latest regulatory and legislative actions affecting water supply, treatment, and distribution.

Association of Boards of Certification

Established in 1972, the Association of Boards of Certification (ABC) is a nonprofit member-driven organization dedicated to protecting public health and the environment by advancing the quality and integrity of environmental certification programs. ABC membership includes almost 100 certifying authorities, representing more than 40 states and nine Canadian provinces as well as several international programs. Existing solely for its members, ABC is the voice for the profession and serves as the conduit for information in an ever-changing industry.

Over 70 certification programs currently test approximately 35,000 operators and laboratory analysts annually through ABC's industry-leading Certification & Testing Services. Over 400,000 water and wastewater operators, laboratory analysts, and backflow prevention assembly testers have taken an ABC exam since the testing program began in 1982.

ABC Vision
Promote integrity in environmental certification throughout the world.

ABC Mission
ABC is dedicated to advancing the quality and integrity of environmental certification programs.

ABC Objectives
- Promote certification as a means of protecting public health, the infrastructure, and the environment
- Promote uniformity of standards and best practices in certification
- Serve as the technical resource for certification entities
- Facilitate the transfer of certification between certifying authorities
- Serve the needs of our members

Preface

This book was developed by the American Water Works Association (AWWA) with the cooperation of the Association of Boards of Certification (ABC). It is based on ABC's Need-to-Know Criteria for four levels of certification for water treatment operators and water distribution operators. These criteria were ratified by the ABC Board of Directors in January 2011. All the questions have been thoroughly reviewed and updated, and many are drawn from the current Principles and Practices of Water Supply Operations (WSO) series books published by AWWA in 2010, as well as the current Safe Drinking Water Act regulations and other timely materials. In addition, this guide contains substantially more additional practice questions and math exercises than previous editions.

All of the questions in the first five chapters of each section and the majority of additional questions have been vetted and ranked by ABC; all questions have been reviewed by experts in the topic area. Ranking refers to the level of knowledge (i.e., comprehension, application, analysis) required for each task by each level of operator.

- *Comprehension* is the most basic level of understanding and remembering. Items written at the comprehension level require examinees to recognize, remember, or identify important ideas.
- Items written at the *application* level require examinees to interpret, calculate, predict, use, or apply information and solve problems.
- Items written at the *analysis* level require examinees to compare, contrast, diagnose, examine, analyze, and relate important concepts.

The level of knowledge is a hierarchy from basic comprehension to analysis. The level of knowledge tested is cumulative. Therefore, tasks identified as application may include questions written at both the application and comprehension levels. Tasks identified as analysis may include questions written at the comprehension, application, and analysis levels.

This guide is intended to give operators practice in answering questions that are similar in format and content to the questions that appear on certification exams. While the questions found on an exam are not duplicated in this guide, similar types of questions are included; they cover relevant areas of study and allow operators to understand what areas they need to study for passing an exam. Operators who have difficulty answering any of the questions in the study guide should consult the reference source provided following the answer to the question.

Tips on Using This Book
1. *Use the study guide with other training materials and courses, not in place of them.* The exam questions in this book are designed to test your knowledge and identify content areas that need more study. Using this book is an excellent way to review your understanding of a topic.

2. *Use the references.* All of the test questions have specific references for further study. These references follow the answers to each practice test.
3. *Follow up on missed questions.* Be systematic about your preparation for the exam. Every time you miss a test question in this study guide, go to the reference and read about it. You'll be more likely to get the next one right.
4. *Don't memorize the questions and answers.* It is unlikely that a study guide question will be duplicated exactly in a certification exam. But you will certainly find many questions that are similar. Therefore, understanding the context of the question is critical. So, learn the basis for each question and you should be able to answer any number of similar questions on the topic.
5. *Check with local/regional certification and training authorities* if you have doubts about questions or answers in this book. There are significant regional and state/provincial differences in treatment plants and distribution systems and in regulatory requirements. By checking with your certification agency, you will know what to emphasize as you prepare for the test.
6. *Practice under test conditions.* Have a friend keep time for you (take 2 hours for a 75-question test). Take the test without books or reference materials (but keep a calculator handy).
7. *Give yourself as much time as you can.* Studying works best when you have the time to review the material several times. Using these tests before you start studying for the test will focus your attention on areas where you need the most work.

ABC Operator Certification Education Requirements

Level	Education	Experience
Level I	High school diploma, GED, or equivalent	1 year of acceptable experience
Level II	High school diploma, GED, or equivalent	3 years of acceptable experience
Level III	High school diploma, GED, or equivalent; 900 contact hours of post-high-school education	4 years of acceptable experience including 2 years of direct responsible charge
Level IV	High school diploma, GED, or equivalent; 1,800 contact hours of post-high-school education	4 years of acceptable experience including 2 years of direct responsible charge
Very Small Water Systems		
	High school diploma, GED, or equivalent. Six contact hours of very small water system education.	Six months of acceptable operating experience of a very small water system or higher utility.

Acknowledgments

The following people served as reviewers for this edition of the *Water Operator Certification Study Guide*. Their service and dedication to the profession is greatly appreciated.

Raymond E. Bordner, President, Florida Water and Pollution Control Operators Association, St. Petersburg, Fla.
Gary J. Coleman, Project Manager, American Water Enterprises, Inc., Voorhees, N.J.
Cindy Cook, Water Certification Officer, California Rural Water Association, Ventura, Calif.
Margaret Doss, Environmental Compliance Manager, Columbia County (Georgia) Water Utility
D. Kim Dyches, Field Services Manager, Utah Division of Drinking Water, Salt Lake City, Utah
Robert Hoyt, Water Filtration Plant Manager, Worcester Dept. of Public Works, Worcester, Mass.
Ken Kerri, Emeritus Professor, Office of Water Programs, California State University–Sacramento
Chuck Kingston, Water Treatment Supervisor, Joint Water Commission, Hillsboro, Ore.
A. Martin Nutt, Training and Certification Officer, Water Operator Licensing Program, Arkansas Department of Health, Little Rock, Ark.
Ray Olson, Operations Manager, Castle Rock Water, Castle Rock, Colo.
Gerald B. Samuel, P.Eng., President, Opertech Consulting Ltd., St. Albert, Alta.
Brian Thorburn, AScT, Senior Operator, EPCOR French Creek, Parksville, B.C.
Michael Wentink, Training and Licensing Coordinator, Office of Drinking Water and Environmental Health, Nebraska DHHS–Division of Public Health

In addition, the following members of the AWWA TEC Education Committee also reviewed questions and provided input into the study guide.

Jerry L. Anderson, Project Manager, CH2M HILL, Louisville, Ky.
D. Scott Borman, General Manager, Benton/Washington Regional Public Water Authority, Rogers, Ark.
Paul Demit, Vice President, CH2M HILL, Atlanta, Ga.
James Malley, Professor, Environmental Research Center, University of New Hampshire, Durham, N.H.
Deborah Metz, Superintendent of Water Quality & Treatment, Greater Cincinnati Water Works, Cincinnati, Ohio

The author would also like to thank and acknowledge his wife, Flora Zhou Giorgi, and the following people for their assistance in making this book possible.

Paul Bishop, Executive Director, ABC
Liz Haigh, Director of Publishing, AWWA
Gay Porter De Nileon, Publications Manager, AWWA
Scott Millard, Manager of Business and Product Development, AWWA
Megan Baker, Project Manager, ABC
Scott Harter, (former) Director of Membership Services, ABC
Martha Ripley Gray, Technical Editor, AWWA

Water Treatment Operator Requirements

Blueprint Area	Level I	Level II	Level III	Level IV
Monitor, Evaluate, and Adjust Treatment Processes	30%	28%	31%	31%
Laboratory Analyses	12%	13%	11%	11%
Comply with Drinking Water Regulations	12%	12%	11%	10%
Operate and Maintain Equipment	27%	26%	24%	25%
Perform Security, Safety, and Administrative Procedures	13%	16%	18%	18%
Evaluate Characteristics of Source Water	6%	5%	5%	5%

Monitor, Evaluate, & Adjust Treatment Processes

	Level I	Level II	Level III	Level IV
Chemical addition				
Chemical pretreatment	Comprehension	Comprehension	Application	Analysis
Chlorine dioxide disinfection	Analysis	Analysis	Analysis	Analysis
Chlorine gas disinfection	Analysis	Analysis	Analysis	Analysis
Corrosion control	Comprehension	Comprehension	Application	Analysis
Fluoridation	Comprehension	Analysis	Analysis	Analysis
Ozone disinfection	Comprehension	Comprehension	Application	Application
pH adjustment	Application	Application	Analysis	Analysis
Sodium hypochlorite disinfection	Analysis	Analysis	Analysis	Analysis
Ultraviolet disinfection	Comprehension	Comprehension	Application	Application
Coagulation and flocculation				
Chemical coagulants	Comprehension	Application	Application	Analysis
Flocculation tanks	Comprehension	Application	Application	Analysis
Rapid mix units	Comprehension	Application	Application	Analysis
Clarification and sedimentation				
Dissolved air flotation	Comprehension	Application	Application	Analysis
Inclined-plate sedimentation	Comprehension	Application	Application	Analysis
Sedimentation basins	Comprehension	Application	Application	Analysis
Tube sedimentation	Comprehension	Application	Application	Analysis
Up-flow solids-contact clarification	Comprehension	Application	Application	Analysis
Filtration				
Cartridge filters	Application	Application	Application	Application
Diatomaceous earth filters	Comprehension	Comprehension	Comprehension	Application
Direct filtration	Comprehension	Application	Application	Analysis
Gravity filtration	Comprehension	Application	Application	Analysis
Membranes (ultrafiltration, nanofiltration, reverse osmosis)	Application	Application	Application	Application

Microscreens	Comprehension	Comprehension	Application	Analysis
Pressure or greensand filtration	Application	Application	Application	Application
Slow sand filters	Comprehension	Application	Application	Analysis
Residuals Disposal				
Discharge to lagoons	N/A	N/A	Comprehension	Comprehension
Discharge to lagoons and then raw water source	N/A	N/A	Comprehension	Comprehension
Discharge to raw water	N/A	N/A	Application	Analysis
Disposal to sanitary sewer	N/A	N/A	Comprehension	Comprehension
Land application	N/A	N/A	Comprehension	Comprehension
Mechanical dewatering	N/A	N/A	Application	Analysis
On-site disposal	N/A	N/A	Comprehension	Comprehension
Solids composting	N/A	N/A	Comprehension	Comprehension
Additional Treatment Tasks				
Aeration	Comprehension	Application	Application	Analysis
Backwash aids	Comprehension	Application	Application	Analysis
Coagulation aids	Comprehension	Application	Application	Analysis
Copper sulfate treatment	Application	Application	Application	Application
Electrodialysis	Comprehension	Comprehension	Comprehension	Application
Filter aids	Comprehension	Application	Application	Analysis
Ion-exchange/softening	Application	Application	Application	Application
Iron manganese/softening	Application	Application	Application	Application
Lime-soda ash softening	Comprehension	Comprehension	Application	Analysis
Packed tower aeration	Comprehension	Comprehension	Comprehension	Comprehension
Powdered activated carbon/GAC	Application	Application	Application	Application

Required Capabilities

Knowledge of:
- Analysis and interpretation
- Basic chemistry
- Chemical properties
- Drinking water treatment concepts
- General electrical principles
- Monitoring requirements
- Normal chemical range
- Physical science
- Principles of measurement
- Proper application of chemicals
- Proper chemical handling and storage

Ability to:
- Adjust chemical feed rates
- Adjust flow patterns
- Adjust process units
- Calculate dosage rates
- Confirm chemical strength
- Diagnose/troubleshoot
- Discriminate between normal and abnormal conditions
- Evaluate facility performance
- Evaluate process units
- Interpret data
- Maintain processes in normal operating condition
- Measure chemical weight/volume
- Perform basic math
- Perform physical measurements
- Perform process control calculations
- Prepare chemicals
- Recognize abnormal analytical results

Monitor, Evaluate, & Adjust Treatment Processes–Chemical Addition

Sample Questions for Level I, Answers on Page 155

1. The two most important factors impacting the effectiveness of chlorination are
    a. pH of the water and the content of foreign substances in the water.
    b. concentration of chlorine and the content of foreign substances in the water.
    c. concentration of chlorine and contact time.
    d. pH and temperature of the water.

2. The treatment process that controls corrosion or scaling is known as
    a. chemical control.
    b. stabilization.
    c. passivation.
    d. corrosion kinetics.

3. Permanganate reactions are highly dependent upon
    a. organics in the water.
    b. pH.
    c. temperature.
    d. alkalinity.

4. It is hardest to kill the organism that causes which one of the following illnesses?
    a. Cholera
    b. Typhoid
    c. Cryptosporidiosis
    d. Infectious hepatitis

5. Disinfection of water wells with free chlorine requires exposure for _____ at a concentration of _____.
    a. 6 to 12 hours; 25 mg/L
    b. 12 to 24 hours; 25 mg/L
    c. 12 to 24 hours; 50 mg/L
    d. 24 to 48 hours; 100 mg/L

Sample Questions for Level II, Answers on Page 155

1. Which index determines the calcium carbonate deposition property of water by calculating the saturation pH, where a negative value indicates corrosive water and a positive value indicates depositing water?

    a. Baylis curve
    b. Langelier saturation index
    c. Marble test
    d. Ryzner index

2. The advantage to using the oxidant ozone is that it

    a. is easily generated using relatively little energy.
    b. is easily fed into the treatment process.
    c. is non-corrosive.
    d. has little pH effect.

3. Pretreatment with chlorine is being eliminated at many water treatment plants because it has been shown to

    a. react with floc and not much with organics, pathogens, or algae, thus it is a waste of resources and money.
    b. react with organics almost exclusively and not much with pathogens or algae, thus it is a waste of resources and money.
    c. sometimes produce disinfection by-products known to be carcinogenic.
    d. react by as much as 95% of its concentration with concrete walls and metal structures before oxidizing pathogens, organics, and algae.

4. CTs are based on

    a. concentration of chlorine, contact time, and pH.
    b. concentration of chlorine, contact time, pH, and temperature.
    c. concentration of chlorine, contact time, pH, and water impurities.
    d. concentration of chlorine, contact time, alkalinity, pH, and temperature.

5. If the natural fluoride content of the raw water is variable, the concentration of the raw water should be measured

    a. every 8 hours.
    b. every 12 hours.
    c. every day.
    d. continuously.

Sample Questions for Level III, Answers on Page 156

1. Which is the primary drawback for facilities that use ultraviolet light to disinfect water?

    a. It does not inactivate all microorganisms
    b. It has the potential to produce trihalomethanes
    c. Dissolved colloids can shield microorganisms from the UV light
    d. There is potential for the light bulbs to be coated with light-obscuring material, preventing the UV light from killing microorganisms

MONITOR, EVALUATE, & ADJUST TREATMENT PROCESSES   7

2. Potassium permanganate is most effective in

   a. color removal.
   b. control of biological growth.
   c. control of trihalomethanes formation potential.
   d. removing iron.

3. Chlorine is advantageous over chloramines in that chlorine

   a. is a much stronger oxidant.
   b. has long history of use.
   c. has simple feeding.
   d. has a persistent residual.

4. Which oxidant has the potential of producing ClO_3 by-products?

   a. Chlorine dioxide
   b. Chlorine
   c. Chloramines
   d. Calcium hypochlorite

5. How thick should the layer of sodium fluoride crystals be maintained in a saturator tank for flows of less than 100 gpm?

   a. 6 inches
   b. 10 inches
   c. 1 foot
   d. 2 feet

Sample Questions for Level IV, Answers on Page 156

1. Which disinfectant would work best against *Cryptosporidium*?

   a. Ozone
   b. Dichloramine
   c. Chlorine dioxide
   d. Hypochlorous acid

2. Which chemical oxidant would be best to use for controlling trihalomethanes formation potential?

   a. Chloramines
   b. Chlorine dioxide
   c. Oxygen
   d. Potassium permanganate

3. Ozone generators

   a. must be supplied with extremely dry air.
   b. are usually plate type generators for large water plants.
   c. will produce about 12% ozone by weight when supplied with air.
   d. will produce about 20% ozone by weight when supplied with oxygen only.

8   WATER OPERATOR CERTIFICATION STUDY GUIDE

4. A conventional treatment plant has raw water with high organic content. Respectively, name the most probable oxidants and disinfectants to use, if the plant applies oxidants/disinfectants at the (1) rapid mix chamber, (2) prefilter, at the (3) clearwell, and (4) clearwell effluent, and a long lasting residual is required.

   a. (1) chlorine; (2) sodium hypochlorite; (3) sodium hypochlorite; (4) chlorine
   b. (1) chloramines; (2) chlorine; (3) chlorine; (4) chloramines
   c. (1) potassium permanganate; (2) chlorine; (3) chlorine; (4) chloramines
   d. (1) hydrogen peroxide; (2) chloramines; (3) chloramines; (4) chloramines

5. If air is used to generate ozone, which percentage of the air is usually converted to ozone?

   a. 1 to 3%
   b. 4%
   c. 5 to 6%
   d. 9 to 11%

Monitor, Evaluate, & Adjust Treatment Processes–Coagulation and Flocculation

Sample Questions for Level I, Answers on Page 156

1. Detention time in flocculation basins are usually designed to provide for

   a. 5 to 15 minutes.
   b. 15 to 45 minutes.
   c. 45 to 60 minutes.
   d. 60 to 90 minutes.

2. Alum works best in a pH range of

   a. less than 4.0.
   b. 4.0 to 5.5.
   c. 5.8 to 7.5.
   d. Greater than 9.0.

3. Which statement is true concerning colloidal particles?

   a. Colloidal particles are so small that gravity has little effect on them
   b. The zeta potential between colloidal particles is balanced by covalent bonding
   c. Electrical phenomenon of colloidal particles predominate and control their behavior
   d. The surface area of colloidal particles is very small compared to their mass

4. Which natural electrical force keeps colloidal particles apart in water treatment?

   a. van der Waals forces
   b. Ionic forces
   c. Zeta potential
   d. Quantum forces

5. The zeta potential measures the number of excess _____ found on the surface of all particulate matter.

   a. electrons
   b. ions
   c. cations
   d. protons

MONITOR, EVALUATE, & ADJUST TREATMENT PROCESSES   9

Sample Questions for Level II, Answers on Page 157

1. Low temperature water can be compensated for when using alum by
   a. increasing the pH.
   b. decreasing the pH.
   c. increasing the alum dosage.
   d. decreasing the alum dosage.

2. Which is the optimal pH range for the removal of particulate matter, when using alum as a coagulant?
   a. 4.5 to 5.7
   b. 5.8 to 6.5
   c. 6.5 to 7.2
   d. 7.3 to 8.1

3. Which forces will pull particles together once they have been destabilized in the coagulation-flocculation process?
   a. van der Waals forces
   b. Zeta potential
   c. Ionic forces
   d. Quantum forces

4. Which is a common mistake that operators make in regards to flocculation units?
   a. Excessive flocculation time
   b. Lack of food-grade NSF-approved grease on the flocculator bearings
   c. Keeping the mixing energy the same in all flocculation units
   d. Too short a flocculation time

5. Ferric sulfate has which advantage over aluminum sulfate (alum)?
   a. Less staining characteristics
   b. Less cost
   c. More dense floc
   d. Not as corrosive

Sample Questions for Level III, Answers on Page 157

1. How much alkalinity as $CaCO_3$ will dry-basis alum consume?
   a. 0.5 mg/L
   b. 0.8 mg/L
   c. 1.2 mg/L
   d. 1.5 mg/L

2. Natural zeolites used for softening that have become exhausted with use are regenerated by immersing them in a strong solution of which chemical?
   a. NaCl
   b. NaOH
   c. HCl
   d. H_2SO_4

3. The zeta potential on a particular sample of water is −2. The degree of coagulation is best described as

   a. poor.
   b. fair.
   c. excellent.
   d. maximum.

4. Which is a disadvantage of using static mixers?

   a. They do not provide good mixing
   b. They are not economical
   c. They increase head loss
   d. They require too much maintenance

5. Which is the usual effective pH range of iron salt coagulants?

   a. 3.5 to 9.0
   b. 6.5 to 8.8
   c. 3.0 to 9.5
   d. 4.2 to 9.0

Sample Questions for Level IV, Answers on Page 158

1. To help avoid short circuiting, which is the minimum recommended number of flocculation basins in a series?

   a. 2
   b. 3
   c. 4
   d. 5

2. Which type of polymer(s) is (are) sometimes formulated with regulated substances?

   a. Polyethylene
   b. Divinylbenzene
   c. Polypropylene and polyethylene
   d. Nonionic and anionic polymers

3. Which is the most probable solution if rotifers are visible in the finished water?

   a. Superchlorinate the water plant
   b. Optimize coagulation, flocculation, and filtration
   c. Use aeration followed by lime softening before the settling process
   d. Use oxygen deprivation

4. The **best** addition for water that is highly colored due to organic matter would be

   a. the addition of lime.
   b. lime addition with increase in the coagulant being used.
   c. a small increase in a nonionic polymer.
   d. the addition of an acid to lower pH before coagulation.

5. If the activation process of silica is not carefully controlled,

   a. the silica could splash due to high heat of reactants.
   b. it could inhibit floc formation.
   c. it could corrode and destroy the metal and rubber in the flocculators.
   d. it could deposit silica on the flocculators and the gears, bringing it eventually to a grinding halt.

Monitor, Evaluate, & Adjust Treatment Processes–Clarification and Sedimentation

Sample Questions for Level I, Answers on Page 158

1. Which device collects the settled water as it leaves the sedimentation basin?

    a. Effluent weir
    b. Effluent flow box
    c. Effluent baffle
    d. Effluent launder

2. In solid-contact basins with fairly constant water quality parameters, how often should the solids concentration be determined?

    a. At least once per week
    b. At least every other day
    c. At least once per month
    d. At least twice per day

3. The definition of decant is

    a. to draw off a liquid layer from a vessel of any size without disturbing any layer(s) above or below.
    b. to draw off the sediment at the bottom of a vessel of any size without disturbing the overlying liquid layer(s).
    c. to remove the precipitate at the bottom of any size vessel.
    d. to draw off the liquid from a vessel of any size without stirring up bottom sediment.

4. How often should sedimentation basins with mechanical sludge removal equipment be drained and inspected?

    a. Twice a year
    b. Once a year
    c. Every other year
    d. Every three years

5. Which is the most important reason to reduce turbidity?

    a. To reduce taste and odor problems
    b. To remove pathogens
    c. To reduce corrosion
    d. To determine the efficiency of coagulation and filtration

Sample Questions for Level II, Answers on Page 159

1. If enteric disease-causing protozoans have been found in the effluent of a water plant, which is the most probable solution?

    a. Where possible, use powdered activated carbon (PAC) throughout water plant; backwashing filters will remove the PAC
    b. Use PAC only in the sedimentation basin; backwashing the filters will remove the PAC
    c. Use multibarrier approach—coagulation, flocculation, sedimentation, and filtration
    d. Superchlorinate the water plant

2. Which is the major cause of short circuiting in a sedimentation basin?

    a. Open basins that are subject to algal growths and thick slime growths on the side of the basin
    b. Basins without a wind break
    c. Poor inlet baffling
    d. Density currents

3. Conventional sedimentation has a _____ removal of *Cryptosporidium* oocysts.

    a. less than 0.5-log
    b. 0.5-log
    c. 1.0-log
    d. 2.0-log

4. In solids-contact basins, the weir loading normally should not exceed _____ of weir length.

    a. 1 gpm/ft
    b. 2 gpm/ft
    c. 5 gpm/ft
    d. 10 gpm/ft

5. Dissolved-air flotation is particularly good for removing

    a. sulfides.
    b. inorganics.
    c. manganese and iron.
    d. algae.

Sample Questions for Level III, Answers on Page 159

1. Which determines whether or not colloidal-sized particles in suspension repel each other, stay in suspension, or agglomerate and eventually settle?

    a. Number of collisions
    b. Flow and temperature of the water
    c. Types of chemical bonding
    d. Magnitude of the charges

2. If the sludge in a sedimentation basin becomes too thick, which could happen?

    a. Gases from decomposition will rise through the settled sludge accelerating normal floc settling
    b. Abundant trihalomethanes and haloacetic acids will form
    c. Solids can become resuspended or taste and odors can develop
    d. The sludge will compact at the bottom of the basin making it very difficult to remove

3. In basins using tube and plate settlers, which parameter must be much better than conventional treatment basins?

    a. Metals must be oxidized before reaching the tubes and plates
    b. Floc rate must be 2 to 3 times slower than conventional basins
    c. Floc rate must be 3 to 4 times faster than conventional basins
    d. Floc must have good settling characteristics

MONITOR, EVALUATE, & ADJUST TREATMENT PROCESSES 13

4. Which type of sedimentation basins have the flow of water admitted at an angle?

   a. Rectangular settling basins
   b. Square settling basins
   c. Center-feed settling basins
   d. Spiral-flow basins

5. At which minimum angle must self-cleaning tube settlers be placed?

   a. 50°
   b. 60°
   c. 65°
   d. 70°

Sample Questions for Level IV, Answers on Page 160

1. If nematodes are interfering with the disinfectant, which is the most probable solution?

   a. Optimize the settling process
   b. Use chloramines
   c. Decrease detention time of finished water in clearwell and tanks
   d. Use oxygen deprivation

2. At which angle should the parallel inclined plates be installed when using the shallow-depth sedimentation method?

   a. 35°
   b. 45°
   c. 50°
   d. 60°

3. Why do solids-contact basins have much shorter detention times than conventional treatment basins?

   a. Because chemical reactions take place throughout the basin
   b. Because the settling zone water moves upward, while at the same time the mixing zone moves upward
   c. Because the gentle upward flow of the water throughout the basin is conducive for producing larger settleable floc
   d. Because of the recycled materials from the sludge blanket, the chemical reactions occur more quickly and completely in the mixing area

4. The pulsating energy in a pulsator clarifier helps to

   a. maintain a uniform sludge blanket layer.
   b. mix the coagulants with the raw water.
   c. mix the coagulant aids with the primary coagulant and water to help in flocculation.
   d. raise the sludge blanket over the weir for wasting.

5. Pulsator clarifiers are used to treat water that is

   a. low in temperature, usually <10°C.
   b. high in color and low in turbidity.
   c. low in color and high in turbidity.
   d. high in organic acids.

Monitor, Evaluate, & Adjust Treatment Processes–Filtration

Sample Questions for Level I, Answers on Page 160

1. Which is the filtration flow rate through a manganese greensand pressure filter at 10°C?

    a. 1 to 2 gpm/ft^2
    b. 2 to 3 gpm/ft^2
    c. 3 to 5 gpm/ft^2
    d. 5 to 8 gpm/ft^2

2. When a filter is ripening,

    a. it is in need of a backwash.
    b. turbidity is just starting to break through.
    c. it is becoming more efficient in particle removal.
    d. it is beginning to grow algae in the filter bed, walls, and troughs.

3. Virgin greensand can be regenerated by soaking the filter bed for several hours in a solution of chlorine containing

    a. 50 mg/L Cl_2.
    b. 75 mg/L Cl_2.
    c. 100 mg/L Cl_2.
    d. 200 mg/L Cl_2.

4. Which role does the action of straining of suspended particles play during filtration?

    a. Minor
    b. Fair
    c. Good
    d. Major

5. The turbidity of settled water before it is applied to the filters (post sedimentation process) should always be kept below

    a. 1 to 2 ntu.
    b. 2 to 4 ntu.
    c. 5 ntu.
    d. 8 to 10 ntu.

Sample Questions for Level II, Answers on Page 160

1. Which is the best process for the removal of turbidity?

    a. Anion exchange
    b. Coagulation, flocculation, sedimentation, and filtration
    c. Chemical oxidation
    d. Granular activated carbon

2. If filter run times between backwashes are long, for example one week, because high quality (low turbidity) water is being applied to the filters, which problem could still arise?

    a. Mudball formation
    b. Air binding and formation of mudballs
    c. Extended backwashing due to media becoming too compacted
    d. Floc breakthrough

3. Gravel displacement in a filter bed from backwash rates with too high of a velocity could eventually cause

    a. compaction of the filter media.
    b. loss of media into the backwash troughs.
    c. a sand boil.
    d. bed shrinkage.

4. Virgin greensand

    a. does not require regeneration.
    b. requires regeneration with potassium permanganate (1 hour soak with 60 grams $KMnO_4/ft^3$).
    c. requires regeneration with manganese dioxide (2 hour soak with 25% by weight solution of MnO_2).
    d. requires regeneration with manganese hydroxide (4 hour soak with 200 grams $Mn(OH)_2/ft^3$).

5. Which conventional treatment step is eliminated by direct filtration?

    a. Oxidation
    b. Aeration
    c. Flocculation
    d. Sedimentation

Sample Questions for Level III, Answers on Page 161

1. If crustaceans have clogged the water treatment plant's filters, which is the most probable solution?

    a. Shut down the filters and physically remove them
    b. Shut down one filter at a time and drain; once the crustaceans have died, physically remove them and then repeat process on the other filters
    c. Backwash filters using a very high concentration of ozone in the water
    d. Use a disinfectant that targets the specific organisms in question

2. Which organism can escape coagulation and thus pass through a granular filter?

    a. *Giardia*
    b. *Entamoeba*
    c. *Cryptosporidium*
    d. *Naegleria*

3. Reverse osmosis membranes will compact faster with

    a. higher iron content.
    b. higher chlorine contact.
    c. higher pressure.
    d. higher pH.

4. Which would immediately occur if newly installed manganese greensand was not skimmed of the fines after backwashing and stratification steps were completed?

    a. Uneven flow through the bed
    b. Cracks would develop in the bed
    c. Mudball formation
    d. Shorter filter runs

5. When a filter is operated at normal flow rates, its ability to trap flocculated particles in suspension is a function of

   a. effective size multiplied by uniformity coefficient.
   b. effective size multiplied by uniformity coefficient divided by media size.
   c. media depth and media size.
   d. media depth and uniformity coefficient.

Sample Questions for Level IV, Answers on Page 161

1. Which is the best solution if iron bacteria are causing corrosion problems in the filters?

   a. Protect the metal parts of the filters by adding zinc orthophosphate
   b. Optimize the settling process
   c. Superchlorinate
   d. Add lime at 50 mg/L to one filter at a time

2. A conventional water treatment plant with dual media filters has very cold water in the winter and warm water in the summer. Which should the operator do to compensate for this temperature change?

   a. Use more coagulants in the summer per million gallons
   b. Sustain the same bed expansion without media loss by reducing or increasing backwash flow rate
   c. Increase summer bed expansion and increase winter backwash flow rates
   d. Increase bed expansion in the winter compared to summer in order to remove turbidity

3. How are reverse osmosis membranes cleaned once they become fouled?

   a. They are soaked in high purity industrial soap for at least 24 hours
   b. They are cleaned with an acid wash
   c. They are cleaned with an acid, then with an industrial soap for 24 hours
   d. They are cleaned first with a high purity industrial soap and then soaked in an acid solution for 3 days

4. Which membrane process is used to treat brackish water or seawater?

   a. Microfiltration
   b. Nanofiltration
   c. Reverse osmosis
   d. Ultrafiltration

5. The amount of reject water from a reverse osmosis unit is dependent on the number of stages in which the membranes are configured and the

   a. feed pressure.
   b. amount of cations.
   c. amount of cations and anions.
   d. pH of the water.

Monitor, Evaluate, and Adjust Treatment Processes–Residuals Disposal

Sample Questions, General, Answers on Page 162

1. In the precipitative softening plant, which percentage of solids sludge is produced?

    a. 1%
    b. 5%
    c. 10%
    d. 30%

2. Which sludge disposal method is most economical for lime–soda ash softening plants?

    a. Disposal into the sewage system
    b. Sand drying beds
    c. Lagoons
    d. Landfill the sludge

3. Current regulations require water treatment wastes to be monitored

    a. daily.
    b. weekly.
    c. monthly.
    d. quarterly.

4. Which process is used to concentrate sludge?

    a. Sand bed
    b. Solar lagoon
    c. Thickener
    d. Centrifuge

5. Which process is used to dewater sludge?

    a. Wash water basin
    b. Sand bed
    c. Thickener
    d. Reclamation basin

Sample Questions for Level III, Answers on Page 162

1. Which is the total concentration of dissolved solids in the wastewater from the regeneration of ion exchange units?

    a. 10,000 to 20,000 mg/L
    b. 20,000 to 30,000 mg/L
    c. 35,000 to 45,000 mg/L
    d. 45,000 to 60,000 mg/L

2. Which is the usual range of percent sludge solids, if the sludge is allowed to accumulate and compact at the bottom of a sedimentation basin?

    a. 2–4%
    b. 4–7%
    c. 7–12%
    d. 12–15%

3. Which should be determined first before an in-ground sedimentation tank is drained?

   a. The solids content
   b. The hazardous metals content
   c. Sludge volume and volume of process area to make sure it will be large enough
   d. Water table level

4. Which sludge dewatering process is best for alum sludges (which are difficult to dewater) when the cakes are very dry, filtrate is clear, and solids capture is very high?

   a. Centrifuge
   b. Vacuum filters
   c. Filter press
   d. Belt filter press

5. Which sludge dewatering process requires a precoat of diatomaceous earth and its use has declined due to other newer methods?

   a. Centrifuge
   b. Vacuum filters
   c. Filter press
   d. Belt filter press

Sample Questions for Level IV, Answers on Page 163

1. Which national law regulates underground disposal of wastes in deep wells?

   a. The Toxic Substances Control Act (TSCA)
   b. The Comprehensive Environmental Response Compensation, and Liability Act (Superfund)
   c. The Resource Conservation and Recovery Act (RCRA)
   d. The Safe Drinking Water Act

2. Which is typically the percentage of solid material of alum sludge?

   a. 0.1 to 2.0%
   b. 2.0 to 5.0%
   c. 4.0 to 7.0%
   d. 5.0 to 15.0%

3. Which is the most troublesome operating problem for sedimentation basins?

   a. Algae and slime growths
   b. Short-circuiting
   c. Density currents
   d. Sludge collection and removal

4. 100,000 gallons of sludge is removed from a sedimentation basin at 0.2 percent (2,000 mg/L) solids and is sent to a thickener where it is expected to be concentrated to 5 percent (50,000 mg/L) before being transferred to a tank truck for hauling. If the tank truck has 4,000 gallons capacity, how many trips will be necessary?

   a. One truckload
   b. Two truckloads
   c. 5 truckloads
   d. 20 truckloads

5. Sludge produced by water drawn from sources with relatively low turbidity

   a. is high in silt and organic content.
   b. is high in liquid content.
   c. can be very gelatinous and difficult to handle.
   d. handles like clay.

Monitor, Evaluate, & Adjust Treatment Processes–Additional Treatment Tasks

Sample Questions for Level I, Answers on Page 163

1. Which precipitates can foul a cation exchange resin?

   a. Sodium chloride and potassium chloride
   b. Chlorate and borate
   c. Sulfates
   d. Iron and manganese

2. Which process works best for sequestering manganese?

   a. Sodium silicate alone
   b. Sodium silicate and chlorine
   c. Polyphosphates alone
   d. Polyphosphates and chlorine

3. When should polyphosphates used for sequestration of iron and manganese from a well be injected into the process?

   a. Right after disinfection
   b. Immediately after aeration to remove unwanted gases
   c. Right after clarification
   d. Right after the water leaves the well

4. Recarbonation is

   a. adding CO_2 to the water.
   b. adding bicarbonate to the water.
   c. adding acid to precipitate the excess lime.
   d. adding caustic soda.

5. In the ion-exchange softening process, once the resin can no longer soften water it must be

   a. renewed.
   b. re-catalyzed.
   c. regenerated.
   d. recharged.

Sample Questions for Level II, Answers on Page 164

1. Ion exchange processes can typically be used for direct groundwater treatment as long as turbidity and _____ levels are not excessive.

   a. calcium carbonate
   b. iron
   c. carbon dioxide
   d. sodium sulfate

2. Softened water has a high pH and a high concentration of $CaCO_3$. Therefore, stabilization is essential in order to prevent the $CaCO_3$ from precipitating out

    a. in household plumbing.
    b. in the clear well.
    c. in the distribution system.
    d. on the filters.

3. Which is the best type of salt to use in the regeneration of ion exchange softener resin?

    a. Fine-grained salt
    b. Block salt
    c. Block or road salt
    d. Rock salt or pellet-type salt

4. Powdered activated carbon is primarily used to control

    a. disinfectant by-products.
    b. organic compounds responsible for tastes and odors.
    c. synthetic organic chemicals.
    d. humic and fulvic acids.

5. Ion exchange will remove

    a. all hardness.
    b. all hardness down to 7.4 mg/L, as $CaCO_3$.
    c. all hardness down to 17.2 mg/L, as $CaCO_3$.
    d. all hardness down to about 25.0 mg/L, as $CaCO_3$.

Sample Questions for Level III, Answers on Page 164

1. It is impossible to produce waters with a hardness of less than _____ when using the lime-soda ash process.

    a. 9 mg/L
    b. 17 mg/L
    c. 25 mg/L
    d. 50 mg/L

2. When added to water for softening purposes, soda ash will do which of the following?

    a. Disinfect the water and kill the vast majority of protozoans, viruses, bacteria, and other multicellular organisms
    b. Raise the pH of the water
    c. Decrease the CO_2 in the water
    d. Add calcium alkalinity to the water

3. Magnetic ion exchange resin has been developed to remove

    a. total organic carbon.
    b. chlorides.
    c. iron and magnesium.
    d. sulfates and sulfides.

MONITOR, EVALUATE, & ADJUST TREATMENT PROCESSES   21

4. Approximately how much carbon is lost during the reactivation process for granular activated carbon?

   a. 5%
   b. 7%
   c. 10%
   d. 15%

5. Which is the most advantageous application point for powdered activated carbon?

   a. Raw water intake
   b. After coagulation
   c. After oxidation with chlorine
   d. In the filters

Sample Questions for Level IV, Answers on Page 165

1. Which is the most effective method for removing tastes and odors?

   a. Coagulation, sedimentation, and filtration
   b. Granular activated carbon
   c. Anion exchange
   d. Lime softening

2. Backwashing rate procedures should be reassessed to determine the cause of granular activated carbon loss if the loss per year exceeds

   a. 2 inches.
   b. 4 inches.
   c. 6 inches.
   d. 8 inches.

3. Which is the most efficient process for the removal of nitrite and nitrate?

   a. Powdered activated carbon
   b. Granular activated carbon
   c. Anion exchange
   d. Cation exchange

4. Which is the main problem if particle agglomeration is occurring in a filter for iron and manganese removal at the interface of the coal layer and the layer below?

   a. Oxidant is too weak
   b. Coagulant dosage is excessive
   c. Coal layer is too fine
   d. Coal layer is too coarse

5. Which is the most effective method for the removal of disinfection by-products?

   a. Reverse osmosis
   b. Lime softening
   c. Ultrafiltration
   d. Granular activated carbon

Laboratory Analyses

	Level I	Level II	Level III	Level IV
Algae identification	Comprehension	Comprehension	Application	Application
Asbestos	Comprehension	Comprehension	Application	Application
Biological	Application	Application	Application	Application
Chemical	Comprehension	Application	Application	Application
Chlorine	Analysis	Analysis	Analysis	Analysis
Coliform bacteria	Application	Application	Application	Analysis
Complete chain-of-custody	Comprehension	Application	Application	Analysis
Corrosivity	Comprehension	Comprehension	Comprehension	Comprehension
Disinfectant by-products (THM/HAA)	Comprehension	Comprehension	Application	Analysis
Dissolved oxygen	Comprehension	Comprehension	Comprehension	Comprehension
Hexavalent chromium	Comprehension	Comprehension	Comprehension	Comprehension
Inorganic minerals	Comprehension	Comprehension	Comprehension	Comprehension
Jar test	Comprehension	Comprehension	Application	Analysis
Langelier Index	Comprehension	Analysis	Analysis	Analysis
Metals	Application	Application	Application	Application
Organics	Comprehension	Comprehension	Analysis	Analysis
pH	Application	Application	Analysis	Analysis
Physical Parameters	Analysis	Analysis	Analysis	Analysis
Radiological parameters	Analysis	Analysis	Analysis	Analysis
Saturation Index	Comprehension	Comprehension	Comprehension	Comprehension
Solids	Comprehension	Comprehension	Comprehension	Comprehension
Streaming Current Analysis	Comprehension	Comprehension	Comprehension	Comprehension

Required Capabilities

Knowledge of:
- Basic chemistry
- Basic laboratory techniques
- Biological science
- Laboratory equipment
- Material Safety Data Sheet
- Monitoring requirements

- Chemical properties
- Data collection Personal protective equipment
- Pesticides
- Physical science
- Principles of measurement
- Proper sampling procedures

- Normal characteristics of water
- Normal chemical range Quality control/quality assurance practices
- Safety procedures
- Standard methods

Ability to:

- Accurately transcribe data
- Communicate in writing
- Communicate verbally
- Determine what information needs to be recorded
- Follow written procedures
- Interpret data

- Measure chemical weight/volume
- Perform basic math
- Perform laboratory calculations
- Perform physical measurements
- Prepare chemicals
- Recognize abnormal analytical results
- Record information

Laboratory Analyses

Sample Questions for Level I, Answers on Page 165

1. Under no circumstances should a composite sample be collected for which type of analysis?

    a. Bacteriological
    b. Total dissolved solids
    c. Alkalinity
    d. Turbidity

2. The number of monthly distribution system chlorine residual samples required is

    a. based on water withdrawal permit limit.
    b. based on system size.
    c. based on population.
    d. different for each state.

3. Which is (are) the ideal indicator for pathogens?

    a. *Salmonella* species
    b. Coliform group bacteria
    c. Gram-negative cocci
    d. Gram-negative coccobacilli

4. When one substance is dissolved in another and will not settle out, which is the product called?

    a. An emulsion
    b. A compound
    c. A suspension
    d. A solution

5. Acids, bases, and salts lacking carbon are

   a. ketones.
   b. aldehydes.
   c. organic compounds.
   d. inorganic compounds.

Sample Questions for Level II, Answers on Page 165

1. Which type of sample should always be collected for determining the presence of coliform bacteria?

   a. Time composite.
   b. Grab sample.
   c. Proportional.
   d. Composite.

2. Samples to be tested for coliforms can be refrigerated for up to _____ hours before analysis, but should be done as soon as possible.

   a. 4
   b. 6
   c. 8
   d. 12

3. When a water sample is acidified, the final pH of the water must be

   a. < 2.0.
   b. < 2.5.
   c. < 3.0.
   d. < 3.5.

4. Which chemical is used to remove residual chlorine from water?

   a. $Na_2S_2O_3$
   b. Na_2SiO_3
   c. Na_2SiF_6
   d. $NaOCl$

5. When a sample is collected, which causes its quality to begin to change?

   a. CO_2
   b. Dissolved gases
   c. Biological activity
   d. pH

Sample Questions for Level III, Answers on Page 166

1. Water that is to be analyzed for inorganic metals should be acidified with

   a. dilute hydrochloric acid.
   b. concentrated hydrochloric acid.
   c. dilute nitric acid.
   d. concentrated nitric acid.

2. A solution used to determine the concentration of another solution is called a

   a. saturated solution.
   b. standardized solution.
   c. concentrated solution.
   d. dilute solution.

3. Which method would you use to concentrate and retrieve low numbers of bacteria from a large quantity of water?

   a. Colilert
   b. Colisure
   c. Membrane filtration
   d. MPN (Most Probable Number)

4. A typical coliform colony in the membrane filter method has the following characteristics:

   a. Blue with lustrous surface sheen
   b. Pink to dark red with green metallic surface sheen
   c. Pink or yellow with lustrous to metallic surface sheen depending on species
   d. Yellow with silver metallic to lustrous surface sheen

5. In the presumptive phase of the Most Probable Number test, how long does it take for the coliforms to produce gas?

   a. 12 to 24 hours
   b. 24 to 48 hours
   c. 24 to 36 hours
   d. 36 to 48 hours

Sample Questions for Level IV, Answers on Page 166

1. Chemical analysis for synthetic organic compounds should not be collected in containers made of

   a. polytetrafluoroethylene.
   b. stainless steel.
   c. polypropylene.
   d. borosilicate.

2. How much acid per 100 mL should be used to preserve a sample for later hardness analyses?

   a. 0.1 mL
   b. 0.2 mL
   c. 0.5 mL
   d. 1.0 mL

3. Conductivity measurements can assist the laboratory analyst in

   a. measuring the electrical strength, which is directly proportional to the number of free electrons.
   b. estimating the concentration of calcium carbonate.
   c. evaluating variations in the concentration of suspended particles.
   d. determining the degree of mineralization of the water.

4. Water that is to be analyzed for inorganic metals should be filtered for _____ before _____.

    a. dissolved metals; analyses
    b. suspended metals; analyses
    c. dissolved metals; preserving
    d. suspended metals; preserving

5. In routine water quality sampling, which one of the following is an early warning sign that conditions are becoming more conducive to sulfate-reducing bacteria?

    a. Increase in ferrous iron
    b. Increase in ferric iron
    c. Dramatic decline in dissolved oxygen
    d. Increase in sulfides

Comply with Drinking Water Regulations

	Level I	Level II	Level III	Level IV
40 CFR 141 Subpart A: General (definitions, coverage, variances and exemptions, siting requirements, and effective dates)	Comprehension	Comprehension	Application	Application
40 CFR 141 Subpart B: Maximum Contaminant Levels (arsenic, nitrate, turbidity)	Comprehension	Comprehension	Application	Application
40 CFR 141 Subpart C: Monitoring and Analytical Requirements (turbidity, coliforms, organic contaminants, organic contaminants)	Comprehension	Comprehension	Application	Application
40 CFR 141 Subpart D: Reporting and Recordkeeping Requirements	Comprehension	Comprehension	Application	Application
40 CFR 141 Subpart E: Special Regulations, Including Monitoring Regulations and Prohibition on Lead Use	Comprehension	Comprehension	Application	Application
40 CFR 141 Subpart F: Maximum Contaminant Level Goals and Maximum Residual Disinfectant Level Goals	Comprehension	Comprehension	Application	Application
40 CFR 141 Subpart G: National Primary Drinking Water Regulations: Maximum Contaminant Levels and Maximum Residual Disinfectant Levels	Comprehension	Comprehension	Application	Application
40 CFR 141 Subpart H: Filtration and Disinfection	Comprehension	Comprehension	Application	Application
40 CFR 141 Subpart I: Control of Lead and Copper	Comprehension	Comprehension	Application	Application
40 CFR 141 Subpart K: Treatment Techniques	Comprehension	Comprehension	Application	Application

40 CFR 141 Subpart L: Disinfectant Residuals, Disinfection Byproducts, and Disinfection Byproduct Precursors	Comprehension	Comprehension	Application	Application
40 CFR 141 Subpart P: Enhanced Filtration and Disinfection Systems Serving 10,000 or More People	Comprehension	Comprehension	Application	Application
40 CFR 141 Subpart Q: Public Notification of Drinking Water Violations	Comprehension	Comprehension	Application	Application
40 CFR 141 Subpart S: Ground Water Rule	Comprehension	Comprehension	Application	Application
40 CFR 141 Subpart T: Enhanced Filtration and Disinfection Systems Serving Fewer Than 10,000 People	Comprehension	Comprehension	Application	Application
40 CFR 141 Subpart U: Initial Distribution System Evaluations	Comprehension	Comprehension	Application	Application
40 CFR 141 Subpart V: Stage 2 Disinfection Byproducts Requirements	Comprehension	Comprehension	Application	Application
40 CFR 141 Subpart W: Enhanced Treatment for *Cryptosporidium*	Comprehension	Comprehension	Application	Application
40 CFR 143: National Secondary Drinking Water Regulations	Comprehension	Comprehension	Application	Application

Required Capabilities

Knowledge of:

- Code of Federal Regulations
- Regulations
- Reporting
- Safe Drinking Water Act

Comply with Drinking Water Regulations

Sample Questions for Level I, Answers on Page 167

1. Which type of violation requires an all clear notification to the public?
    a. Tier I
    b. Tier II
    c. Tier III
    d. Tier IV

2. How many coliform positive samples are allowed each month for water systems that collect less than 40 samples per month?

   a. 1
   b. 2
   c. 3
   d. 4

3. If a water sample tests positive for coliforms, the system must collect a set of repeat samples within

   a. 18 hours.
   b. 24 hours.
   c. 36 hours.
   d. 48 hours.

4. A public water system has at least _____ service connections or serves _____ or more persons _____ or more days each year.

   a. 15; 15; 50
   b. 25; 25; 50
   c. 15; 25; 60
   d. 25; 50; 60

5. Where are the sampling points located for required sampling of coliform bacteria in a community water system?

   a. At consumers' faucets
   b. 75% at locations representative of population distribution and 25% at the farthest points in the distribution system
   c. At points where water enters the distribution system
   d. At representative points within the distribution system

Sample Questions for Level II, Answers on Page 167

1. Water being served to the public for a population greater than 3,300 must not have a disinfectant residual entering the distribution system below _____ for more than 4 hours.

   a. 0.1 mg/L
   b. 0.2 mg/L
   c. 0.3 mg/L
   d. 0.4 mg/L

2. Which is the Action Level for copper?

   a. 0.5 mg/L
   b. 1.0 mg/L
   c. 1.3 mg/L
   d. 1.8 mg/L

3. Where are the sampling point(s) located for required sampling of turbidity in a community water system?

    a. Representative points within the distribution system
    b. 75% at locations representative of population distribution and 25% at the farthest points in the distribution system
    c. Point(s) where water enters the distribution system, including all filter effluents if a surface water treatment plant
    d. Effluents of all filters if a surface water treatment plant, entry points to the distribution system and at locations representative of population distribution

4. Under the Surface Water Treatment Rule, disinfection residuals must be collected at the same location in the distribution system as

    a. coliform samples.
    b. total trihalomethanes.
    c. disinfection by-products.
    d. alkalinity, conductivity, and pH for corrosion studies.

5. Which is the Maximum Contaminant Level for haloacetic acids (HAA5)?

    a. 0.040 mg/L
    b. 0.060 mg/L
    c. 0.080 mg/L
    d. 0.100 mg/L

Sample Questions for Level III, Answers on Page 168

1. If a water system collects at least 40 samples per month for the analyses of total coliforms, which percent of total coliform positive samples are acceptable for the system to remain in compliance with the maximum contaminant level for total coliforms?

    a. No more than 2%
    b. No more than 3%
    c. No more than 4%
    d. No more than 5%

2. Water systems are required to achieve at least _____ removal and/or inactivation of viruses between a point where the raw water is not subject to recontamination by surface water runoff and a point downstream before or at the first customer.

    a. 2 log
    b. 2.5 log
    c. 3 log
    d. 4 log

3. Where are the sampling points located for required sampling of organics (except trihalomethanes) in a community water system?

    a. Representative points within the distribution system
    b. 75% at locations representative of population distribution and 25% at the farthest points in the distribution system
    c. Entry points to the distribution system
    d. Entry points to the distribution system and representative points within the distribution system

4. Where are the sampling point(s) located for required sampling of natural radionuclides in a community water system?

   a. Consumer's faucet
   b. Representative points within the distribution system
   c. Each entry point to the system
   d. 75% at locations representative of population distribution and 25% at the farthest points in the distribution system

5. Continuous chlorine residual monitoring is required where the water enters the distribution system under the Surface Water Treatment Rule when the

   a. population served is > 3,300 people.
   b. population served is > 10,000 people.
   c. number of taps is > 1,000.
   d. number of taps is > 2,500.

Sample Questions for Level IV, Answers on Page 168

1. A community water system must post the Consumer Confidence Report on a publicly accessible Web site if it serves more than

   a. 10,000 people.
   b. 25,000 people.
   c. 50,000 people.
   d. 100,000 people.

2. Public Water Systems that cannot meet the required removal of TOC can comply if the source-water TOC level is _____, calculated quarterly as a running annual average.

   a. <1.0 mg/L
   b. <2.0 mg/L
   c. <2.5 mg/L
   d. <5.0 mg/L

3. To avoid filtration, one criterion is that the fecal coliform concentration must be equal to or less than _____ in representative samples of the source water immediately before the first or only point of disinfectant application in at least _____ of the measurements made for the previous 6 months that the system served water to the public on an ongoing basis.

   a. 5/100 mL; 95%
   b. 10/100 mL; 99%
   c. 20/100 mL; 90%
   d. 25/100 mL; 90%

4. Water systems may be allowed by their state to monitor annually, if the state has determined that the system is reliable and consistently below the maximum contaminant level for organic compounds. When must the water system collect these annual samples?

   a. During the fall when plant debris starts to increase in surface waters
   b. During the spring when plant debris has had time to decay and thus increase organic levels in surface waters
   c. During the month that previously yielded the highest analytical result
   d. During the quarter that previously yielded the highest analytical result

5. Water treatment operators must strike a balance between TOC removal and the

   a. Stage 1 Disinfectants/Disinfection By-products Rule.
   b. Stage 2 Disinfectants/Disinfection By-products Rule.
   c. Filter Backwash Recycle Rule.
   d. Lead and Copper Rule.

Operate and Maintain Equipment

	Level I	Level II	Level III	Level IV
Evaluate Operation of Equipment				
Check speed of equipment	Comprehension	Application	Application	Analysis
Inspect equipment for abnormal conditions	Comprehension	Application	Application	Analysis
Measure temperature of equipment	Comprehension	Application	Application	Analysis
Read charts	Application	Application	Application	Analysis
Read meters	Application	Application	Application	Analysis
Read pressure gauges	Application	Application	Application	Analysis
Operate Equipment				
Blowers and compressors	Application	Application	Application	Application
Chemical feeders	Analysis	Analysis	Analysis	Analysis
Computers (SCADA systems, HMI, etc.)	Application	Application	Application	Application
Drives	Application	Application	Application	Application
Electronic testing equipment	Application	Application	Application	Application
Engines	Application	Application	Application	Application
Gates	Application	Application	Application	Application
Generators	Application	Application	Application	Application
Hand tools	Application	Application	Application	Application
Hydrants	Application	Application	Application	Application
Hydraulic equipment	Application	Application	Application	Application
Instrumentation	Application	Application	Application	Application
Motors	Application	Application	Application	Application
Pneumatic equipment	Application	Application	Application	Application
Power tools	Application	Application	Application	Application
Pumps	Application	Application	Application	Application
Valves	Application	Application	Application	Application

Perform Maintenance				
Backflow prevention devices	Application	Application	Application	Analysis
Blowers and compressors	Application	Application	Application	Application
Bulk chemical storage systems	Application	Application	Application	Analysis
Calibration of chemical feeders	Application	Application	Application	Analysis
Chemical feeders	Application	Application	Application	Application
Drives	Comprehension	Application	Application	Application
Electrical grounding	Comprehension	Application	Application	Application
Engines	Comprehension	Application	Application	Application
Gates	N/A	N/A	N/A	Comprehension
Generators	Comprehension	Comprehension	Comprehension	Comprehension
Hydrants	N/A	N/A	N/A	Comprehension
Hydraulic equipment	N/A	N/A	N/A	Comprehension
Instrumentation	Application	Application	Application	Application
Lock-out/tag-out	Application	Application	Application	Application
Motors	Application	Application	Application	Application
Off-gas equipment	Comprehension	Comprehension	Comprehension	Comprehension
Pipes	Comprehension	Comprehension	Comprehension	Comprehension
Pneumatic equipment	Application	Application	Application	Application
Pumps	Comprehension	Application	Application	Application
Treatment units	Application	Application	Application	Application
Valves	Application	Application	Application	Analysis

Required Capabilities

Knowledge of:
- Facility operation and maintenance
- Function of tools
- General electrical principles
- HVAC equipment
- Hydraulic principles
- Internal combustion engines
- Lubricant and fluid characteristics
- Mechanical equipment
- Mechanical principles
- Operation and maintenance practices
- Personal protective equipment
- Pneumatics
- Process control instrumentation
- Proper lifting procedures
- Start-up and shutdown procedures
- Storage

Ability to:
- Adjust equipment
- Assign work to proper trade
- Differentiate between preventative and corrective maintenance

- Calibrate equipment
- Communicate in writing
- Communicate verbally
- Diagnose/troubleshoot Operate safety equipment
- Order spare parts
- Organize information
- Perform general maintenance
- Perform general repairs
- Perform physical measurements
- Discriminate between normal and abnormal conditions
- Evaluate operation of equipment
- Monitor equipment
- Recognize unsafe work conditions
- Record information
- Report findings
- Translate technical language into common terminology
- Use hand tools

Operate and Maintain Equipment

Sample Questions for Level I, Answers on Page 169

1. The main purpose of mechanical seals is to
   a. keep lubrication in and dirt and other foreign materials out.
   b. control water leakage from the stuffing box.
   c. keep contamination from entering or leaving.
   d. save on costs, as they last longer than packing.

2. The purpose of packing is to
   a. keep oil or graphite on the shaft.
   b. control water leakage along the pump's shaft.
   c. prevent water leakage from the pump shaft.
   d. help prevent shaft from warping.

3. Which part of a pump houses the packing or mechanical seal?
   a. The shroud
   b. The stuffing box
   c. The volute
   d. The casing head

4. Which is the only type of pump that can be operated against a closed valve?
   a. Vertical turbine pump
   b. Centrifugal pump
   c. Axial-flow pump
   d. Mixed-flow pump

5. Which is a principal problem associated with preliminary treatment screening?
   a. Broken screens
   b. Excessive downtime due to numerous shear pins being broken
   c. Clogging
   d. Chain comes out of foot sprocket

Sample Questions for Level II, Answers on Page 169

1. Which device applies an even pressure to the packing such that it compresses tight around the pump shaft?

    a. Lantern ring
    b. Mechanical seal
    c. Packing gland
    d. Seal cage

2. When using four packing rings, the rings should be staggered at

    a. 45 degrees.
    b. 90 degrees.
    c. 120 degrees.
    d. 180 degrees.

3. How is velocity head expressed mathematically?

    a. V/g
    b. V^2/g
    c. $V^2/2g$
    d. $V^2/32.2 \text{ ft}/\text{sec}^2$

4. Which type of centrifugal pump impeller is used for pumping medium-sized solids?

    a. Open
    b. Closed
    c. Semi-open
    d. Radial

5. The shaft's main function is to transmit _____ from the motor to the impeller.

    a. centrifugal force
    b. torque
    c. kinetic energy
    d. thrust

Sample Questions for Level III, Answers on Page 170

1. A single chlorine cylinder is delivering 48 lb/d of chlorine to the water process, causing the cylinder to form a little frost. Which would be the best solution to this problem?

    a. Install a fan to improve air circulation
    b. Heat the cylinder immediately below the valve with heat tape
    c. Heat the valve only
    d. Add another cylinder and feed from both

2. The pressure-reducing and shutoff valve on a vaporizer will shut off when there is a/an

    a. loss of electrical power.
    b. high water level.
    c. high water temperature.
    d. over-pressurization of the vaporization system.

3. Particle counters use the principle of light

    a. scattering.
    b. reflection.
    c. refraction.
    d. blockage.

4. How much time does it usually take to slake lime in the detention-time lime slaker?

    a. 20 to 30 minutes
    b. 30 to 45 minutes
    c. 45 to 60 minutes
    d. 60 to 75 minutes

5. At which temperature should the slaking process be maintained?

    a. 120°F
    b. 135°F
    c. 150°F
    d. 160°F or higher

Sample Questions for Level IV, Answers on Page 170

1. When external seal water is used as a water source, which should its pressure be in relation to the volute pressure?

    a. 3 to 10 psi higher
    b. 10 to 15 psi higher
    c. 20 to 25 psi higher
    d. 25 to 30 psi higher

2. Under which conditions are regenerative turbine pumps used?

    a. High head conditions
    b. Low head conditions
    c. High volume output
    d. Low suction conditions

3. Approximately, which percentage of flow increase occurs when a two-flight screw pump is changed to a three-flight screw pump, all other things being equal?

    a. 10%
    b. 20%
    c. 33%
    d. 40%

4. The arrangement of pumps in series is primarily used to increase the

    a. flow, usually by 70-80%.
    b. discharge head.
    c. flow, but keep discharge head the same.
    d. pump efficiency.

5. How much will a flexible disk coupling compensate for in angular movement?

    a. 1.0 degree
    b. 2.0 degrees
    c. 3.2 degrees
    d. 3.7 degrees

Perform Security, Safety, and Administrative Procedures

	Level I	Level II	Level III	Level IV
Write/Complete Reports (State/Provincial)	Comprehension	Application	Application	Analysis
Write/Complete Reports (State/Provincial)				
Administer safety program	Comprehension	Comprehension	Comprehension	Comprehension
Develop budget	N/A	N/A	Comprehension	Comprehension
Respond to Complaints	Analysis	Analysis	Analysis	Analysis
Respond to Emergencies				
Facility upset	Application	Application	Application	Application
Natural disasters	Application	Application	Application	Application
Major spill response	Comprehension	Application	Application	Analysis
System contamination	Analysis	Analysis	Analysis	Analysis
Safety Procedures				
Calibration of atmospheric testing devices	Application	Application	Application	Application
Chemical hazards and chemical spill response	Application	Application	Application	Application
Confined space entry	Analysis	Analysis	Analysis	Analysis
General safety and health	Analysis	Analysis	Analysis	Analysis
Pathogens	Application	Application	Application	Application
Personal protective equipment	Analysis	Analysis	Analysis	Analysis
Record Information				
Compliance	Application	Application	Analysis	Analysis
Corrective actions	Application	Application	Analysis	Analysis
Customer complaints	Application	Application	Application	Application
Facility Operation	Application	Application	Application	Application
Laboratory	Comprehension	Application	Application	Analysis
Maintenance	Application	Application	Application	Analysis

Required Capabilities

Knowledge of:
- Arbitration procedures
- Building codes
- Disciplinary procedures
- Emergency plans
- Legislative process
- Local codes and ordinances
- Material Safety Data Sheet
- Personal protective equipment
- Potential causes of disasters in facility
- Potential impact of disasters on facility
- Principles of finance
- Principles of management
- Principles of public relations
- Principles of supervision
- Proper chemical handling and storage
- Proper lifting procedures
- Public administration procedures
- Recordkeeping policies
- Regulations
- Reporting requirements
- Retrieval
- Risk management
- Safety procedures
- Safety regulations

Ability to:
- Assess likelihood of disaster occurring
- Communicate in writing
- Communicate verbally
- Conduct meetings
- Conduct training programs
- Coordinate emergency response with other organizations
- Demonstrate safe work habits
- Determine what information needs to be recorded
- Develop a budget
- Develop a public relations campaign
- Develop a staffing plan
- Develop a work unit
- Evaluate employee performance
- Evaluate promotional materials
- Evaluate proposals
- Generate capital plans
- Generate long- and short-term plans
- Generate written safety procedures
- Identify potential safety hazards
- Negotiate contracts
- Operate safety equipment
- Perform impact assessments
- Prepare proposals
- Recognize unsafe work conditions
- Report findings
- Select safety equipment

Perform Security, Safety, and Administrative Procedures

Sample Questions for Level I, Answers on Page 171

1. Which is the most important thing to consider in automatic startup after a power failure?

    a. Operator safety
    b. Power surges that could trip a breaker
    c. Importance to process
    d. Importance to water quality

2. At which range of temperatures will fusible plugs on chlorine cylinders melt?

    a. 147 to 152°F
    b. 155 to 159°F
    c. 157 to 162°F
    d. 167 to 171°F

PERFORM SECURITY, SAFETY, AND ADMINISTRATIVE PROCEDURES   43

3. When a chlorine cylinder or container is changed, a new gasket should be used
    a. every time.
    b. after 2 uses.
    c. after 3 uses.
    d. after 4 or 5 uses.

4. Which is the best type of gasket to use when connecting a chlorine cylinder or container to the chlorine feed system?
    a. Fiber gasket
    b. Aluminum gasket
    c. Copper gasket
    d. Lead gasket

5. Which part of the human body is the most vulnerable to allowing microorganisms to enter it?
    a. Stomach area where most of the organs are located
    b. Heart
    c. Eyes
    d. Throat and neck area

Sample Questions for Level II, Answers on Page 171

1. Which will occur if dry alum and quicklime are mixed together?
    a. Highly explosive hydrogen gas will be released.
    b. Dust may be released which may cause an explosion.
    c. A coagulated gel will form.
    d. Nothing, as they are neutral towards each other.

2. Chlorine gas
    a. will burn in the presence of oxygen and moisture.
    b. will explode in the presence of oxygen and an ignition source.
    c. will support combustion.
    d. will conduct electricity.

3. Chlorine storage rooms should
    a. have sealed walls and doors that open inward.
    b. be fitted with chlorine-resistant power exhaust fans ducted out at ceiling level.
    c. have lights and fans on the inside and wired to the same switch.
    d. have a window in the door so an operator can look into the room to detect any abnormal conditions.

4. Before an authorized entrant enters a permitted confined space,
    a. he or she must don SCUBA gear.
    b. the internal atmosphere of the space must be tested to measure oxygen content, presence of flammable gases and vapors, and potentially toxic air contaminants.
    c. verify the availability of rescue services.
    d. ensure there is adequate lighting in the space to conduct the necessary work.

5. Fluoride injection should be as far away as possible from any chemical that contains

   a. chlorine.
   b. aluminum (from alum).
   c. calcium.
   d. sodium (for example, soda ash).

Sample Questions for Level III, Answers on Page 172

1. Water treatment personnel should only use self-contained breathing apparatus equipment that has been approved by

   a. the Occupational Safety and Health Administration (OSHA).
   b. their state's Department of Public Health.
   c. the National Institute of Occupational Safety and Health (NIOSH).
   d. the American Standards and Testing Methods.

2. If a substantial chlorine leak incident occurs, which agency should be called for actual hands-on assistance?

   a. The Occupational Safety and Health Administration (OSHA)
   b. The Chemical Transportation Emergency Center
   c. The Transportation Emergency Institute
   d. The Chlorine Institute

3. In regards to safety, wet activated carbon will remove which from the air?

   a. Organic gases and hydrogen sulfide
   b. Oxygen
   c. Carbon dioxide
   d. Carbon monoxide

4. An employee's average airborne exposure in any 8-hour shift in a 40-hour workweek that should not be exceeded is called

   a. Short Term Exposure Limit (STEL).
   b. Time-Weighted Average (TWA).
   c. Threshold Limit Value (TLV).
   d. Recommended Exposure Limits (REL).

5. Sites are required to do a site assessment under the process safety management (PSM) regulations (OSHA) if the facility in a single process has more than how many pounds of chlorine?

   a. 1,000 lb
   b. 1,500 lb
   c. 2,000 lb
   d. 4,000 lb

Sample Questions for Level IV, Answers on Page 172

1. Which factor is used to express the relative volatility of a substance?

   a. Chaitin's constant
   b. Feigenbaum constant
   c. Henry's constant
   d. Boyle's Law

2. Ammonia gas can be fatal to humans at concentrations as low as

   a. 500 ppm.
   b. 900 ppm.
   c. 1,400 ppm.
   d. 2,000 ppm.

3. Which emergency kit is for chlorine tank cars?

   a. Kit A
   b. Kit B
   c. Kit C
   d. Kit D

4. Each year approximately which percentage of confined space deaths are would-be rescuers?

   a. 25%
   b. 50%
   c. 67%
   d. 75%

5. When a water source is destroyed or damaged due to some disaster that causes total contamination, which would be a short-term goal?

   a. Install a dual potable-nonpotable water system
   b. Construct a new raw water source
   c. Clean up the source of contamination caused by the disaster
   d. Draw water from an adjoining system or systems

Evaluate Characteristics of Source Water

	Level I	Level II	Level III	Level IV
Algae control	Comprehension	Comprehension	Comprehension	Application
Bacteriological	Application	Analysis	Analysis	Analysis
Biological	Comprehension	Comprehension	Application	Application
Chemical	Comprehension	Comprehension	Application	Application
Chemical treatment (copper sulfate)	Application	Application	Application	Analysis
Identify and evaluate potential sources of source water contamination	Comprehension	Application	Analysis	Analysis
Monitor, evaluate, and adjust source water	Comprehension	Application	Analysis	Analysis
Physical	Comprehension	Comprehension	Application	Application
Stratification control	Comprehension	Comprehension	Application	Analysis

Required Capabilities

Knowledge of:
- Contaminants
- Hydrology
- Normal characteristics of water
- Watershed protection

Ability to:
- Communicate in writing
- Communicate orally
- Discriminate between normal and abnormal conditions

Evaluate Characteristics of Source Water

Sample Questions for Level I, Answers on Page 172

1. The organisms used to indicate the likelihood that pathogenic bacteria may be present are
   a. *Salmonella* bacteria.
   b. enteric viruses.
   c. coliform bacteria.
   d. *Pseudomonas* group bacteria.

2. Which adverse effects does the secondary contaminant manganese have?

    a. Unappealing to drink, undesirable taste, and possible indication of corrosion
    b. Discolored laundry and changed taste of water, coffee, tea, and other beverages
    c. Undesirable metallic taste and possible indication of corrosion
    d. Added total dissolved solids and scale, indication of sewage contamination and tastes

3. In general, which causes an increase in taste and odors when natural waters lack dissolved oxygen?

    a. Increase in algae growth
    b. pH reduction causing an increase in iron and manganese going into solution
    c. Anaerobic decomposition
    d. Oxidation of organic material

4. A physical property, as opposed to a chemical property, important to water treatment is

    a. turbidity.
    b. pH.
    c. dissolved solids.
    d. electrical conductivity.

5. Which are the two principal chemicals that cause water hardness?

    a. Aluminum and iron
    b. Aluminum and calcium
    c. Iron and manganese
    d. Calcium and magnesium

Sample Questions for Level II, Answers on Page 173

1. Temporary hardness is caused by

    a. magnesium sulfate.
    b. calcium nitrate.
    c. magnesium chloride.
    d. calcium bicarbonate.

2. Which type of algae would most likely be found in nutrient-rich waters?

    a. Blue-green algae
    b. Green algae
    c. Diatoms
    d. Brown algae

3. The quantity of dissolved oxygen in water is a function of

    a. pH, alkalinity, temperature, and total dissolved solids.
    b. temperature and alkalinity.
    c. pH and temperature.
    d. temperature, pressure, and salinity.

4. Most of the carbon dioxide in groundwater originates from

   a. weathering of rocks.
   b. biological oxidation of organic matter.
   c. animal respiration.
   d. plant respiration.

5. At which temperature does the maximum density of water occur?

   a. −0.4°C
   b. 0.0°C
   c. 0.4°C
   d. 4.0°C

Sample Questions for Level III, Answers on Page 173

1. When water freezes it expands by _____ of its original volume.

   a. one-twentieth
   b. one-fortieth
   c. one-ninth
   d. one-seventh

2. Which type of polar bonds are between the hydrogen atoms and the oxygen atom in water?

   a. Ionic
   b. Covalent
   c. van der Waals
   d. Hydrogen bonding

3. Which would be the most probable solution to control algae in the source water if the algae were clogging the filters at the water plant?

   a. Use activated carbon
   b. Decrease oxygen levels
   c. Backwash filters more frequently
   d. Control nutrients

4. Total dissolved minerals in rainwater are typically close to

   a. 0.05 mg/L.
   b. 1.00 mg/L.
   c. 10.00 mg/L.
   d. 40.00 mg/L.

5. Because manganese is only slightly soluble, it is rarely found in surface waters above a concentration of

   a. 1.0 mg/L.
   b. 2.0 mg/L.
   c. 2.5 mg/L.
   d. 3.0 mg/L.

Sample Questions for Level IV, Answers on Page 174

1. Where does the microbial growth that affects groundwater sources and thus the treatment for removing iron and manganese occur?

    a. Anaerobic zone
    b. Aerobic zone
    c. Reduction zone
    d. Where aerobic and anaerobic conditions meet

2. Which conditions will +Eh values of water support?

    a. Anaerobic activities
    b. Aerobic activities
    c. Low pH values
    d. Organic acid production

3. Radon is a radioactive decay daughter of

    a. uranium 235.
    b. uranium 238.
    c. radium 226.
    d. radium 228.

4. Which algal genera are most likely to cause taste-and-odor problems when abundant in the water?

    a. *Volvox*
    b. *Chlorella*
    c. *Oocystis*
    d. *Spirogyra*

5. Shellfish grown in polluted waters have been known to contain the organism that causes

    a. dysentery.
    b. typhoid.
    c. infectious hepatitis.
    d. gastroenteritis.

Additional Water Treatment Operator Practice Questions

Monitor, Evaluate, and Adjust Treatment Processes

Chemical Addition

Answers on Page 175

1. Which is the principal scale-forming substance in water?

    a. Zinc orthophosphate
    b. Sodium carbonate
    c. Calcium
    d. Calcium carbonate

2. Which oxidant should be fed as early as possible in the treatment process to allow for complete reaction before it enters the distribution system?

    a. Ozone
    b. Chlorine dioxide
    c. Chloramines
    d. Potassium permanganate

3. The flexible copper tubing that connects a 150-lb chlorine cylinder to a manifold should be rated at

    a. 100 psi.
    b. 150 psi.
    c. 250 psi.
    d. 500 psi.

4. The buildup of corrosion products is a process known as

    a. deposition.
    b. electrochemical deposition.
    c. physiochemical deposition.
    d. tuberculation.

5. Potassium permanganate

    a. usually has a dosage range from 10 to 15 mg/L.
    b. oxidizes to manganese dioxide, which can give the water a purple color.
    c. should be fed before chlorine.
    d. should be used with powdered activated carbon (PAC).

6. Low values for which water characteristic may require the addition of lime, caustic soda, or sodium bicarbonate?

    a. Turbidity
    b. Water temperature
    c. pH
    d. Alkalinity

7. Which chemical could cause pink water problems?

    a. Chloramines
    b. Potassium permanganate
    c. Bromine
    d. Iodine

8. The most frequent in situ well treatment is the application of a(n)

    a. chemical to adjust pH of well water that is either too acidic or too alkaline.
    b. phosphate inhibitor to prevent well pump corrosion.
    c. oxidant to reduce iron bacteria.
    d. coagulant for high turbidity well water.

9. Which chemical is slaked lime?

    a. $Ca(OH)_2$
    b. $Ca(HCO_3)_2$
    c. $CaCO_3$
    d. $Ca(H_2O)_2$

10. Calcium sulfate ($CaSO_4$) is also known as

    a. gypsum.
    b. limestone.
    c. dolomite.
    d. talc.

11. Which color does the oxidant potassium permanganate turn at its end point?

    a. Reddish-brown
    b. Red
    c. Pink
    d. Yellow

12. The simplest chlorine gas feeder for controlling chlorine gas flow is

    a. constant differential pressure.
    b. variable differential pressure.
    c. sonic flow.
    d. induction mixer.

13. Which is usually used to disinfect the gravel packing of a groundwater well?

    a. Chlorine dioxide
    b. Powder of calcium hypochlorite tablet
    c. Sodium hypochlorite
    d. Chloramines

14. Which is true regarding chlorine one-ton containers?

    a. Full one-ton containers weigh at least 4,500 pounds
    b. Full one-ton containers weigh about 2,750 pounds
    c. The chlorine containers have three fusible plugs on the valve end and two on the opposite end
    d. One-ton containers are equipped with two valves, one for liquid and one for gas

15. Teeth are likely to become pitted when the fluoride concentration in drinking water goes above which amount?

    a. 3 mg/L
    b. 4 mg/L
    c. 5 mg/L
    d. 6 mg/L

16. How many pounds of ammonia are contained in a full 1-ton cylinder?

    a. 800
    b. 1,000
    c. 1,500
    d. 2,000

17. High pH favors the formation of

    a. haloacetic acids.
    b. total trihalomethanes.
    c. both haloacetic acids and total trihalomethanes.
    d. neither haloacetic acids nor total trihalomethanes.

18. How often is it recommended to check the fluoride feed system?

    a. At least once every 24 hours
    b. Every 12 hours
    c. Every 8 hours
    d. Every hour

19. Low pH favors the formation of

    a. haloacetic acids.
    b. total trihalomethanes.
    c. both haloacetic acids and total trihalomethanes.
    d. neither haloacetic acids or total trihalomethanes.

20. Dental fluorosis is

    a. dental cavities.
    b. mottling of the teeth.
    c. fluoride dental rinse.
    d. coating of fluoride on the enamel of teeth.

21. One volume of liquid chlorine will expand to approximately how many volumes of gas?

    a. 440
    b. 460
    c. 480
    d. 484

22. In general, how long does ozone last when it is dissolved in water for disinfection purposes?

    a. A few seconds to a few minutes
    b. 5 to 10 minutes
    c. 10 to 15 minutes
    d. 15 to 30 minutes

23. Sodium fluoride

    a. was once called silly acid.
    b. is a straw yellow color.
    c. has a solution pH of about 1.0 pH units.
    d. is odorless.

24. Which does the saturation point of calcium carbonate primarily depend upon?

    a. pH
    b. Temperature
    c. Concentration of calcium
    d. TDS

25. Which will increase pH, hardness, and alkalinity?

    a. Sodium hydroxide
    b. Zinc orthophosphate
    c. Lime
    d. Sodium zinc phosphate

26. How often should the fluoride concentration of treated water be measured?

    a. Every 8 hours
    b. Every 12 hours
    c. Every day
    d. Continuously

27. Which is the pH of a saturated solution of sodium fluoride?

    a. 6.2
    b. 7.0
    c. 7.8
    d. 8.0 to 8.4

28. Which is the most effective wavelength when using ultraviolet light for disinfection?

    a. 2,235 angstroms
    b. 2,650 angstroms
    c. 2,875 angstroms
    d. 2,965 angstroms

29. Which type of chlorine gas feeder is most commonly used?

    a. Pressure
    b. Combination water and pressure
    c. Vacuum
    d. Combination pressure and vacuum

30. Which substance is a depolarizer?

    a. Carbonates
    b. Nitrates
    c. Polyphosphates
    d. Silicates

31. Which composes the majority of scale in pipe?

    a. $Ca(HCO_3)_2$
    b. $CaCO_3$
    c. $MgCO_3$
    d. $CaSO_4$

32. The specific gravity standard for gases is

    a. water vapor at 100°C.
    b. air.
    c. oxygen.
    d. nitrogen.

33. Which chemical oxidant is most effective in reducing the concentration of taste-and-odor compounds caused by actinomycetes species?

    a. Chloramines
    b. Peroxone
    c. Chloride dioxide
    d. Potassium permanganate

34. Which oxidant is a bluish toxic gas with a pungent odor?

    a. Bromine
    b. Iodine
    c. Ozone
    d. Potassium permanganate

35. Past the breakpoint chlorination point, which percentage of the total chlorine residual should be free chlorine?

    a. 75 to 85%
    b. 85 to 90%
    c. 90 to 95%
    d. 100%

36. Sodium hypochlorite has a pH range of

    a. 6.0 to 7.5 pH units.
    b. 7.5 to 8.5 pH units.
    c. 8.5 to 9.5 pH units.
    d. 9.0 to 11.0 pH units.

37. Where should the fluoride injection point be located?

    a. Right after flocculation
    b. After sedimentation, but before lime softening
    c. Before filtration
    d. After water has received complete treatment

38. The corrosion process can be accelerated by certain bacterial organisms because they produce which chemical?

    a. N_2
    b. CO_2
    c. $MgCO_3$
    d. $CaCO_3$

39. Which chemical would be best to use for corrosion inhibition if the water is very low in alkalinity and calcium concentration?

    a. Sodium carbonate
    b. Sodium bicarbonate
    c. Sodium hydroxide
    d. Polyphosphates

40. If the temperature drops below _____ a 50% solution of caustic soda will begin to crystallize.

    a. 44°F
    b. 48°F
    c. 54°F
    d. 58°F

41. When chlorine has destroyed all reducing compounds, any chlorine remaining will react with

    a. nitrite and form chloramines.
    b. nitrates and form chloramines.
    c. ammonia and form chloramines.
    d. organics and form aromatics.

42. Low values for which water characteristic usually cause poor coagulation, flocculation, and settling characteristics?

    a. Water temperature
    b. Turbidity
    c. Alkalinity
    d. pH

43. Polyphosphates would most likely give the best results at

    a. high alkalinity levels.
    b. high water temperatures.
    c. low pH values.
    d. high pH values.

44. Which is the most effective disinfectant for inactivating bacteria?

    a. Hypochlorite ion
    b. Hypochlorous acid
    c. Chloramines
    d. Bromine

45. Which chemical oxidant is the weakest?

    a. Potassium permanganate
    b. Ozone
    c. Chloramines
    d. Chlorine dioxide

46. Which is the density of liquid chlorine?

    a. 1.25
    b. 1.50
    c. 2.50
    d. 3.50

47. Which device(s) uniformly disperse(s) the chlorine solution into the main flow of water?

    a. Injectors
    b. Pressure regulating valve
    c. Diffusers
    d. Effluent nozzles

48. The pressure in a chlorine cylinder depends on the

    a. amount of chlorine in the cylinder.
    b. temperature of the chlorine liquid.
    c. vacuum placed on the regulator.
    d. amount of gas being withdrawn.

49. Sodium bicarbonate needs to be stored in a cool, dry place because it decomposes rapidly as the temperature nears

    a. 80°F.
    b. 100°F.
    c. 115°F.
    d. 125°F.

50. In the breakpoint chlorination curve, in which zone do inorganic chlorine demand-causing compounds consume free chlorine?

    a. Zone 1
    b. Zone 2
    c. Zone 3
    d. Zone 4

51. Which is the optimal pH level for sulfate-forming reactions with chlorine?

    a. 6
    b. 8
    c. 9
    d. 11

52. Liquid chlorine is which color?

    a. Yellow
    b. Dark green
    c. Amber
    d. Light brown

53. When chlorine gas is under a vacuum reliquefaction will not occur until the temperature drops below

    a. −20°F.
    b. −30°F.
    c. −38°F.
    d. −48°F.

54. Which method can be used to control scale and corrosion?

    a. Softening
    b. pH and alkalinity adjustment with lime
    c. Sequestering
    d. Chelation

55. Which is the chemical formula for hydrated lime?

   a. CaO
   b. $CaCO_3$
   c. $Ca(OH)_2$
   d. $CaMg(CO_3)_2$

56. Once iron bacteria become established in a well, shock chlorination will be required. The shock treatment is required

   a. at increasingly shorter intervals.
   b. at increasingly longer intervals.
   c. at very regular intervals.
   d. sporadically, depending on many physical water quality factors and specifically on pumping rates.

57. Which should accompany shock treatment to maximize its effectiveness for a well that has been fouled by iron bacteria?

   a. Constant addition of more chlorine as it is consumed, as determined by chlorine wet tests
   b. Air surging, valved surge blocks, or jetting
   c. Slow addition of surfactants
   d. 1 to 5% acid solution added with the chlorine

58. Which should the concentration of chlorine be just before the pump is started (for surging and moving the disinfectant into the aquifer and well) in order to sufficiently disinfect a public water supply well?

   a. 10 mg/L
   b. 25 mg/L
   c. 50 mg/L
   d. 100 mg/L

59. Which compound will decrease in solubility as the temperature rises?

   a. Slaked lime
   b. Calcium carbonate
   c. Magnesium hydroxide
   d. Magnesium oxalate

60. Which is the most probable solution if nitrifying bacteria are increasing the demand for the disinfectant being used?

   a. Optimize nutrient control at the water plant
   b. Superchlorinate reservoirs and storage tanks
   c. Increase detention time in reservoirs and the distribution system
   d. Use powdered activated carbon in the filters

61. When air is used to generate ozone, an air dryer should be used to reduce the production of

   a. nitrous oxides.
   b. hydrogen gas (H_2).
   c. water.
   d. ammonia (NH_3).

ADDITIONAL WATER TREATMENT OPERATOR PRACTICE QUESTIONS   59

62. Which shows the correct relative strengths of chlorine species as disinfectant?

    a. $HOCl > OCl > NH_2Cl > HCl$
    b. $OCl > HOCl > NH_2Cl > HCl$
    c. $NH_2Cl > HCl > HOCl > OCl$
    d. $HCl > HOCl > OCl > NH_2Cl$

63. When ultraviolet radiation is used to inactivate microorganisms, which are the optimum wavelengths?

    a. 100 to 200 nanometers (nm)
    b. 250 to 265 nm
    c. 330 to 360 nm
    d. 400 to 500 nm

64. Regarding the breakpoint chlorination curve, the chlorine dosage must go beyond which point in the curve to eliminate tastes and odors?

    a. Point 2
    b. Point 3
    c. Point 4
    d. Point 5

65. Which chemical will decrease alkalinity?

    a. Silicates
    b. Carbon dioxide
    c. Sodium hydroxide
    d. Sodium bicarbonate

66. Which oxidant would cause the most corrosion and scaling problems?

    a. Ozone
    b. Chlorine dioxide
    c. Chloramines
    d. Oxygen

67. Which metal, if in contact with cast iron, would become the most active (anode)?

    a. Aluminum
    b. Zinc
    c. Cadmium
    d. Mild steel

68. Low alkalinity water becomes quite corrosive when it is heavily saturated with

    a. lime.
    b. soda ash.
    c. oxygen.
    d. sodium bicarbonate.

69. Immediately before breakpoint chlorination occurs, which chemical species are destroyed?

    a. Aromatic hydrocarbons
    b. Aromatic hydrocarbons and aliphatic hydrocarbons
    c. Ammonia and chloroorganics
    d. Chloroorganics and chloramines

70. Which is the most effective disinfectant for inactivating viruses?

    a. Hypochlorite ion
    b. Diatomic iodine
    c. Chloramines
    d. Chlorine dioxide

71. Which is the most effective disinfectant for inactivating protozoan cysts?

    a. Chloramines
    b. Hypochlorous acid
    c. Ozone
    d. Bromine

72. Which chemical oxidant would be most effective in removing color and for controlling tastes and odors?

    a. Ozone
    b. Potassium permanganate
    c. Chloramines
    d. Chlorine

73. Which chemical oxidant is the weakest?

    a. Oxygen
    b. Ozone
    c. Potassium permanganate
    d. Chlorine dioxide

74. Why can hypochlorous acid (HOCl) penetrate bacterial surfaces better than hypochlorite ion (OCl$^-$)?

    a. Because it is a proton donor
    b. Because it is an acid
    c. Because it is a smaller molecule than OCl$^-$ despite having a hydrogen atom; the hydrogen atom helps compact the molecule
    d. Because it is neutral

75. Which chlorine compound will produce the least detectable chloronous taste and odor?

    a. Hypochlorous acid
    b. Monochloramine
    c. Dichloramine
    d. Nitrogen trichloride

76. Which type of pumps would be the most economical to use if large volumes of sodium hypochlorite were required?

    a. Chemical feed pumps using piston rods
    b. Chemical feed pumps using plunger drives
    c. Centrifugal pumps
    d. Solenoid-operated pumps

77. A disadvantage to using chloramines is

    a. their vulnerability to nitrification.
    b. that they cannot penetrate biofilms in the distribution system.
    c. their propensity to form trihalomethanes.
    d. that their residual persistence in the distribution system is low.

ADDITIONAL WATER TREATMENT OPERATOR PRACTICE QUESTIONS 61

78. Where is the best place to add corrosive inhibitors such that disinfection can take place under more advantageous conditions at a conventional water treatment plant?

   a. Rapid mix basin
   b. Pre-filter
   c. Pre-clearwell
   d. Post clearwell

79. Which method can be used to control scaling but is never used to control corrosion?

   a. Chelation
   b. Controlled $CaCO_3$ scaling
   c. Polyphosphate addition
   d. pH and alkalinity adjustment with lime

80. How much more soluble is ozone in water than oxygen?

   a. 7.6 times
   b. About 10.0 times
   c. 12.0 times
   d. 20.0 times

81. Which is the most effective ratio of hydrogen peroxide to ozone?

   a. 1:2
   b. 2:5
   c. 1:3
   d. 1:4

82. Which is the chlorine demand of nitrite?

   a. 2.0 parts chlorine to 1 part nitrite
   b. 2.5 parts chlorine to 1 part nitrite
   c. 4.0 parts chlorine to 1 part nitrite
   d. 5.0 parts chlorine to 1 part nitrite

83. A sodium hypochlorite solution at room temperature will lose which percentage of its available chlorine content per month?

   a. 1 to 2%
   b. 2 to 4%
   c. 4 to 6%
   d. 6 to 8%

84. When supplied with air, ozone generators will produce about

   a. 2% ozone.
   b. 5% ozone.
   c. 7% ozone.
   d. 12% ozone.

85. Which method for chlorine dioxide generation is most likely to have substantial amounts of chlorine in solution?

   a. Add chlorine to water, then this water to sodium citrate and sodium chlorate
   b. Add chlorine to water, then this water to hydrochloric acid and sodium chlorate
   c. Inject chlorine gas under vacuum into a stream of chlorite solution
   d. Add hydrochloric acid to a chlorite solution

86. Which chemical used for stabilization of potable water is a bacterial nutrient?

    a. Polyphosphate
    b. Lime
    c. Soda ash
    d. Sodium bicarbonate

87. Once formed, hypochlorous acid instantaneously establishes equilibrium as follows:

    $$HOCl \leftrightarrow H^+ + OCl^-$$
    hypochlorous acid, hypoclorite ion

    This equilibrium reaction at 20°C has which characteristic?

    a. It contains 50% of HOCl and OCl⁻ at a pH of 7.3
    b. The reaction depends on concentration
    c. It responds slowly to pH changes
    d. It is completely reversible

88. Ultraviolet light used in disinfection is generated by mercury or

    a. germanium.
    b. antimony.
    c. cesium.
    d. cadmium.

89. The production of earthy–musty compounds, geosmin, and 2-methylisoborneol (MIB) by cyanobacteria are difficult to remove except when using

    a. lime.
    b. ozone.
    c. aeration.
    d. filtration.

90. Which chemical oxidant would be most effective for removing synthetic organics?

    a. Oxygen
    b. Chlorine
    c. Chloramines
    d. Chlorine dioxide

91. The advantage to using the oxidant oxygen is that it

    a. is non-corrosive.
    b. has no by-products.
    c. causes small amount of scaling problems.
    d. is a strong oxidant.

92. Which method for chlorine dioxide generation is best for large-scale application?

    a. Add chlorine to water, then this water to sodium citrate and sodium chlorate
    b. Add chlorine to water, then this water to hydrochloric acid and sodium chlorate
    c. Inject chlorine gas under vacuum into a stream of chlorite solution
    d. Add hydrochloric acid to a chlorite solution

93. Which is the commercial purity of fluorosilicic acid?

    a. 20 to 30%
    b. 20 to 40%
    c. 30 to 40%
    d. 35 to 48%

94. Which type of fluoride chemical should be used in down flow saturators?

    a. Tablets of sodium fluorosilicate
    b. Granular sodium silicofluoride
    c. Crystalline sodium fluoride
    d. Powdered sodium fluoride

95. The decomposition products of sodium hypochlorite into chlorate and O_2 would be caused by which factor?

    a. Temperature
    b. Metallic impurities
    c. UV light
    d. pH

96. Which are the only two metals that can contact sodium hypochlorite; all other metals must be eliminated from storage vessels, piping, valves, and feed equipment?

    a. Titanium and tantalum
    b. Titanium and germanium
    c. Vanadium and germanium
    d. Vanadium and selenium

97. Sites that require chlorine vaporizers typically have feed rates that exceed

    a. 450 lb/d.
    b. 800 lb/d.
    c. 1,000 lb/d.
    d. 2,000 lb/d.

98. The deterioration of elastomeres by chloramines is enhanced by

    a. higher pH.
    b. lower pH.
    c. higher temperatures.
    d. lower temperatures.

99. Most water treatment systems can limit nitrification by

    a. maintaining a pH of 7.0 to 7.5.
    b. maintaining a chlorine to ammonia-nitrogen ratio between 4.5:1 and 5.0:1.
    c. limiting excess free ammonia to below 0.5 mg/L nitrogen.
    d. maintaining temperature between 18 and 28°C.

100. If the percentage of chlorine delivered to a water treatment plant in liquid form is 12.5% chlorine, which chemical is it most likely to be?

    a. Calcium hypochlorite
    b. Sodium hypochlorite
    c. Chlorine
    d. Chlorine dioxide

Coagulation and Flocculation

Answers on page 183

1. How many steps are in the coagulation process?

    a. 3
    b. 4
    c. 5
    d. 6

2. Polymers used as flocculants are

    a. water-soluble inorganic electrolytes.
    b. water-insoluble organic polyelectrolytes.
    c. water-insoluble inorganic electrolytes.
    d. water-soluble organic polyelectrolytes.

3. Which is the zeta potential?

    a. Electrical charge on a suspended particle
    b. Electrical resistance of a suspended particle
    c. Electrical potential of a suspended particle
    d. Electrical charge of a suspended particle as it relates electrochemically to the coagulate being used

4. All colloids are

    a. sols.
    b. hydrophobic.
    c. electrically charged.
    d. emulsoids.

5. Colloidal particle movement and aggregation caused by thermal energy is called

    a. orthokinetic flocculation.
    b. parakinetic dynamism.
    c. Brownian movement.
    d. thermodynamic flocculation.

6. When used with alum, which chemical improves coagulation?

    a. Ferric chloride
    b. Ferric sulfate
    c. Sodium aluminate
    d. Aluminum sulfate

7. The coagulant alum reacts with natural alkalinity in the water forming

    a. aluminum hydroxide.
    b. aluminum carbonate.
    c. aluminum bicarbonate.
    d. aluminum oxide.

8. When ferric coagulants react with natural alkalinity in the water they form

    a. ferric hydroxide.
    b. ferric carbonate.
    c. ferric bicarbonate.
    d. ferric oxide.

9. Detention time in mechanical mixers usually ranges from
    a. 5 to 15 seconds.
    b. 15 to 45 seconds.
    c. 1 to 2 minutes.
    d. 2 to 4 minutes.

10. The aluminum ions in alum neutralize the negatively charged particles in the raw water in about
    a. 1 to 2 seconds.
    b. 3 to 5 seconds.
    c. 5 to 8 seconds.
    d. 5 to 12 seconds.

11. Suspended and colloidal solids found in natural waters are usually
    a. positively charged.
    b. negatively charged.
    c. neutral.
    d. polyionic.

12. After the oxidation treatment, H_2S in the water turns a milky-blue color because of elemental sulfur (solid sulfur). Which is the best procedure to remove this sulfur?
    a. Adsorption
    b. Sedimentation
    c. Coagulation and filtration
    d. Lime-soda ash softening

13. Particles will stay in suspension as long as the
    a. ionic forces are stronger than the van der Waals forces.
    b. zeta potential is stronger than the van der Waals forces.
    c. van der Waals forces are stronger than the ionic forces.
    d. van der Waals forces are stronger than the zeta potential.

14. The most probable effective ratio of the coagulants aluminum sulfate to hydrated lime is
    a. 1:1 aluminum sulfate to hydrated lime.
    b. 3:1 aluminum sulfate to hydrated lime.
    c. 5:1 aluminum sulfate to hydrated lime.
    d. 6:1 aluminum sulfate to hydrated lime.

15. Which should the "tip speed" of the flocculator be in water, assuming the water is not extremely cold?
    a. 0.2 ft/sec
    b. 0.5 ft/sec
    c. 1.0 ft/sec
    d. 2.0 ft/sec

16. Coagulants consisting of trivalent ions are how much more effective than bivalent ions?
    a. 50 to 60 times
    b. 100 to 120 times
    c. 200 to 250 times
    d. 1,000 times

17. When tip velocities of the impellers reach speeds greater than _____ localized shearing of floc occurs.

    a. 6.0 fps
    b. 8.0 fps
    c. 10.0 fps
    d. 10.5 fps

18. In the flash mixing process, which type of rapid mixing system has the most significant head loss?

    a. Mechanical mixers
    b. Static mixers
    c. Pumps and conduits
    d. Baffled chambers

19. When these types of chemicals are used, flash mixing becomes less critical, but thorough mixing (flocculation) remains very important.

    a. Ferric or ferrous chloride
    b. Aluminum sulfates
    c. Lime, soda ash, or caustic soda
    d. Polymers

20. Weighing agents are used primarily to treat water that is

    a. low in color and mineral content and high in turbidity.
    b. high in color and mineral content and low in turbidity.
    c. low in turbidity and mineral content and high in color.
    d. high in turbidity and mineral content and low in color.

21. The flash mixing process that has the best control features are

    a. mechanical mixers.
    b. static mixers.
    c. pumps and conduits.
    d. baffled chambers.

22. The zeta potential on a particular sample of water is +2. The degree of coagulation is best described as

    a. poor.
    b. fair.
    c. excellent.
    d. maximum.

23. The most effective ratio of the coagulant ferric sulfate to the oxidant chlorine is

    a. 2:1 ferric sulfate to chlorine.
    b. 4:1 ferric sulfate to chlorine.
    c. 5:1 ferric sulfate to chlorine.
    d. 8:1 ferric sulfate to chlorine.

24. Mixing of a coagulant is least influenced by flow when using

    a. a static mixer.
    b. a baffled chamber.
    c. a pump (adding coagulant on suction side).
    d. pipe grids that have been perforated.

25. When alum or ferric coagulants are used to treat water, the reactions are governed by all the following. Which one is critical to floc formation?

    a. Initial application at point of highest mixing intensity
    b. Temperature and ratio of turbidity to coagulant
    c. Alkalinity
    d. Amount of turbidity

Clarification and Sedimentation

Answers on page 185

1. Sedimentation depends on

    a. size of flocculated particles.
    b. gravity.
    c. the depth of the basin.
    d. the type of coagulant used.

2. Most clarifiers are designed to have a theoretical detention time of about _____ when the plant is at its maximum flow.

    a. 1 hour
    b. 2 hours
    c. 3 hours
    d. 4 hours

3. Which percentage of source water turbidity do sedimentation basins typically remove?

    a. 80% or more
    b. 90% or more
    c. 95% or more
    d. 96% or more

4. How many zones are defined in all sedimentation basins?

    a. 3
    b. 4
    c. 5
    d. 6

5. Which sedimentation zone is particularly affected by the other zones?

    a. Influent zone
    b. Settling zone
    c. Effluent zone
    d. Sludge zone

6. The material that accumulates on the surface of a dissolved-air flotation unit is called

    a. scum.
    b. sludge.
    c. float.
    d. floc cake.

7. At what rate are overflow rates from outlet weirs of conventional basins commonly designed not to exceed?

    a. 10,000 gpd/ft
    b. 20,000 gpd/ft
    c. 25,000 gpd/ft
    d. 30,000 gpd/ft

8. Name the two major areas in solids-contact clarifiers.

    a. Reaction and settling zones
    b. Reaction and mixing zones
    c. Mixing and settling zones
    d. Sludge and reaction zones

9. Lamella plates are usually inclined

    a. 20.0 to 30.0 degrees.
    b. 33.7 degrees.
    c. 55.0 degrees.
    d. 67.7 degrees.

10. Which high rate process has the objective of producing pin-sized floc?

    a. Tube settlers
    b. Superpulsators
    c. Actiflo process
    d. Dissolved air flotation

11. Slime growth in uncovered sedimentation basins can be eliminated by coating the walls with

    a. copper sulfate and lime.
    b. copper sulfate.
    c. lime and potassium permanganate.
    d. copper sulfate and potassium permanganate.

12. Which is the most probable result of high winds on a sedimentation basin?

    a. Sheared floc
    b. Short-circuiting
    c. Density currents
    d. Floc will be partially resuspended and it will be more difficult to settle the floc

13. How often is the vent valve opened to release the vacuum that produces pulsating energy in a pulsator clarifier?

    a. Every 5 to 10 seconds
    b. Every 40 to 50 seconds
    c. Every 2 to 3 minutes
    d. Every 4 to 6 minutes

22. Which is the angle of inclination from the horizontal for the plates in a superpulsator clarifier?

    a. 30°
    b. 33°
    c. 45°
    d. 60°

23. Which high rate process has the highest surface overflow rate?

    a. Tube settlers
    b. Superpulsators
    c. Actiflo process
    d. Dissolved air flotation

24. How high can surface overflow rates for the Actiflo process reach?

    a. 8 gpm/ft^2
    b. 12 gpm/ft^2
    c. 16 gpm/ft^2
    d. 20 gpm/ft^2

25. Which is the best process for the removal of radium (226 and 228)?

    a. Anion exchange
    b. Coagulation, sedimentation, and filtration
    c. Aeration and stripping
    d. Chemical oxidation

Filtration

Answers on page 186

1. Manganese greensand filters can be regenerated by using

    a. a surface wash and an air-water backwash.
    b. brine water during backwashing.
    c. potassium permanganate solution during backwashing.
    d. first a brine solution during the first backwashing cycle followed by potassium permanganate solution for the second backwash cycle.

2. Which material is manganese greensand and which is the coating?

    a. Quartz sand coated with manganese hydroxide [Mn(OH)$_2$]
    b. Garnet sand coated with manganese dioxide [MnO$_2$]
    c. Ilmenite sand coated with manganese hydroxide
    d. Glauconite sand coated with manganese dioxide

3. Which is the layer of solids and biological growth that forms on the top of a slow sand filter?

    a. Biosolids film
    b. Bio-carbonated scale layer
    c. Schmutzdecke
    d. Saprophytic layer

14. Typically, the time between backwashing of Trident contact adsorption clarifiers averages

    a. 4 to 8 hours.
    b. 16 to 24 hours.
    c. 1 to 2 days.
    d. 3 to 5 days.

15. Which will result if the influent water to a sedimentation basin contains more suspended solids than the water in the basin?

    a. Too much sludge will settle at the inlet of the basin
    b. Poor floc formation
    c. Septic conditions will began to occur
    d. Density current

16. Horizontal tube settlers

    a. are actually inclined about 5 degrees.
    b. work independently with the downstream filter(s).
    c. require jet sprayers to clean the collected sludge out of the tubes.
    d. have high maintenance requirements due to frequent cleaning.

17. Which high rate process uses microsand?

    a. Tube settlers
    b. Superpulsators
    c. Actiflo process
    d. Dissolved air flotation

18. Which high rate process applies a vacuum?

    a. Tube settlers
    b. Superpulsators
    c. Actiflo process
    d. Dissolved air flotation

19. Dissolved air flotation has a _____ removal of *Cryptosporidium* oocysts.

    a. 0.5-log
    b. 1.0-log
    c. 1.5-log
    d. 2.0-log

20. In dissolved-air flotation units, mechanical rubber scrappers travel over the tank surface and push the float over a ramp called a

    a. trap.
    b. beach.
    c. launder.
    d. loading plate.

21. Which is the most effective method of reducing nematode concentrations in surface water treatment plants?

    a. Coagulation
    b. Free chlorine
    c. Rapid sand filters
    d. Settling

4. Which filter type utilizes pressure for filtration?

   a. Diatomaceous earth filter
   b. High-rate filter
   c. Rapid sand filter
   d. Deep-bed monomedium filter

5. Slow sand filter runs can be as long as

   a. 3 months.
   b. 4 months.
   c. 6 months.
   d. 1 year.

6. Gravity rapid sand filters can have head losses as high as

   a. 6 ft.
   b. 8 ft.
   c. 10 ft.
   d. 12 ft.

7. Which group of algae is usually associated with "blinding filters"?

   a. Green algae
   b. Blue-green algae
   c. Pigmented flagellates
   d. Diatoms

8. Depending on water temperature, which is the typical backwashing flow rate for a manganese greensand filter bed?

   a. 7 to 8 gpm/ft^2
   b. 8 to 10 gpm/ft^2
   c. 10 to 12 gpm/ft^2
   d. 12 to 14 gpm/ft^2

9. The typical filter run time for a high-rate filter system is

   a. 8 hours.
   b. 24 hours.
   c. 48 hours.
   d. 72 hours.

10. Below are four membrane technologies. Which is the correct sequence from larger to smaller pore sizes?

    a. Microfiltration, reverse osmosis, ultrafiltration, and nanofiltration
    b. Microfiltration, ultrafiltration, nanofiltration, and reverse osmosis
    c. Reverse osmosis, ultrafiltration, microfiltration, and nanofiltration
    d. Ultrafiltration, microfiltration, reverse osmosis, and nanofiltration

11. Optimum water quality in manganese greensand filters can be attained by backwashing and regenerating the greensand before the effluent manganese level goes above

    a. 0.02 mg/L.
    b. 0.05 mg/L.
    c. 0.10 mg/L.
    d. 0.12 mg/L.

12. Granular media filters should remove suspended solids including iron and manganese greater than

    a. 10 µm.
    b. 15 µm.
    c. 20 µm.
    d. 25 µm.

13. Which filter media material is given an abrasive number?

    a. Garnet
    b. Activated carbon
    c. Sand
    d. Greensand

14. Which is the typical filtration rate for high-rate filters?

    a. 0.5 to 2.0 gpm/ft^2
    b. 3.0 to 12.0 gpm/ft^2
    c. 15.0 to 20.0 gpm/ft^2
    d. >25.0 gpm/ft^2

15. After backwashing and stratifying newly installed manganese greensand, approximately which percentage of the greensand should be skimmed off?

    a. 3%
    b. 5%
    c. 8 to 9%
    d. 12 to 15%

16. The media in a slow sand filter is

    a. stratified fine.
    b. stratified medium.
    c. stratified coarse.
    d. unstratified.

17. Clumping of filter media or ion exchange resin is caused by

    a. electrostatic charges.
    b. media or resin breakdown, respectively.
    c. temperature and pH of the clay in the water.
    d. iron and manganese oxides reacting with the coagulant.

18. Backwash water is which percentage of the totaled filtered water in a conventional rapid sand filter system?

    a. 1%
    b. 2 to 4%
    c. 5 to 6 %
    d. 6 to 10%

19. Why does anthracite always stay on top of sand during backwashing?

    a. Because the grain sizes are smaller
    b. Because it is lighter and less dense
    c. Because of its shape
    d. Because of its porosity

ADDITIONAL WATER TREATMENT OPERATOR PRACTICE QUESTIONS 73

20. The length of filter runs for manganese greensand filters can be increased by
    a. adding a high molecular weight polymer filter aid.
    b. keeping the pH above 9.0 and lowering it after filtration.
    c. keeping the pH below 7.3 and raising it after filtration.
    d. adding a layer of anthracite above the greensand.

21. Which is the most common type of valves used on filters?
    a. Gate
    b. Globe
    c. Butterfly
    d. Ball

22. Manganese greensand removes manganese and iron by a combination of
    a. adsorption and oxidation.
    b. adsorption and absorption.
    c. adsorption and precipitate screening.
    d. absorption and precipitate screening.

23. Which is the best means to remove *Giardia lamblia*?
    a. Coagulation, flocculation, and filtration
    b. Ion exchange
    c. Powdered activated carbon
    d. Activated alumina

24. Which is the best method for removing trihalomethanes precursors?
    a. Reverse osmosis
    b. Powdered activated carbon
    c. Lime softening
    d. Aeration and stripping

25. Reverse osmosis membranes will compact faster
    a. with higher iron content.
    b. with higher temperature.
    c. with higher chlorine.
    d. if the pH is greater than 8.5.

26. Which is the best method for the removal of arsenic (+3)?
    a. Reverse osmosis
    b. Coagulation, sedimentation, and filtration
    c. Cation exchange
    d. Activated alumina

27. Which is the filtration rate for filters with granular activated carbon as a medium?
    a. 0.5 to 1.0 gpm/ft^2
    b. 2.0 gpm/ft^2
    c. 4.0 gpm/ft^2
    d. 5.0 to 6.0 gpm/ft^2

28. When manganese greensand filters are run in the continuous regeneration mode [continuous regeneration of the $MnO_2(s)$ surfaces], the free chlorine residual in the filter effluent should be kept at

    a. 0.50 mg/L.
    b. 1.00 mg/L.
    c. 1.20 mg/L.
    d. 1.75 mg/L.

29. The sorption kinetics and sorption capacity of Mn (II) by $MnOx(s)$-coated filter media increase with

    a. increasing solution pH.
    b. decreasing alkalinity.
    c. increasing alkalinity.
    d. decreasing temperature.

30. Direct filtration is used to treat raw water that has average turbidities

    a. below 10 ntu.
    b. up to 25 ntu.
    c. 40 to 50 ntu.
    d. above 50 ntu.

31. Diatomaceous earth filters

    a. have a relatively high installation cost.
    b. have relatively high operating costs.
    c. are used only for water with low turbidity.
    d. produce very little backwash sludge.

32. Besides regenerating manganese greensand beds, potassium permanganate will also regenerate

    a. anthracite filter beds.
    b. granular activated carbon beds.
    c. pyrolusite filter beds.
    d. polystyrene upflow clarifier media beds.

33. Why is there a limited effect of organic absorption when granular activated carbon (GAC) is used as a filter medium above the sand layer?

    a. GAC has a low porosity
    b. GAC repels much of the organics due to the attached coagulants
    c. Backwashing does not remove the organics absorbed over time nor does it regenerate the GAC
    d. The water being filtered has limited contact time with the GAC

34. An advantage of a variable declining-rate filtration system is that

    a. it does not require a loss-of-head indicator.
    b. it does not require a rate-of-flow controller.
    c. individual filters keep a constant filtration rate.
    d. flow rate changes do not impact the system.

35. Which is the most probable solution if chironomids can be seen in the finished water?

    a. Increase chlorination and backwash
    b. Isolate distribution area, superchlorinate, flush mains and customer lines
    c. Add powdered activated carbon to the filters, then backwash
    d. Use chloramines, or, if already using, temporarily increase by 100%

Residuals Disposal

Answers on page 189

1. Which should be determined first before an in-ground sedimentation tank is drained?

    a. The solids content
    b. The hazardous metals content
    c. Sludge volume and volume of process area to make sure it will be large enough
    d. Water table level

2. Which is the recommended number of solar drying lagoons?

    a. 2
    b. 3
    c. 4
    d. 5

3. Which sludge dewatering process has a low consumption of energy, is easy to operate, has small land requirements, and can produce a dry filter cake of between 35 and 40% solids?

    a. Centrifuge
    b. Vacuum filters
    c. Filter press
    d. Belt filter press

4. Which sludge dewatering process has the advantage of varying the density of the sludge cake from a thickened liquid slurry to a dry cake?

    a. Centrifuge
    b. Vacuum filters
    c. Filter press
    d. Belt filter press

5. Which sludge dewatering process is best for alum sludges (which are difficult to dewater) when the cakes are very dry, filtrate is clear, and solids capture is very high?

    a. Centrifuge
    b. Vacuum filters
    c. Filter press
    d. Belt filter press

6. Which sludge dewatering process requires a precoat of diatomaceous earth and its use has declined due to other newer methods?

   a. Centrifuge
   b. Vacuum filters
   c. Filter press
   d. Belt filter press

Additional Treatment Tasks

Answers on page 190

1. Softener resin is regenerated by passing a sufficient volume and concentration of which through the resin bed?

   a. Caustic soda
   b. Dilute hydrochloric acid
   c. Brine
   d. Sodium sulfate

2. Sequestration is a process in which iron and manganese are

   a. precipitated with a chemical, but not filtered out of the water.
   b. kept in solution by certain chemicals.
   c. oxidized and thus made insoluble.
   d. reduced and thus made insoluble.

3. Which chemical is often used with lime softening?

   a. Ferric chloride
   b. Ferric sulfate
   c. Sodium aluminate
   d. Sodium silicate

4. If sequestration of iron and manganese is used, the chlorine residual in the water system should always be maintained at or above

   a. 0.2 mg/L.
   b. 0.4 mg/L.
   c. 0.5 mg/L.
   d. 1.0 mg/L.

5. Which must be added for the magnesium carbonate hardness in the softening process?

   a. Lime
   b. Two times as much lime
   c. Soda ash
   d. Lime and soda ash

6. Which must be added for magnesium noncarbonated hardness in the softening process?

   a. Lime
   b. 2 times as much lime
   c. Soda ash
   d. Lime and soda ash

7. Once the ion-exchange resin can no longer remove hardness, it is said to be
   a. wasted.
   b. consumed.
   c. expended.
   d. exhausted.

8. Which of the following is the most effective method for the removal of hardness?
   a. Granular activated carbon
   b. Lime softening
   c. Anion exchange
   d. Coagulation, sedimentation, and filtration

9. When should polyphosphates be added to the process to control iron and manganese?
   a. 20 to 30 minutes before chlorine addition
   b. A few minutes before chlorine application
   c. At the same time as chlorine application
   d. 20 to 30 minutes after chlorine addition

10. The major advantage for using activated silica is that it will
    a. reduce sensitivity to pH changes.
    b. inhibit bacterial growth.
    c. increase alkalinity.
    d. strengthen the floc.

11. Which chemical is used to stabilize water after it has been lime-softened?
    a. Sulfuric acid
    b. Carbon dioxide
    c. Sodium hydroxide
    d. Zinc orthophosphate

12. Sodium polyphosphate
    a. will produce a protective coating.
    b. will increase pH and alkalinity.
    c. is a sequestering agent.
    d. is a coagulant aid.

13. Usually, raw water can be softened with lime alone when it contains
    a. mostly calcium hardness.
    b. mostly magnesium hardness.
    c. very little carbonate hardness.
    d. little or no noncarbonated hardness.

14. In the lime-soda ash softening process, the soda ash dosage is based
    a. only on carbonate and magnesium hardness to be removed.
    b. on carbonate hardness and carbon dioxide to be removed.
    c. primarily on noncarbonated and magnesium hardness.
    d. only on the amount of noncarbonated hardness to be removed.

15. Powdered activated carbon is usually added to the treatment process

    a. before the normal coagulation-flocculation step.
    b. before the sedimentation basin.
    c. after the sedimentation basin.
    d. in the filters.

16. Softening resin exchanges what for the calcium and magnesium that it removes from the water?

    a. Sodium
    b. Hydroxide
    c. Potassium
    d. Hydrogen ions (H^+)

17. Sequestration of iron and manganese prevents them from being

    a. oxidized and made insoluble.
    b. reduced and made insoluble.
    c. reduced and made soluble.
    d. utilized by bacteria.

18. Ion exchange processes can typically be used for direct groundwater treatment as long as turbidity and _____ levels are not excessive.

    a. calcium carbonate
    b. iron
    c. carbon dioxide
    d. sodium sulfate

19. Which color is oxidized iron?

    a. Yellow
    b. Brown
    c. Red
    d. Dark brown

20. Recarbonation after lime-soda ash softening is required to remove calcium carbonate that is in suspension to prevent

    a. deposition of the calcium carbonate in the clear well and distribution system.
    b. scale formation in distribution piping.
    c. customer complaints of turbidity and color.
    d. cementation of filter media.

21. Which chemical will cause corrosion if overfeeding occurs?

    a. Sodium hydroxide
    b. Carbon dioxide
    c. Sodium hexametaphosphate
    d. Tetrasodium polyphosphate

22. Which would be the best chemical treatment or method to use to effectively reduce the concentration of taste-and-odor compounds caused by actinomycetes species?

    a. Activated carbon adsorption
    b. Chlorine
    c. Potassium permanganate
    d. Chlorine dioxide

23. The process of decomposing a substance by electricity is called

   a. cathodic corrosion.
   b. galvanic corrosion.
   c. electrolysis.
   d. ionization.

24. Sequestration is effective only for groundwater that has relatively low levels of dissolved iron and manganese and

   a. no dissolved oxygen.
   b. no carbonate hardness.
   c. is slightly acidic.
   d. is slightly basic.

25. Which treatment procedure should be followed after a well has been shock chlorinated?

   a. An alkaline neutralization solution should be added
   b. Oxygen gas should be bubbled throughout the well water
   c. A dissolution-dispersion step using acidification and surfactants
   d. Three well volumes should be purged to flush the well and reduce the chlorine concentration

26. Which type of hardness usually calls for a two-stage softening process?

   a. Calcium hardness above 100 mg/L
   b. Magnesium hardness
   c. Noncarbonated hardness
   d. Carbonate hardness

27. The activated alumina process is sensitive to

   a. humic substances.
   b. pH.
   c. high total dissolved solids.
   d. fluoride.

28. How much will the alkalinity of the treated water increase for each mg/L of lime added?

   a. 0.42 mg/L
   b. 0.64 mg/L
   c. 1.00 mg/L
   d. 1.28 mg/L

29. When should core samples be collected from filters using granular activated carbon?

   a. Every six months
   b. At the time of installation and every six months thereafter
   c. Every year after installation
   d. At the time of installation and every year thereafter

30. Which of the following is the best process to remove lead?

   a. Activated alumina
   b. Anion exchange
   c. Chemical oxidation
   d. Lime softening

31. The ion exchange softening process does not alter which properties of the water?

    a. Chloride and bromide concentrations
    b. pH and alkalinity
    c. Heavy metals concentrations
    d. Nitrogen and phosphorus concentrations

32. A flocculant is required for iron removal if most of the iron passes through a _____ filter membrane.

    a. 5 µm
    b. 10 µm
    c. 15 µm
    d. 20 µm

33. Which would be the most probable solution to control algae in the water plant if the algae were clogging the filters?

    a. Add copper sulfate before the filters
    b. Use activated carbon
    c. Increase backwashing rate and frequency
    d. Add a long chain polymer filter aid

34. Activated silica is prepared by the water treatment operator. Which is the actual chemical that is delivered to the water treatment plant?

    a. Na_2SiO_3
    b. Ca_2SiO_4
    c. Al_2SiO_5
    d. K_2SiO_3

35. Raw water requires both lime and soda ash when there are nominal amounts of which type of hardness in the water?

    a. Magnesium hardness
    b. Iron hardness
    c. Calcium hardness
    d. Manganese hardness

36. The double-stage process in excess-lime softening

    a. causes greater removal of iron.
    b. accomplishes greater removal of magnesium.
    c. eliminates the need for soda ash for removing noncarbonated hardness.
    d. removes the recarbonation requirement.

37. The reaction $CO_2 + Ca(OH)_2 \rightarrow CaCO_3 + H_2O$ is complete when the pH reaches

    a. 7.6.
    b. 8.0.
    c. 8.3.
    d. 8.5.

38. Trihalomethanes can form along with other disinfectant by-products due to lime, excess lime, or lime soda ash softening of water because of the

    a. temperatures produced in the process increasing chemical reactions.
    b. high pH that is required to operate the process.
    c. coagulants used along with the softening process.
    d. gases produced in the process reacting with organics acids in the water.

39. When backwashing an ion exchange unit, how much should the bed expand from water introduced from the bottom of the ion exchange column?

    a. 25 to 35%
    b. 35 to 50%
    c. 50 to 75%
    d. 75 to 85%

40. Which bed expansion during backwashing must be achieved for filters using granular activated carbon?

    a. 20%
    b. 25%
    c. 30%
    d. 50%

41. Which type of aerator causes aeration to occur in splash areas and can be used to oxidize iron and to partially reduce dissolved gases?

    a. Slat-and-coke-tray aerator
    b. Packed tower
    c. Cascade aerator
    d. Cone aerator

42. Which type of aerator is much more efficient in removing dissolved gases and for oxidizing iron and manganese?

    a. Packed tower
    b. Slat-and-coke-tray aerator
    c. Cascade aerator
    d. Draft aerator

43. The lime-soda ash softening process is enhanced best with which other application?

    a. Increase alkalinity with caustic soda
    b. Aeration
    c. Solids-contact unit
    d. Oxidation with chlorine

44. After first stage recarbonation, which is almost completely dissolved?

    a. Calcium crystals
    b. Calcium carbonate crystals
    c. Magnesium crystals
    d. Magnesium carbonate crystals

45. Which are the exchange capacities of natural zeolites?

    a. 1 to 2 kilograins/ft^3
    b. 3 to 5 kilograins/ft^3
    c. 5 to 8 kilograins/ft^3
    d. 9 to 12 kilograins/ft^3

46. Which best describes the purpose of granular activated carbon?

    a. Particulate removal and particulate adsorption
    b. Particulate screening and particulate absorption
    c. Particulate absorption and particulate removal
    d. Particulate adsorption and particulate absorption

47. Which is the best process for removing trihalomethanes?

    a. Activated alumina
    b. Coagulation, sedimentation, and filtration
    c. Anion exchange
    d. Aeration and stripping

48. The most common cation exchange resins are

    a. weak-base resins.
    b. strong-base resins.
    c. weak-acid resins.
    d. strong-acid resins.

49. Sequestration of iron and manganese is usually not recommended for waters where the concentration of iron, manganese, or both exceeds

    a. 1.0 mg/L.
    b. 2.0 mg/L.
    c. 3.0 mg/L.
    d. 5.0 mg/L.

50. Which coagulant aids use extremely small dosages?

    a. Polyelectrolytes
    b. Activated alumina
    c. Weighing agents
    d. Aluminum chloro-hydrate (ACH)

51. The design parameter for predicting granular activated carbon (GAC) performance is called

    a. empty-bed contact time.
    b. GAC saturation point.
    c. GAC potential.
    d. GAC saturation potential.

52. Which is the most effective method for the removal of synthetic organic chemicals?

    a. Lime softening
    b. Granular activated carbon
    c. Aeration and stripping
    d. Chemical oxidation

53. Which is the most effective process for removing radon?

    a. Reverse osmosis
    b. Aeration and stripping
    c. Activated alumina
    d. Coagulation, sedimentation, and filtration

54. Lime softening in the presence of calcium and magnesium will form which magnesium precipitate?

    a. $MgSO_4$
    b. $MgCO_3$
    c. $Mg(OH)_2$
    d. $MgCO_2$

55. The cartridge filters that are sometimes used immediately between reverse osmosis units should be replaced when the difference in pressure between the inlet and outlet gauges reaches

    a. 10.
    b. 15.
    c. 25.
    d. 30.

56. Which is the problem if a sequestering agent does not suppress color and/or turbidity appears?

    a. Chlorine is being fed too far ahead of the sequestering agent
    b. Chlorine is being fed too close to the sequestering agent
    c. Lime or soda ash is being fed too far ahead of the sequestering agent
    d. Lime or soda ash is being fed too close to the sequestering agent

57. The performance of sodium silicate in the sequestration of iron is reduced with

    a. increasing temperatures.
    b. low pH values.
    c. high pH values.
    d. increasing levels of calcium and magnesium.

58. Chose the most effective means of removing toxins produced by cyanobacteria.

    a. Oxidation with potassium permanganate followed by lime-softening
    b. Aeration and oxidation with chlorine
    c. Conventional treatment—coagulation-flocculation, sedimentation, filtration, and disinfection
    d. Activated carbon and ozone

59. Biological treatment of water using microbes has which disadvantage?

    a. Increased chlorine demand
    b. Effluent water conducive to microbial growth in the distribution system
    c. Risk of introduction of pathogenic microorganisms into the finished water if not properly designed
    d. Production of tastes and odors in the finished water if not properly maintained

60. The concentration of solids in the lower level of a sludge blanket in a softening facility's solid contact basin should be

    a. 2 to 7%.
    b. 5 to 15%.
    c. 15 to 20%.
    d. 25 to 30%.

61. Activated silica for anion removal is most efficient when the

    a. temperature is above 12°C.
    b. total dissolved solids are below 300 mg/L.
    c. total dissolved solids are below 500 mg/L.
    d. pH is lower than 8.2.

62. In the ion exchange softening process, how much will the total dissolved solids increase for each 1 mg/L of calcium removed and replaced by sodium?

    a. 0.15 mg/L
    b. 0.28 mg/L
    c. 0.40 mg/L
    d. 0.56 mg/L

63. In ion exchange, a bumping order based on valence exists. Which would be the preferred ion?

    a. Aluminum
    b. Calcium
    c. Sodium
    d. Hydrogen

64. Which is the most probable cause of tastes and odors persisting or getting worse when powdered activated carbon (PAC) is used?

    a. PAC is added at the inlet
    b. PAC is added during flocculation
    c. PAC and chlorine are added too close to each other
    d. PAC is added to water with too high an alkalinity

65. Which is the normal air-to-water ratio at peak flow in a packed tower aerator?

    a. 10:1
    b. 15:1
    c. 25:1
    d. 35:1

66. Precipitation softening water treatment facilities that do not recarbonate need to

    a. monitor filters for carbonate deposition.
    b. add more coagulants.
    c. add more coagulants and disinfectants.
    d. have longer sedimentation detention times.

67. After granular activated carbon (GAC) has all adsorption sites filled, which option(s) should be taken?

    a. Replace or recatalyze the GAC
    b. Recharge the GAC
    c. Reactivate followed by recharging the adsorption sites on the GAC
    d. Replace or reactivate the GAC

68. Which is the most efficient process for removing organic mercury?

    a. Coagulation, sedimentation, and filtration
    b. Lime softening
    c. Activated alumina
    d. Granular activated carbon

69. Which is the most efficient process for the removal of nitrate?

    a. Strong base anion exchange
    b. Granular activated carbon
    c. Chemical oxidation
    d. Coagulation, sedimentation, and filtration

70. When activated carbon is used to remove natural organic matter, it should be used

    a. after filtration.
    b. after settling.
    c. before chlorine addition.
    d. in conjunction with a lime slurry.

71. The cross section of the (repeating) layers in an electrodialysis membrane would be as follows:

    a. Cation permeable membrane, nonionic permeable membrane, cation permeable membrane
    b. Cation permeable membrane, nonionic permeable membrane, anion permeable membrane
    c. Cation permeable membrane, plastic spacer, cation permeable membrane
    d. Cation permeable membrane, plastic spacer, anion permeable membrane

72. An electrodialysis system will operate at temperatures as high as

    a. 78°F.
    b. 88°F.
    c. 100°F.
    d. 110°F.

73. In the operation of an electrodialysis system, the most commonly encountered problem is

    a. leaky membranes.
    b. broken membranes.
    c. scaling or membrane fouling.
    d. the associated pumping equipment.

74. Which catalytic agent transforms chromium-3 into chromium-6 in the distribution system?

    a. High pH
    b. Phosphates
    c. Organics
    d. Oxidants

75. Which is the best method for analysis of hexavalent chromium (chromium-6)?

    a. Ion chromatography
    b. Atomic absorption
    c. Sulfuric acid-nitric acid digestion
    d. Titration using 0.02 N silver nitrate

Additional Practice Questions–Laboratory Analyses

Answers on page 196

1. Turbidity analyses on samples should be measured

    a. as soon as possible.
    b. within 2 hours.
    c. within 4 hours.
    d. within 8 hours.

2. Which sample quantity should be collected for a hardness analysis?

    a. 100 mL
    b. 250 mL
    c. 500 mL
    d. 1,000 mL

3. Which is the minimum size turbidity sample that should be collected?

    a. 10 mL
    b. 25 mL
    c. 100 mL
    d. 250 mL

4. Which sample should not be collected in a glass container?

    a. Haloacetic acid
    b. Fluoride
    c. Trihalomethanes
    d. Volatile organic compounds

5. Bacteriological sample storage refrigerator temperatures should range from

    a. −5 to 5°C.
    b. 0 to 2°C.
    c. 1 to 5°C.
    d. 5 to 8°C.

6. How long can a preserved sample that is to be analyzed for manganese content be stored?

    a. 24 hours
    b. 48 hours
    c. 7 days
    d. 6 months

7. Samples to be analyzed for manganese are preserved with

    a. nitric acid.
    b. citric acid.
    c. sulfuric acid.
    d. phosphoric acid.

8. In the total coliform Colisure test, if coliforms are present, the special medium will turn

    a. yellow to brown.
    b. red to purple.
    c. yellow.
    d. brown.

9. Which is the chemical formula for ammonia?

    a. NH
    b. NH_2
    c. NH_3
    d. NH_4

10. Electrically charged atoms such as Na^+ and Cl^- are called

    a. cations.
    b. ions.
    c. radicals.
    d. compounds.

11. How often should a softening treatment plant analyze for hardness?

    a. Every 4 hours
    b. Every 8 hours
    c. Every 12 hours
    d. Daily

12. Which test would be used to determine total alkalinity?

    a. Marble test
    b. Phenolphthalein test
    c. Methyl orange test
    d. Violet-green test

13. Which method is used to calculate the calcium carbonate stability of water?

    a. Marble test
    b. Langelier Index
    c. Alkalinity test
    d. Calcium carbonate solubility index

14. Who is responsible for sample preservation?

    a. The sampler
    b. The laboratory technician
    c. The laboratory analyst
    d. The laboratory supervisor

15. If hardness samples need to be stored, which preservative should be used?

    a. Sulfuric acid
    b. Nitric acid
    c. Hydrochloric acid
    d. Acetic acid

16. How long can samples be held for fluoride analysis?

    a. 24 hours
    b. 5 days
    c. 7 days
    d. 1 month

17. Which water quality parameter can be properly determined by collecting a composite sample?

    a. Zinc
    b. pH
    c. Chlorine residual
    d. Dissolved gases

18. Water that is to be analyzed for inorganic metals should be acidified to a pH that is

    a. <2.0 pH units.
    b. <3.0 pH units.
    c. <4.0 pH units.
    d. <4.5 pH units.

19. Nematodes are

    a. small snails.
    b. insects.
    c. roundworms.
    d. pelecypods.

20. An actinomycetes bacterial colony on an agar plate would have a

    a. smooth dull rounded appearance.
    b. smooth, shiny mucous-like appearance.
    c. dull powdery appearance with fuzzy border.
    d. smooth mucous-like appearance with an irregular border.

21. Which level of threshold odor number (TON) will result in customer complaints?

    a. 5 TON
    b. 7 TON
    c. 8 TON
    d. 10 TON

22. Suspended solids that are referred to as settleable solids will settle unaided to the bottom of a sedimentation basin within

    a. 2 hours.
    b. 4 hours.
    c. 6 hours.
    d. 8 hours.

23. Which would be considered very hard water measured in mg/L as $CaCO_3$ for the Sawyer, Briggs, or Ficke classification?

    a. >120 mg/L
    b. >160 mg/L
    c. >180 mg/L
    d. >300 mg/L

24. Water that is too soft will cause

    a. calcium scale.
    b. carbonate scale.
    c. soap scum.
    d. water spots.

25. Why is it important to measure total organic carbon (TOC)?

    a. TOC correlates with production of disinfection by-products
    b. TOC indicates the relative amounts of bacterial contamination
    c. TOC indicates the relative amounts of microbial activity
    d. TOC increases biofouling on the filters

26. Which element is found in all acids?

    a. Hydrogen
    b. Carbon
    c. Oxygen
    d. Chlorine

27. When a base reacts with an oxide of a nonmetal it will produce

    a. a salt.
    b. an acid.
    c. a hydroxide.
    d. a binary acid.

28. When anhydrides react with water they produce

    a. bases and oxides.
    b. acids and hydroxides.
    c. oxides and acids.
    d. acids or bases.

29. Oxides of nonmetals in water will

    a. cause acidity.
    b. cause alkalinity.
    c. produce amphiprotic compounds.
    d. produce neutral compounds.

30. Which formula is correct for the Langelier Index?

    a. LI = – log(alkalinity) + TDS/log($CaCO_3$)
    b. LI = pH – pHs
    c. LI = Alkalinity + TDS – log(pH)
    d. LI = Temperature + calcium content – log(alkalinity)

31. *Giardia lamblia* species are

    a. viruses.
    b. bacteria.
    c. protozoans.
    d. metazoans.

32. The specific gravity of water is based on using a temperature of

    a. 0°C.
    b. 4°C.
    c. 20°C.
    d. 25°C.

33. Jar testing procedures are applicable to

    a. conventional treatment.
    b. sludge blanket clarifiers.
    c. upflow clarifiers.
    d. conventional softening.

34. The incubation time for the membrane filter method is

    a. 18 hours.
    b. 24 hours.
    c. 36 hours.
    d. 48 hours.

35. Which type of glassware is used in dispensing solutions during titrations?

    a. Mohr pipette
    b. Burette
    c. Volumetric pipette
    d. Kjeldahl flask

36. Samples confirmed in the presence-absence test must also be tested for

    a. *Salmonella.*
    b. *Shigella.*
    c. *Escherichia coli.*
    d. *Giardia lamblia.*

37. Which should the heterotrophic plate count value be in water that has been properly treated?

    a. <10 colonies per mL
    b. <100 colonies per mL
    c. <300 colonies per mL
    d. <500 colonies per mL

38. Why should a jar test procedure be done as quickly as possible after collecting the raw water sample?

    a. Changes in dissolved gases will significantly affect results
    b. Changes in pH will significantly affect results
    c. Changes in temperature will significantly affect results
    d. Changes in alkalinity will significantly affect results

39. Which is the preferred method of hardness analysis?

    a. Titration with 0.50 normal hydrochloric acid solution
    b. Titration with 0.02 normal sulfuric acid solution with methyl blue indicator
    c. Titration with ethylenediaminetetraacetic acid
    d. Titration with acetic acid

40. Which method is used for fluoride analysis?

    a. Acid titration
    b. Caustic titration
    c. SPADNS
    d. EDTA

41. When EDTA is added to a sample being tested for hardness, the end point causes the dye in the sample to change to

    a. red.
    b. yellow.
    c. blue.
    d. pink.

42. Which type of pipette has a single ring near the top?

    a. Mohr pipette
    b. Transfer pipette
    c. Koch pipette
    d. Standard pipette

43. Titration for alkalinity analyses uses which one of the following to determine the end point?

    a. pH meter
    b. Colorimeter
    c. Methylene red
    d. Methylene di-amine

44. The number of grains per gallon (gpg) can be converted to mg/L hardness by

    a. multiplying by 17.12.
    b. multiplying by 17.72.
    c. dividing by 17.12.
    d. dividing by 17.72.

45. Hardness samples that have been acidified can be stored for

    a. 2 days.
    b. 5 days.
    c. 7 days.
    d. 28 days.

46. Which type of bottle should be used and how should it be cleaned for taste and odor samples?

    a. Plastic, cleaned with detergent and rinsed with de-ionized water
    b. New glass bottles only
    c. Glass bottles, cleaned with detergent and rinsed with distilled water
    d. New glass or plastic bottles only, rinsed with de-ionized water

47. After a sample to be analyzed for manganese has been acidified, its pH must be

    a. <2 pH units.
    b. <3 pH units.
    c. <4 pH units.
    d. <5 pH units.

48. Conductivity (as it pertains to water) is

    a. the measure of ions in the water and is indirectly proportional to the number of free electrons in solution.
    b. the measure of the positive and negative metals ions in water.
    c. the measure of a solution's ionic strength and an indirect measure of the total dissolved solids.
    d. the measure of the electrical strength, and is directly proportional to the number of free electrons in the water.

49. When should a water treatment plant take action on odors in finished water?

    a. 1 TON
    b. 3 TON
    c. 5 TON
    d. 8 TON

50. Dissolved metals in a water sample are considered to be metals that are

    a. unacidified and pass through a 0.25 μm membrane.
    b. acidified and pass through a 0.25 μm membrane.
    c. unacidified and pass through 0.45 μm membrane.
    d. acidified and pass through 0.45 μm membrane.

51. *Salmonella* species are

    a. bacteria.
    b. viruses.
    c. protozoans.
    d. flagellates.

52. *Standard Methods* defines QA as

    a. producing data that is precise and unbiased.
    b. a set of analytical procedures.
    c. a set of regulations promulgated by the USEPA.
    d. a set of USEPA procedures and regulations.

53. Atoms that have the same atomic number, but a different mass number are called

    a. isotopes.
    b. radioactive.
    c. radicals.
    d. isomers.

54. The oxidation number of an element is also called its _____ number.

    a. ion
    b. valence
    c. radical
    d. isotope

55. When oxidation occurs

    a. electrons are lost by the species being oxidized.
    b. electrons are gained by the species being oxidized.
    c. electrons are shared with the oxygen atom.
    d. oxygen is always added to the species being oxidized.

56. When reduction occurs

    a. electrons are lost by the species being reduced.
    b. electrons are gained by the species being reduced.
    c. electrons are shared with the oxygen atom.
    d. oxygen is always lost to the species being reduced.

57. A double replacement reaction is the same as

    a. a composition reaction.
    b. a decomposition reaction.
    c. an ionic reaction.
    d. a redox reaction.

58. Which is the strongest oxidizing agent?

    a. $Cl_2 + 2e^-$
    b. $Cu^{++} + 2e^-$
    c. $Al^{+3} + 3e^-$
    d. $Sn^{+4} + 2e^-$

59. Which will show Brownian motion?

    a. Solute
    b. Colloids
    c. Solvent
    d. Electrolyte

60. Molarity is the number of

    a. gram equivalent weights of solute per liter of solution.
    b. gram atomic weight per oxidation number.
    c. moles of solute per kilogram of solvent.
    d. moles of solute per liter of solution.

61. Which would increase in solubility as the temperature decreases?

    a. Magnesium carbonate
    b. Chlorine gas
    c. Iron hydroxide
    d. Sulfuric acid

62. Crystalline compounds that contain water are called

    a. hygroscopic crystals.
    b. hydroxides.
    c. hydrous crystals.
    d. hydrates.

63. Which is it called when water molecules associate with ions?

    a. Ionic hydration
    b. Ionic dissociation
    c. Hydration of ions
    d. Index of water solubility

64. If a hydrocarbon has one or more of its hydrogen atoms replaced with a hydroxyl (OH^-) group it is called

    a. a ketone.
    b. an ester.
    c. an aldehyde.
    d. an alcohol.

65. When is water considered to be in equilibrium?

    a. $pH - pH_s = 0$
    b. Calcium carbonate is slightly soluble
    c. $pH - pH_s > 0$
    d. $pH - pH_s < 0$

66. How are total calcium analyses reported?

    a. As percent hardness
    b. As percent Ca
    c. In mg/L as $CaCO_3$
    d. In mg/L as Ca

67. Which is the most probable solution if midges can be seen in the finished water?

    a. Isolate distribution area, superchlorinate, flush mains and customer lines
    b. Use chloramines, or, if already using, temporarily increase by 100%
    c. Increase chlorination and backwash
    d. Add powdered activated carbon to the filters, then backwash

68. Typically, which constituent would account for the least amount of total dissolved solids (TDS)?

    a. Chloride
    b. Calcium
    c. Dissolved organic matter
    d. Bicarbonates

69. Which algal genera in the Chlorophyta Phylum are filamentous and can form tangled mats in ponds?

    a. *Oocystis*
    b. *Rhizoclonium*
    c. *Actinastrum*
    d. *Volvox*

70. If phenolphthalein alkalinity equals total alkalinity, which species is present?

    a. OH^-
    b. NO_3^-
    c. HCO_3^-
    d. CO_3^{-2}

71. If phenolphthalein alkalinity is zero, which species is present?

    a. OH^-
    b. NO_3^-
    c. HCO_3^-
    d. CO_3^{-2}

72. In general, for every 10 electrical conductivity units there are approximately how many mg/L of dissolved solids?

    a. 2 to 3 mg/L
    b. 6 to 7 mg/L
    c. 12 to 13 mg/L
    d. 16 to 18 mg/L

73. Ionic compounds are formed when

    a. an acid is added to a mixture.
    b. an acid is added to a compound.
    c. a transfer of electrons occurs.
    d. electrons are shared.

74. The measure of attraction that atoms have for each other when they share electrons is called

    a. ionization energy.
    b. van der Waals forces.
    c. polarization.
    d. electronegativity.

75. Which genera of cyanobacteria release a variety of odorous organic sulfur compounds, especially after dying?

    a. *Microcystis*
    b. *Phormidium*
    c. *Anabaena*
    d. *Oscillatoria*

76. If two combining atoms have an electronegativity difference that is more than 1.6,

    a. ionic bonding will occur.
    b. polarization will occur.
    c. covalent bonding will occur.
    d. they will repel each other.

77. Which sulfur bacteria are photosynthetic?

    a. *Chromatium*
    b. *Desulfotomaculum*
    c. *Thiotrix*
    d. *Beggiatoa*

78. Which is the quantitative relationship of chemical reactants and their products known as?

    a. Mass balance
    b. Mass ratio
    c. Stoichiometry
    d. A chemical reaction

79. If a water molecule has both of its hydrogen atoms replaced by alkyl groups, it is called

    a. a ketone.
    b. an ester.
    c. an aldehyde.
    d. an alcohol.

80. In general, an electrical conductivity change of 10 units represents approximately _____ of dissolved solids.

    a. 1.0 mg/L
    b. 2.0 to 2.5 mg/L
    c. 6.0 to 7.0 mg/L
    d. 10.0 mg/L

81. Which is the first level of TDS that is considered unfit for human consumption?

    a. 1,000 mg/L
    b. 2,500 mg/L
    c. 4,000 mg/L
    d. 5,000 mg/L

82. Which sulfur bacteria are photosynthetic?

    a. *Beggiatoa*
    b. *Thiobacillus*
    c. *Chlorobium*
    d. *Desulfuromonas*

83. Which chemical is used in desiccators to remove moisture from the air in the desiccator?

    a. Calcium oxide
    b. Sodium chloride
    c. Calcium sulfate
    d. Sodium oxalate

84. Which laboratory incubator is most useful for total coliform and HPC (Heterotrophic Plate Count) analyses?

    a. Water-bath incubator
    b. Dry-heat incubator
    c. Autoclave
    d. Low-temperature incubator

85. Which type of incubator is used when reactions must be performed with mixtures or reagents at a specific temperature?

    a. Low-temperature dessicators
    b. Dry-heat dessicators
    c. Water-bath
    d. Low-heat desiccators

86. Which type of special photometer is used to analyze for heavy metals in water?

    a. Electrophotometer
    b. Magnetophotometer
    c. Spectrophotometer
    d. Atomic absorption spectrophotometer

87. In the Multiple-Tube Fermentation method, which type of culture media is used in the confirmed test?

    a. Lactose agar
    b. Green lactose bile broth
    c. Eosin methylene red agar
    d. Glucose bile tubes

88. All positive samples in the presence-absence test must be confirmed using

    a. brilliant green bile tubes.
    b. eosin methylene red agar.
    c. lauryl tryptose broth.
    d. eosin 12-amino acid broth.

89. The bicarbonate alkalinity equals the total alkalinity if the pH is

    a. 4.3.
    b. 8.3.
    c. 4.3 or more.
    d. 8.3 or less.

90. Which formula is correct for determining the pHs (saturation)?

    a. pH_s = Temperature – log(alkalinity) + TDS/log($CaCO_3$)
    b. pH_s = TDS + log(Ca^{+2}) – log[OH^-] + pH/log(marble test result)
    c. pH_s = Alkalinity + TDS + temperature – log(pH)
    d. pH_s = Temperature + TDS – log(Ca^{+2}) – log(alkalinity)

Additional Practice Questions – Comply with Drinking Water Regulations

Answers on page 203

1. How is the term "lead free" defined, when used with respect to solders and flux?

    a. Containing not more than 0.01% lead
    b. Containing not more than 0.02% lead
    c. Containing not more than 0.04% lead
    d. Containing not more than 0.05% lead

2. What is the maximum contaminant level goal for *Giardia lamblia*?

   a. Zero
   b. 1 cyst per liter
   c. 2 cysts per liter
   d. 5 cysts per liter

3. Which is the maximum residual disinfection level goal for chlorine?

   a. 3 mg/L
   b. 4 mg/L
   c. 5 mg/L
   d. 8 mg/L

4. If a water system collects at least 40 samples per month for the analyses of total coliforms, which percent of total coliform positive samples are acceptable for the system to remain in compliance with the maximum contaminant level for total coliforms?

   a. No more than 2%
   b. No more than 3%
   c. No more than 4%
   d. No more than 5%

5. If a water system collects less than 40 samples per month for the analyses of total coliforms, how many samples can be total coliform positive and the system still remain in compliance with the maximum contaminant level for total coliforms?

   a. Zero
   b. 1
   c. 2
   d. 3

6. Which is the maximum contaminant level for bromate?

   a. 0.010 mg/L
   b. 0.020 mg/L
   c. 0.050 mg/L
   d. 0.10 mg/L

7. Which is the best available technology (BAT) for removing combined radium-226 and radium-228 from source water?

   a. Enhanced coagulation
   b. Greensand filtration
   c. Lime softening
   d. Aeration

8. Which log removal must water systems achieve for the removal and/or inactivation of *Giardia lamblia* cysts between a point where the raw water is not subject to recontamination by surface water runoff and a point downstream before or at the first customer?

   a. At least 2-log removal
   b. At least 2.5-log removal
   c. At least 3-log removal
   d. At least 4-log removal

ADDITIONAL WATER TREATMENT OPERATOR PRACTICE QUESTIONS   99

9. Water systems are required to achieve at least _____ removal and/or inactivation of viruses between a point where the raw water is not subject to recontamination by surface water runoff and a point downstream before or at the first customer.

   a. 2 log
   b. 2.5 log
   c. 3 log
   d. 4 log

10. One criterion that a public water system must meet to avoid filtration is for the turbidity level not to exceed _____ in representative samples of the source water immediately before the first or only point of disinfectant application in at least 90% of the measurements made for the previous 6 months that the system served water to the public on an ongoing basis.

    a. 0.3 ntu
    b. 0.5 ntu
    c. 1.0 ntu
    d. 5.0 ntu

11. A public water system that does not provide filtration must have a residual disinfection concentration in the distribution system, measured as total chlorine, combined chlorine, or chlorine dioxide, which cannot be undetectable in more than _____ of the samples each month, for any two consecutive months that the system serves water to the public.

    a. 1%
    b. 2%
    c. 3%
    d. 5%

12. For water systems using conventional treatment or direct filtration, the turbidity level of representative samples of a system's filtered water must be less than or equal to _____ in at least _____ of the samples collected each month, according to the Enhanced Surface Water Treatment Rule.

    a. 0.3 ntu; 90%
    b. 0.3 ntu; 95%
    c. 0.5 ntu; 95%
    d. 1.0 ntu; 95%

13. Which are the turbidity requirements for slow sand filtration?

    a. <0.5 ntu in 95% of the measurements collected each month
    b. <1.0 ntu in 95% of the measurements collected each month
    c. ≤0.5 ntu in 95% of the measurements collected each month
    d. ≤1.0 ntu in 95% of the measurements collected each month

14. Which are the turbidity requirements for diatomaceous earth filtration?

    a. <0.5 ntu in 95% of the measurements collected each month
    b. <1.0 ntu in 95% of the measurements collected each month
    c. ≤0.5 ntu in 95% of the measurements collected each month
    d. ≤1.0 ntu in 95% of the measurements collected each month

15. Measurements for pH, turbidity, temperature, and residual disinfectant concentrations must be performed by a person approved by the

    a. County Public Health Department.
    b. National Sanitary Foundation.
    c. State.
    d. US Environmental Protection Agency.

16. Who certifies laboratories for the measurement of total coliforms, fecal coliforms, and heterotrophic plate counts?

    a. American Standard Testing Methodology
    b. ASTM International or NSF International
    c. American Water Works Association
    d. USEPA or the State

17. Which is the maximum holding time for fecal coliforms using the Fecal Coliform Procedure?

    a. 8 hours
    b. 12 hours
    c. 24 hours
    d. 36 hours

18. Samples to be analyzed for total coliforms by the total coliform Fermentation Technique, must be kept during transit at a temperature below

    a. 4°C.
    b. 8°C.
    c. 10°C.
    d. 12°C.

19. For water systems that collect inorganic samples more frequently than annually (for antimony, arsenic, barium, beryllium, cadmium, chromium, cyanide, fluoride, mercury, nickel, selenium, and thallium), compliance with the maximum contaminant levels is determined by

    a. a running annual average at any sampling point.
    b. individual samples each time at any sampling point.
    c. an average of all samples at any sampling point.
    d. an average of the last two samples at any sampling point.

20. How should a distribution operator preserve a sample to be analyzed for asbestos?

    a. Preserve with HNO_3
    b. Preserve with NaOH
    c. Preserve with H_2SO_4
    d. Preserve by keeping it at 4°C

21. How should a distribution operator preserve a sample to be analyzed for nitrate–nitrite?

    a. Preserve with HNO_3
    b. Preserve with NaOH
    c. Preserve with H_2SO_4
    d. Preserve by keeping it at 4°C

22. Which is the appropriate holding time for nitrate, if the sample is chlorinated and not acidified?

    a. 24 hours
    b. 48 hours
    c. 72 hours
    d. 14 days

23. Water systems may be allowed by their state to monitor annually, if the state has determined that the system is reliable and consistently below the maximum contaminant level for organic compounds. When must the water system collect these annual samples?

    a. During the fall when plant debris starts to increase in surface waters
    b. During the spring when plant debris has had time to decay and thus increase organic levels in surface waters
    c. During the month which previously yielded the highest analytical result
    d. During the quarter which previously yielded the highest analytical result

24. How many consecutive annual samples with no detection results for an organic contaminant must a water system achieve before it may apply to the state for a waiver?

    a. 3
    b. 4
    c. 5
    d. 6

25. If a groundwater system detects one or more organic compounds in the chloroethylene family (e.g., trichloroethylene, tetrachloroethylene and trans-1,2-dichloroethylene), which constituents(s) will the water system have to monitor quarterly?

    a. Ethylene
    b. Ethylene and xylene
    c. Vinyl chloride
    d. Perchlorate

26. If the state allows a water system to composite samples that are to be analyzed for organic compounds {§141.61 (a)}, which entity must composite the samples and what is the holding time?

    a. Laboratory, 3 days
    b. Laboratory, 14 days
    c. Sampler, 3 days
    d. Sampler, 14 days

27. A follow-up sample is required if the concentration in the composite sample is greater than or equal to _____ for any contaminant listed in the National Primary Drinking Water Regulations, §141.61 (a) – Organic Chemicals.

    a. 0.0001 mg/L
    b. 0.0005 mg/L
    c. 0.001 mg/L
    d. 0.005 mg/L

28. Which analytical test can be performed by any person acceptable to the State?

    a. Nitrate
    b. Alkalinity
    c. Arsenic
    d. Toluene

29. If a water supplier fails to comply with any national primary drinking water regulation, including monitoring requirements, the water supplier must report to the state the failure to comply within

    a. 24 hours.
    b. 48 hours.
    c. 10 days.
    d. 1 month.

30. When is it acceptable for a water supplier not to report analytical results to the state?

    a. When all constituents analyzed are below the detection limit
    b. When all constituents analyzed are below the maximum contaminant levels
    c. When a state laboratory performs the analyses
    d. When the samples have been waived by the state

31. How long does a public water system have for completing and submitting to the primacy agency that it has fully complied with a public notification requirement?

    a. 5 days
    b. 10 days
    c. 14 days
    d. 21 days

32. When must a water supplier monitor for unregulated contaminants?

    a. When the water supplier serves more than 3,300 people
    b. When the water supplier serves more than 10,000 people
    c. When the supplier is a public water system and the state has asked them to participate
    d. When the water supplier wishes to participate

33. If a water system is required to monitor unregulated contaminants, it must collect the samples in one continuous

    a. 12-month period.
    b. 24-month period.
    c. 36-month period.
    d. 60-month period.

34. If a water system does not collect a sample for unregulated contaminant monitoring according to federal instructions for a listed contaminant, the deviation must be reported to the UCMR sampling coordinator or state within

    a. 7 days.
    b. 10 days.
    c. 14 days.
    d. 30 days.

35. How often must a public water system using surface water in whole or in part collect and analyze one sample per water plant at the entry to the distribution system for sodium concentration levels?

    a. Once per month
    b. Quarterly
    c. Annually
    d. At least every three years

36. How often must a public water system using groundwater collect and analyze one sample per water plant at the entry to the distribution system for sodium concentration levels?

    a. Once per month
    b. Quarterly
    c. Annually
    d. At least every three years

37. The prohibition on the use of lead for any plumbing in a residential or nonresidential facility connected to a public water system that monitors water for human consumption does not apply to

    a. repair of cast iron pipes using leaded joints.
    b. repair of copper tubing.
    c. the use of solders which have between 2% and 5% lead.
    d. the use of flux which has between 2% and 3% lead.

38. Which is the sample size for lead and copper analyses?

    a. 100 mL
    b. 250 mL
    c. 500 mL
    d. 1,000 mL

39. The first draw samples from taps for lead and copper analyses shall have stood motionless in the plumbing system pipes of each sampling site for at least

    a. 4 hours.
    b. 5 hours.
    c. 6 hours.
    d. 8 hours.

40. Lead and copper samples collected by customers at their taps must be acidified within

    a. 3 days.
    b. 5 days.
    c. 10 days.
    d. 14 days.

41. Which is not an acceptable method of sampling for lead and copper analyses lead service lines?

    a. Tapping directly into the lead service line
    b. At the tap after flushing the volume of water between the tap and the lead service line
    c. At the tap after the water has remained motionless for at least 4 hours
    d. At the tap after allowing the water to run until there is a significant change in temperature

42. When can a large water system serving more than 100,000 people, reduce the number of sampling sites for lead and copper?
    a. When the system meets the lead and copper action level for 2 consecutive six-month monitoring periods
    b. When the system meets the lead and copper action level for 3 consecutive six-month monitoring periods
    c. When the system meets the lead and copper action level for 2 consecutive annual monitoring periods
    d. When the system meets the lead and copper action level for 3 consecutive annual monitoring periods

43. When can a water system that meets the lead and copper action levels for two consecutive six-month periods reduce the number of samples and reduce the frequency of sampling so that it only is required to sample every 3 years?
    a. When the 90th percentile for lead is ≤0.003 mg/L and for copper is ≤0.50 mg/L
    b. When the 90th percentile for lead is ≤0.004 mg/L and for copper is ≤0.60 mg/L
    c. When the 90th percentile for lead is ≤0.005 mg/L and for copper is ≤0.65 mg/L
    d. When the 90th percentile for lead is ≤0.007 mg/L a and for copper is ≤0.70 mg/L

44. A small water system has met certain criteria and has been granted a full waiver, and thus a reduced frequency for monitoring lead and copper. How often does it now have to monitor?
    a. At least once every 3 years
    b. At least once every 5 years
    c. At least once every 9 years
    d. At least once every 12 years

45. All water systems that exceed the lead and copper action level shall monitor water quality parameters. Which sample criteria shall be followed if exceptions are not granted?
    a. 1 sample for each water quality parameter, per entry point to the distribution system during each monitoring period
    b. 2 samples for each water quality parameter, per entry point to the distribution system during each monitoring period
    c. 3 samples for each water quality parameter, per entry point to the distribution system during each monitoring period
    d. 3 samples for each water quality parameter, per entry point to the distribution system during each monitoring period, and one blind sample for each water quality parameter

46. When a water system is required to monitor water quality parameters because it exceeded the lead and copper action level, which parameter must be measured every two weeks at each entry point to the distribution system?
    a. Alkalinity
    b. Disinfectant residual
    c. pH
    d. Orthophosphate

47. When a water system is required to monitor water quality parameters because it exceeded the lead and copper action level and the system uses a corrosion inhibitor, which parameter must be measured every two weeks at each entry point to the distribution system?

    a. Orthophosphate
    b. Total alkalinity
    c. Calcium carbonate
    d. Dosage rate of the chemical corrosion inhibitor

48. Water systems are required to monitor for water quality parameters when they exceed the lead and copper action levels. If these systems maintain the range of values for the water quality parameters that reflect optimal corrosion control treatment during each of two consecutive six-month monitoring periods, then, although they must still monitor at the entry point(s) to the distribution system, they are allowed to

    a. reduce the number of sites sampled.
    b. reduce the frequency of sampling from 6-month periods to annually.
    c. eliminate all water quality parameters monitored, except pH and alkalinity.
    d. eliminate all water quality parameters monitored, except pH, and they must record in mg/L the dosage for any chemical alkalinity adjustment.

49. If a water system that monitors for water quality parameters because it exceeded the lead and copper action levels maintains the range of values for the water quality parameters reflecting optimal corrosion control treatment during three consecutive years of monitoring the system may

    a. reduce the number of sites sampled.
    b. reduce the frequency of sampling from 6-month periods to annually.
    c. eliminate all water quality parameters monitored.
    d. eliminate all water quality parameters monitored, except pH and alkalinity.

50. Groundwater systems that have failed to meet the lead and copper action levels shall take a minimum of

    a. one sample at representative sampling points throughout the distribution system.
    b. two samples at representative sampling points throughout the distribution system.
    c. one sample at every entry point to the distribution system that is representative of each well after treatment.
    d. two samples at every entry point to the distribution system that is representative of each well after treatment.

51. Surface water systems that have failed to meet the lead and copper action levels shall take a minimum of

    a. one sample at representative sampling points throughout the distribution system.
    b. two samples at representative sampling points throughout the distribution system.
    c. one sample after any application of treatment or in the distribution system at a point which is representative of each source after treatment.
    d. two samples after any application of treatment or in the distribution system at a point which is representative of each source after treatment.

52. Water systems monitoring for lead and copper and water quality parameters must report all the results within the first _____ after each monitoring period.

    a. 7 days
    b. 10 days
    c. 14 days
    d. 21 days

53. Any water system subject to the NPDWR Subpart I, Control of Lead and Copper, shall retain on its premises original records of all sampling data and analysis, reports, surveys, letters evaluations, schedules, State determinations, and any other information for no fewer than

    a. 10 years.
    b. 12 years.
    c. 15 years.
    d. 20 years.

54. Which test must a water utility perform to determine monitoring frequency and specific radionuclide analyses?

    a. Extraction and concentration of the radioactive compounds
    b. ICP mass spectroscopy
    c. Gross alpha particle activity scan
    d. Anodic stripping voltametry

55. Under the Stage 1 Disinfectants and Disinfection By-products Rule, jar testing may be conducted as part of the enhanced coagulation requirements to

    a. determine alternative total organic carbon removal requirements for a particular plant.
    b. determine optimum flash mix conditions.
    c. evaluate alternative coagulation chemicals.
    d. determine turbidity and color removal.

56. The USEPA recommends water systems using a surface water source to collect samples for hexavalent chromium (chromium-6) on a

    a. monthly basis.
    b. quarterly basis.
    c. semi-annual basis.
    d. annual basis.

57. The USEPA recommends water systems using a groundwater source(s) to collect samples for hexavalent chromium (chromium-6) on a

    a. monthly basis.
    b. quarterly basis.
    c. semi-annual basis.
    d. annual basis.

58. Which is the National Primary Drinking Water Regulations maximum contaminant level for total chromium that includes hexavalent chromium (chromium-6)?

    a. 0.001 mg/L
    b. 0.050 mg/L
    c. 0.1 mg/L
    d. 0.5 mg/L

59. A water system using conventional filtration treatment does not have to use enhanced coagulation to achieve the total organic carbon (TOC) percent removal if the total trihalomethanes (TTHM) and the sum of haloacetic acids (HAA5) running annual averages are

    a. ≤0.020 mg/L and ≤0.010 mg/L for TTHM and HAA5, respectively.
    b. ≤0.030 mg/L and ≤0.020 mg/L for TTHM and HAA5, respectively.
    c. ≤0.040 mg/L and ≤0.030 mg/L for TTHM and HAA5, respectively.
    d. ≤0.050 mg/L and ≤0.040 mg/L for TTHM and HAA5, respectively.

60. A water system that uses enhanced softening, but cannot achieve the required total organic carbon (TOC) removals, can still comply by using results from lowering the treated water alkalinity to less than _____ as CaCO3 measured monthly and calculated quarterly as a running annual average.

    a. <60 mg/L
    b. <80 mg/L
    c. <100 mg/L
    d. <120 mg/L

61. A water system that detects nitrate at levels above _____, but below the MCL, must include in the Consumer Confidence Report a short informational statement about the impacts of nitrate to children.

    a. 0.5 mg/L
    b. 1.0 mg/L
    c. 2.0 mg/L
    d. 5.0 mg/L

62. How long must a water system monitor the CT 99.9% values daily to determine the total logs of inactivation for each day of operation?

    a. 6 months
    b. 12 months
    c. 2 years
    d. 3 years

63. Water system parameters for calculating CT (pH, temperature, residual disinfectant concentration, and disinfectant contact time) must be determined

    a. when the disinfectant is at its lowest point for each day the system operates.
    b. when the disinfectant is at its lowest point and the pH at its highest for each day the system operates.
    c. when the disinfectant is at its lowest point and the temperature is at its highest for each day the system operates.
    d. each day during peak hourly flows.

64. A water system that calculates the logs of inactivation for *Giardia lamblia* must also calculate the logs of inactivation for viruses, if it uses which type of disinfectant?

    a. Sodium hypochlorite
    b. Calcium hypochlorite
    c. Chlorine dioxide
    d. Chloramines

65. If a water system changes its disinfection sampling point, disinfection type, process, or any other modification identified by the state, it must

   a. perform a sanitary survey.
   b. develop a disinfection benchmark.
   c. develop a disinfection profile.
   d. start over or go back to routine monitoring of disinfection by-products.

66. Filtered water systems can avoid source water monitoring for *Cryptosporidium* and *E. coli* if the system provides a total of at least _____ of treatment for *Cryptosporidium*.

   a. 3-log
   b. 4-log
   c. 5-log
   d. 5.5-log

67. A system must begin a second round of source water monitoring no later than _____ following initial bin classification (average source water *Cryptosporidium* concentration) or determination of the mean *Cryptosporidium* level, as applicable.

   a. 3 years
   b. 5 years
   c. 6 years
   d. 9 years

68. Unfiltered water systems with a mean *Cryptosporidium* level greater than _____ must provide at least a 3-log *Cryptosporidium* inactivation.

   a. 0.01 oocysts/L
   b. 0.05 oocysts/L
   c. 0.10 oocysts/L
   d. 0.50 oocysts/L

69. If a water supplier fails to comply with any National Primary Drinking Water Regulation, including monitoring requirements, the water supplier must report to the state the failure to comply within

   a. 24 hours.
   b. 48 hours.
   c. 10 days.
   d. 1 month.

70. If benzene, toluene, ethylbenzene, or xylene is detected in water from groundwater well(s) at a level exceeding 0.0005 mg/L, how often must the system monitor at each sampling point where the contaminant(s) were detected?

   a. Monthly for a minimum of 6 months
   b. Quarterly for a minimum of 2 quarters
   c. Biannually for 2 years for a minimum of 4 times
   d. Yearly for 3 years

71. How many consecutive annual samples with no detection results for an organic contaminant must a water system achieve before it may apply to the state for a waiver?

   a. 3
   b. 4
   c. 5
   d. 6

72. How soon must a small water system notify the state in writing after it becomes aware of lead-containing and/or copper-containing material in the system?

    a. Within 7 days
    b. Within 14 days
    c. Within 30 days
    d. Within 60 days

73. A public water system that uses surface water or groundwater under the direct influence of surface water and provides filtration treatment is required by the primacy agency to report an acute violation of the filtered water turbidity standard if any turbidity measurement collected during the month exceeds

    a. 0.3 ntu.
    b. 2.0 ntu.
    c. 3.0 ntu.
    d. 5.0 ntu.

74. The federal Safe Drinking Water Act requires a water system to provide a plant schematic showing the origin of all flows that are recycled. Which other requirement for this recycle provision must be shown on the schematic?

    a. The hydraulic conveyance used to transport the recycled flows
    b. The chemical injection points for treating the recycled flows
    c. All water quality sampling points for the recycled flows
    d. All maximum turbidity reads for all recycled flows each month the water system recycled flows

Additional Practice Questions – Operate and Maintain Equipment

Answers on page 210

1. The most common solution feeder used for liquid coagulants and coagulant aids is the

    a. metering pump.
    b. peristaltic pump.
    c. progressive cavity pump.
    d. centrifugal pump.

2. Fixed screens or bar screens usually have openings ranging from

    a. 0.25 to 0.50 inches.
    b. 0.50 to 1.00 inches.
    c. 1.00 to 3.00 inches.
    d. 3.00 to 5.00 inches.

3. Which is the simplest instrument for measuring levels?

    a. Pressure-activated level transmitters
    b. Bubbler units
    c. Float-type sensor
    d. Sonic type indicators

4. Electricity is the study of the movement that effects large numbers of

   a. electrons.
   b. neutrons.
   c. neutrinos.
   d. positrons.

5. The bearings are protected from water leaking out of the stuffing box by the

   a. shaft sleeve.
   b. lantern ring.
   c. slinger ring.
   d. packing gland.

6. The purpose of the lantern ring is to

   a. distribute lubrication to the packing.
   b. prevent water from getting into the stuffing box.
   c. seal the low-pressure zone from the high-pressure zone.
   d. hold the shaft in place.

7. How often should chlorine supply lines be replaced?

   a. Every 6 months
   b. Every year
   c. Every 2 years
   d. Every 3 years

8. Which is the most common well operational problem?

   a. Plugging of the pump with sediment
   b. Pump failure
   c. Plugging of the well screen
   d. Corrosion of the well's metal parts

9. Why are roller bearings used?

   a. To increase torque
   b. To increase thrust
   c. To reduce shaft erosion corrosion
   d. To reduce friction and heat

10. Which type of pump is typically used for dispensing chemical solutions?

    a. Centrifugal
    b. Diaphragm
    c. Peripheral
    d. Pneumatic ejector

11. Bar screen assemblies are installed in a waterway at a(n) _____ angle from the horizontal.

    a. 40 to 60°
    b. 60 to 80°
    c. 80 to 90°
    d. 90 to 120°

12. Which disadvantage do peristaltic pumps have?

    a. The are expensive
    b. They cannot handle many non-flammable liquids
    c. They are not very precise
    d. They require periodic flexible tubing replacement

13. Which is the purpose of the vacuum in the line to the chlorine injector?

    a. To apply suction so chlorine gas is fed to the system
    b. To shut off chlorine gas flow in case of a leak
    c. To adjust chlorine gas flow by applying variable vacuum pressures
    d. To apply vacuum when one cylinder or bank of cylinders becomes empty and a switchover to a full container or containers is required

14. Which device creates the vacuum that is required to operate the chlorinator?

    a. Injector
    b. Regulating diaphragm
    c. Pressure regulating valve
    d. Vacuum valve

15. Which is the maximum chlorine gas withdrawal rate from a one-ton container at 70°F?

    a. 200 lb/d
    b. 300 lb/d
    c. 400 lb/d
    d. 450 lb/d

16. When opening a chlorine cylinder valve,

    a. pull or tug at the wrench.
    b. use a wrench that is 12 inches long for added leverage.
    c. use common bleach to check for any chlorine leaks.
    d. open only one turn to permit maximum withdrawal rate.

17. Which type of pump is generally used for feeding a fluoride solution because it is accurate and will operate against pressure?

    a. Centrifugal pump
    b. Diaphragm pump
    c. Radial flow velocity pump
    d. Axial flow velocity pump

18. Auxiliary equipment for fluoridation includes

    a. a compound meter to measure solution makeup water for flow to the saturator.
    b. a flow meter for measuring total flow such that pacing can be provided.
    c. a day tank designed to hold 5 to 7 days of supply for safety.
    d. a required extension hopper that is always located next to the primary hopper.

19. Screen openings in raw water intakes are designed for flows of

    a. 1 ft/sec.
    b. 2 ft/sec.
    c. 3 ft/sec.
    d. 4 ft/sec.

20. Which type of flow meter should be used for chlorinators, which require very low flows?

    a. Rotameter
    b. Loss of head meter
    c. Positive displacement meter
    d. Magnetic meter

21. Aspirators should always be equipped with

    a. an atmospheric vacuum breaker.
    b. a water seal.
    c. an inert gas source.
    d. noncollapsible tubing.

22. If chlorine is added before the water enters the microstrainer, which of the following will occur?

    a. Algae will be killed, making it easier to clean off the mesh
    b. Iron will be precipitated, eliminating soluble iron from adhering to the mesh
    c. Tastes and odors caused by algae will increase significantly
    d. Free chlorine may corrode the mesh

23. Which chlorine gas feeder can also feed liquids?

    a. Constant differential pressure
    b. Variable differential pressure
    c. Sonic flow
    d. Induction mixer

24. The voltage rating on a fuse should be _____ the voltage of the circuit on which it is applied.

    a. slightly larger than
    b. equal to
    c. equal to or greater than
    d. slightly lower than

25. In a centrifugal pump, which part physically separates the high pressure zone in the volute from the low pressure zone at the impeller's "eye"?

    a. Wear rings
    b. Packing gland
    c. Shaft sleeves
    d. Slinger ring

26. When removing the volute case at the gasket from a pump, which tool(s) should be used?

    a. Hammer and screwdrivers
    b. Chisels
    c. Soft-faced hammer
    d. Screwdrivers or chisels

27. How should a pump's shaft be stored when not in use?

    a. Stored vertically, on end
    b. Stored horizontally in foam or other soft material
    c. Stored in a shaft box and coated with a light oil to prevent corrosion
    d. Stored horizontally in a long box on foam that is saturated with a light oil

28. Which should be used to clean an old bearing?

    a. Solvent
    b. Mild detergent
    c. Kerosene
    d. Ethyl alcohol

29. How often should amperage and voltage tests be performed on three-phase motors?

    a. Monthly
    b. At least every three months
    c. At least every six months
    d. At least once per year

30. When should a mechanical seal be replaced?

    a. Before any leakage occurs
    b. When it first starts to leak
    c. When it leaks a small amount
    d. When leakage becomes excessive

31. Which type of lubrication is best used on pump bearings for light to moderate, high speed, horizontally-shafted pumps?

    a. Water
    b. Oil
    c. Vegetable based grease
    d. NSF approved animal based grease

32. Which is a principal problem associated with preliminary treatment screening?

    a. Corrosion
    b. Chain comes out of foot sprocket
    c. Foot shaft freezes up due to debris at bottom
    d. Spray pipes and nozzles clog

33. Which is the formula for kinetic head?

    a. $2Vg / 2h$
    b. $V^2g / 2h$
    c. $V^2 / 2g$
    d. $V / 2g$

34. Which type of centrifugal pump can have water enter from two sides (double suction)?

    a. End suction pump
    b. Split-case pump
    c. Vertical turbine pump
    d. Close coupled pump

35. Which type of pump has the ability to rotate its discharge 360°?

    a. End suction pump
    b. Split case pump
    c. Vertical pump
    d. Jet pump

36. A split case pump has three impellers. Which type of multistage pump is this?

    a. One stage
    b. Two stage
    c. Three stage
    d. Six stage

37. A load, perpendicular to a pump shaft, applied to a bearing is called

    a. shear load.
    b. thrust load.
    c. radial load.
    d. axial.

38. Which type of coupling is used when the coupled shafts are of two different sizes?

    a. Split coupling
    b. Flexible disc coupling
    c. Flange coupling
    d. Jaw coupling

39. Excessive up-thrust on line shaft turbines is caused by

    a. misalignment of shaft.
    b. pumping too much water.
    c. an impeller that is too small or worn.
    d. a vertical clearance between bottom of pump bowl and impeller that is too small.

40. Besides providing the enclosure for the impeller, the volute case is also cast and machined to provide a seat for the

    a. lantern ring.
    b. impeller rings.
    c. mechanical seal.
    d. slinger ring.

41. Which is a type of rigid coupling?

    a. Chain coupling
    b. Split coupling
    c. Gear coupling
    d. Jaw coupling

42. The bearings that maintain the axial positioning of a shaft are called

    a. sleeve bearings.
    b. line bearings.
    c. thrust bearings.
    d. rolling bearings.

43. Which type of bearing can operate well against radial loads, but can only withstand very low thrust loads?

    a. Self-aligning, spherical roller bearings
    b. Single-row, tapered roller bearings
    c. Self-aligning double-row ball bearings
    d. Angular contact bearings

44. A pump fails to prime because the suction lift is too high. Which is the best solution?

    a. Inspect, clean, or repair the priming unit
    b. Open suction piping air bleed-off valve
    c. Check external water seal unit
    d. Re-evaluate the pump's requirements and correct the condition

45. Which type of pump has little chance of losing its prime, but has the disadvantage of having a lack of access for maintenance or repair?

    a. Recessed impeller or vortex pump
    b. Turbine pump
    c. Submersible pump
    d. Positive displacement pump

46. Which type of pump adds energy to the fluid flow intermittently?

    a. Recessed impeller or vortex pump
    b. Turbine pump
    c. Submersible pump
    d. Positive displacement pump

47. Diaphragm pumps do not work very well if they have to lift liquids more than about

    a. 4 feet.
    b. 8 feet.
    c. 12 feet.
    d. 20 feet.

48. The streaming current detector is based on what principle?

    a. Zeta potential
    b. Particle reaction kinetic principle
    c. Electromotive principle
    d. Chemical thermodynamic principle

49. Which does the streaming current detector actually measure for determining the effectiveness of the coagulant chemical?

    a. Charge density
    b. Electrical resistance
    c. Electrical potential
    d. Net particle charge of water

50. Which is the normal rotational speed of a microstrainer?

    a. 1 rpm
    b. 2 rpm
    c. 5 rpm
    d. 10 rpm

51. A low vacuum condition on a chlorinator can be caused by

    a. a plugged chlorine supply line.
    b. a failed injector.
    c. an empty chlorine container.
    d. a switchover valve that failed.

52. A high vacuum condition on a chlorinator can be caused by

    a. the solenoid valve being stuck open.
    b. a failed injector.
    c. an empty chlorine container.
    d. a stuck diaphragm regulator.

53. Which metal has the highest corrosion potential?

    a. Aluminum
    b. Copper
    c. Cast iron
    d. Brass

54. Which flow measuring device is mechanical?

    a. Weir
    b. Flume
    c. Venturi
    d. Loss of head meter

55. Which is the last step in well construction?

    a. Placement of the well seal
    b. Disinfection
    c. Installation of the pump
    d. Pump test to confirm capacity

56. Aspiration is the removal of

    a. air from a container by suction.
    b. fluid from a container by suction.
    c. oxygen from water by an exothermic reaction.
    d. oxygen from water by biological or chemical action.

57. Which type of chlorine gas feeding equipment requires an external instrument and is often called closed-loop control?

    a. Automatic residual control gas feeder
    b. Compound-loop control gas feeder
    c. Automatic proportioning control gas feeder
    d. Semiautomatic control gas feeder

58. Which is the best material to use for fittings, lighting, and ventilation systems in a room that houses an ammonia hydroxide tank?

    a. Aluminum
    b. Stainless steel
    c. Brass
    d. Monel

59. Wire size is selected according to which criteria?

    a. Distance from energy source to equipment
    b. Number of resistors and winding for all equipment on that circuit
    c. Number of amperes to be carried
    d. Number of volts to be carried

60. Field experience has shown that streaming current detectors require very intensive maintenance at water plants utilizing treatment chemicals containing

    a. alum.
    b. iron salts.
    c. short chain polymers.
    d. cationic polymers.

61. Sleeve bearings that are used in a vertical pump's column may be lubricated by

    a. lithium.
    b. unsealed grease packed bearings.
    c. sealed grease packed bearings.
    d. oil drip system.

62. Grease that is water resistant is made with

    a. graphite.
    b. animal fat.
    c. calcium.
    d. potassium phosphate.

63. What is the normal temperature for centrifugal pump motor bearings?

    a. 160°F
    b. 180°F
    c. 195°F
    d. 200°F

64. Where is scale most likely to form in a hypochlorinator?

    a. In the push rod
    b. In the plunger
    c. In the suction and discharge hoses
    d. In the spring

65. Which will a flexible diaphragm coupling compensate for in parallel misalignment?

    a. 1/32 of an inch
    b. 1/16 of an inch
    c. 1/8 of an inch
    d. 1/4 of an inch

66. The booster pump used to provide water pressure necessary for a chlorine system's injector to operate properly is usually a

    a. low-head, low-capacity positive-displacement pump.
    b. high-head, high-capacity positive-displacement pump.
    c. high-head, low-capacity centrifugal-type pump.
    d. low-head, high-capacity centrifugal-type pump.

67. Which type of continuous control will maintain control action when signal is lost?

    a. Floating proportional control
    b. Proportional control
    c. Proportional plus reset control
    d. Proportional plus reset plus derivative control

68. Which method for measuring water levels has a distinct disadvantage where freezing occurs?

   a. Pressure-activated level transmitters
   b. Bubbler units
   c. Float-type sensor
   d. Sonic type indicators

69. Which is the general test period to confirm design capacity for a public water supply well that is in an unconfined aquifer?

   a. At least 24 hours
   b. At least 48 hours
   c. At least 72 hours
   d. At least 96 hours

70. Because vaporizers operate under pressure, they must be designed to meet the _____ code.

   a. American Institute of Physics
   b. ASTM International
   c. American Society of Mechanical Engineers
   d. Building Officials and Code Administrators of America

71. Which temperature range of superheat is considered sufficient, when comparing the off-gas temperature from the vaporizer to its position on the chlorine or ammonia vapor pressure curves?

   a. 5 to 10°C
   b. 10 to 20°C
   c. 20 to 30°C
   d. 30 to 40°C

72. In vaporizer systems, expansion chambers must be installed in a liquid line between any two valves to protect against over-pressurizing the liquid line. The expansion chamber must be sized to protect which percentage of the line volume?

   a. 15%
   b. 20%
   c. 25%
   d. 50%

73. Volumetric feeders are

   a. usually more difficult to operate than gravimetric feeders.
   b. more expensive to purchase than gravimetric feeders.
   c. less expensive to maintain than gravimetric feeders.
   d. more accurate than gravimetric feeders.

74. Which type of continuous control requires meters, transmitters, and transducers?

   a. Feedback control
   b. Feedforward control
   c. Floating control
   d. Proportional control

75. Which type of continuous control has a controller that constantly tries to change the output if there is deviation from the set point?

   a. Proportional plus reset plus derivative control
   b. Proportional control
   c. Floating proportional control
   d. Proportional plus reset control

Additional Practice Questions – Perform Security, Safety, and Administrative Procedures

Answers on page 216

1. Which gas is commonly called swamp gas?

   a. Hydrogen sulfide
   b. Methane
   c. Carbon monoxide
   d. Radon

2. Which statement concerning contact with chlorine is true?

   a. If chlorine contacts the eyes, flush for 15 minutes with water, then neutralize with appropriate electrolytes that are safe for the eyes
   b. Flush the eyes and give a sedative to the person that contacted chlorine, as it usually leads to excited behavior
   c. Apply an appropriate ointment to the area of the skin that liquid chlorine came in contact with
   d. Chlorine inhalation may lead to delayed reactions such as pulmonary edema

3. If sodium hypochlorite comes into contact with the eyes, it should be immediately flushed with water

   a. for at least 15 minutes.
   b. for at least 20 minutes.
   c. for at least 30 minutes.
   d. until the burning sensation stops.

4. Manganese causes a

   a. brown stain.
   b. yellow stain.
   c. black stain.
   d. yellow stain for mn^{+2} and black stain for mn^{+3}.

5. When is the typical minimal demand on a public water system?

   a. 12 midnight
   b. 3 a.m.
   c. 7 a.m.
   d. 12 noon

6. When is the typical peak demand on a public water system?

    a. 12 midnight
    b. 10 a.m.
    c. 3 p.m.
    d. 7 p.m.

7. Which water quality complaint is the most common for most utilities?

    a. Appearance of the water
    b. Taste and odors
    c. Stained laundry and plumbing fixtures
    d. Illness caused by the water

8. Which is true regarding the storage and handling of 150 lb chlorine cylinders?

    a. Chlorine cylinder storage facilities should be made of brick or block material and only have one secured door for protection
    b. Chlorine cylinders stored outside should be stored on elevated racks to help prevent corrosion
    c. Lifting a chlorine cylinder should be done by putting a chain through the holes in the protective hood
    d. Chlorine cylinders should be moved by a hand truck with the restraining chain three-fourths of the way up the cylinder

9. Chemical splash goggles (no face shield) are required PPE when working with potential exposure to 3 to 20% sodium hypochlorite solutions at temperatures below 100° F only when

    a. handling the material.
    b. an initial line break occurs.
    c. inspecting the dome and no product is flowing.
    d. loading that is remotely activated.

10. In permit entry confined space, who is responsible for knowing the conditions within that confined space?

    a. Standby attendant
    b. Entry supervisor
    c. Authorized entrant
    d. Authorized attendant

11. Currently in the United States, which is the most frequently diagnosed waterborne disease?

    a. Gastroenteritis
    b. Giardiasis
    c. Legionnaires disease
    d. Dysentery

12. Infectious hepatitis is caused by a type of

    a. mycobacterium.
    b. basiomycetes.
    c. virus.
    d. *Gonidium.*

13. The largest percentage of reported cases for the transmission of the *Giardia* disease is

    a. water.
    b. food.
    c. swimming in infested waters.
    d. person-to-person contact.

14. Self-contained breathing apparatus (SCBA) units

    a. are very different than the units used by SCUBA divers.
    b. are fitted with a low-air-pressure alarm that sounds, alerting the wearer to leave the contaminated site.
    c. should not be stored in storage rooms far away from the chlorine location, but two units should be stored in the chlorine feed room and two in the chlorine room.
    d. should have all straps rolled up so they can properly fit in their cases.

15. At which concentrations is ammonia gas first detectable by humans?

    a. 2 to 7 ppm
    b. 7 to 15 ppm
    c. 10 to 15 ppm
    d. 20 to 50 ppm

16. In permit entry confined space, who is responsible for knowing the behavioral effects of exposure?

    a. Authorized entrant, entry supervisor and the standby attendant
    b. Entry supervisor and authorized attendant
    c. Authorized attendant
    d. Standby attendant

17. In humans, *Salmonella* will cause

    a. hemochromatosis.
    b. cholera.
    c. gastroenteritis.
    d. dysentery.

18. Which disease has a high mortality rate?

    a. Cholera
    b. Infectious hepatitis
    c. Giardiasis
    d. Cryptosporidiosis

19. Hypochlorite solutions are

    a. highly volatile.
    b. highly acidic.
    c. highly viscous.
    d. soluble in organic solvents.

20. Introducing water into a strange tank containing ammonia vapors can cause

    a. a rapid exothermic reaction.
    b. a rapid endothermic reaction.
    c. the tank to collapse.
    d. an explosion.

21. Which metal reacts with wet chlorine to form an acid?

    a. Platinum
    b. Silver
    c. Copper
    d. Tantalum

22. The IDLH (Immediately Dangerous to Life and Health) represents the maximum concentration from which, if respiratory equipment failed, one could not escape within _____ without a respirator and without experiencing any escape impairing or irreversible health effects.

    a. 10 minutes
    b. 20 minutes
    c. 30 minutes
    d. 60 minutes

23. A benefit of water conservation would be

    a. reduced demand on supply source.
    b. loss of revenue for the utility.
    c. possible stimulation of water service growth.
    d. difficulty dealing with drought conditions.

24. Tastes and odors become more noticeable when the

    a. pH increases.
    b. pH decreases.
    c. temperature increases.
    d. dissolved oxygen increases.

25. At which concentration level can most customers notice iron tastes?

    a. 0.1 mg/L
    b. 0.5 mg/L
    c. 1.0 mg/L
    d. 2.0 mg/L

26. When the iron content of drinking water is too high, coffee can turn very dark. This is caused by

    a. iron being oxidized more readily in the coffee, that is, iron reacting with oxygen.
    b. iron reacting with chlorogenic acids in the coffee.
    c. iron reacting with benzene in the coffee.
    d. iron reacting with tannic acid in the coffee.

27. The principal source of excess lead and copper at the customer's tap is from

    a. transmission lines.
    b. distribution appurtenances.
    c. household plumbing systems.
    d. raw groundwater or surface water.

28. Which emergency kit is for chlorine ton containers?

    a. Kit A
    b. Kit B
    c. Kit C
    d. Kit D

29. Gaseous ammonia may be fatal when it reaches levels of

    a. 400 to 700 ppm.
    b. 800 to 1,100 ppm.
    c. 1,100 to 1,700 ppm.
    d. 2,000 to 3,000 ppm.

30. Which is the size range of *Cryptosporidium*?

    a. 4 to 6 µm
    b. 6 to 8 µm
    c. 10 to 12 µm
    d. 12 to 18 µm

31. Which species causes typhoid?

    a. *Salmonella*
    b. *Shigella*
    c. *Klebsiella*
    d. *Pseudomonas*

32. Which chemical, when it contacts moisture, will produce a cake that is both difficult and hazardous to handle?

    a. Quicklime
    b. Calcium oxide
    c. Calcium hydroxide
    d. Soda ash

33. Which is the highest recommended stacking height for bags of powdered activated carbon and granular activated carbon?

    a. 5 ft
    b. 6 ft
    c. 8 ft
    d. 10 ft

34. How many fusible plugs do 1-ton ammonia cylinders have?

    a. 0
    b. 2
    c. 4
    d. 6

35. Ammonia gas produces general discomfort in humans at concentrations of

    a. 10 to 20 ppm.
    b. 65 to 80 ppm.
    c. 150 to 200 ppm.
    d. 300 to 350 ppm.

36. Prolonged *Entamoeba histolytica* disease can cause amoebic abscesses which usually occur in the

    a. liver.
    b. urinary tract.
    c. small intestine.
    d. kidney.

37. Which is the size range for viruses?

    a. 10 to 25 nanometers
    b. 300 to 900 nanometers
    c. 0.12 to 1.25 µm
    d. 0.75 to 6.30 µm

38. Which is the current OSHA-PEL level for ammonia?

    a. 10
    b. 35
    c. 50
    d. 75

39. Which is the working strength of sodium hydroxide solution in a chlorine neutralization tank?

    a. 15%
    b. 20%
    c. 25%
    d. 30%

40. Which mixture may be produced if ammonia hydroxide is accidentally unloaded into a sodium hypochlorite tank?

    a. Cyanide
    b. Strychnine
    c. An explosive mixture of nitrogen trichloride
    d. An explosive mixture of nitroglycerin

41. Which is the goal for the reduction in water consumption in a Stage III drought or emergency?

    a. 25 to 30%
    b. 35 to 40%
    c. 40 to 45%
    d. 45 to 50%

42. Which is the goal for the reduction in water consumption in a Stage IV drought or emergency?

    a. More than 40%
    b. More than 50%
    c. More than 60%
    d. More than 75%

43. Which national law regulates the storage, transportation, treatment, and disposal of solid and hazardous wastes?

    a. The Toxic Substances Control Act (TSCA)
    b. The Comprehensive Environmental Response, Compensation, and Liability Act (Superfund)
    c. The Resource Conservation and Recovery Act (RCRA)
    d. The Safe Drinking Water Act

44. The minimum standards for specific chemical substance have requirements that are itemized in

    a. OSHA Safety Standards.
    b. *Code of Federal Regulations*.
    c. NSF International Standards.
    d. AWWA Standards.

45. Which is the rationale for establishing the three categories of public water systems?

    a. Billing purposes
    b. Federal funding
    c. State and local funding
    d. Exposure differences to contaminants

46. The American Water Works Association recommends that water treatment chemicals be purchased in quantities that maintain a minimum supply at all times of

    a. 14 days.
    b. 15 days.
    c. 20 days.
    d. 30 days.

47. The Norwalk virus causes

    a. liver disease.
    b. severe kidney infection.
    c. stomach flu symptoms.
    d. lung lesions.

48. When ingested *Giardia* cysts mature and multiply, which problem do they cause?

    a. Interfere with nutrient absorption
    b. Destroy the lining of the stomach
    c. Secrete a toxin, which causes diarrhea
    d. Kill millions of cells in the large intestine, which causes diarrhea

49. Microsporidiosis spores range in size from _____ in diameter.

    a. 0.1 to 1.0 µm
    b. 2.0 to 3.0 µm
    c. 5.0 to 10.0 µm
    d. 25.0 to 75.0 µm

50. Which emergency kit is for chlorine tank trucks?

    a. Kit A
    b. Kit B
    c. Kit C
    d. Kit D

51. Which plan is a mutual aid program for chlorine incidents that occur during transportation or at user locations?

    a. North American Chlorine Emergency Plan
    b. Chemical Transportation Emergency Plan
    c. Chlorine Institute Plan
    d. Transportation Emergency Assistance Plan

52. Which is the discharge concentration limit for chlorine vapor from a neutralization system?

    a. 1 ppm
    b. 2 ppm
    c. 5 ppm
    d. 10 ppm

53. Sites using chlorine in a single process are required by the US Environmental Protection Agency to write a risk management plan (RMP), if chlorine exceeds

    a. 2,000 lb.
    b. 2,500 lb.
    c. 4,000 lb.
    d. 6,000 lb.

54. When a water source is destroyed or damaged due to some disaster that causes total contamination, which would be a long term goal?

    a. Clean up the source of contamination caused by the disaster
    b. Draw water from an adjoining system or systems
    c. Supply customers with bottled water
    d. Commence conservation or rationing

55. Water utilities that make a sewer system available to customers that previously had a private septic system can expect the water use to increase by

    a. 20 to 25%.
    b. 30 to 50%.
    c. 50 to 100%.
    d. more than 150%.

56. Which is a potential problem for water conservation?

    a. Reduced costs
    b. Possible stimulation of water service growth
    c. Energy savings
    d. Reduction of wastewater flow

57. The injection of water into the ground for aquifer recharge is subject to federal and state approval. This approval is under the rules of which program?

    a. US Environmental Protection Agency
    b. Resource Conservation and Recovery Act
    c. Federal Groundwater Rules
    d. Federal Underground Injection Control Program

58. The "rate of reasonable sharing" of water is usually called the

    a. Riparian doctrine.
    b. Legal use doctrine.
    c. Priority use doctrine.
    d. Beneficial use doctrine.

59. When a landowner can "capture" all the water under only their land, it is called

    a. correlative rights.
    b. absolute ownership.
    c. reasonable use.
    d. appropriation-permit systems.

60. When a landowner has the right to groundwater as long as it does not harm a neighbor, it is called

    a. correlative rights.
    b. absolute ownership.
    c. reasonable use.
    d. appropriation-permit systems.

61. Which rule applies when prorating the available water supply based on the overlying size of the land parcel?

    a. Correlative rights
    b. Absolute ownership
    c. Reasonable use
    d. Appropriation-permit systems

62. Which rule applies when the most important issue is priority, where water rights are based on who has used the water for the longest time?

    a. Correlative rights
    b. Absolute ownership
    c. Reasonable use
    d. Appropriation-permit systems

63. The US Environmental Protection Agency estimates that the radioactivity in potable water contributes about which percentage of the annual dose received by an average person?

    a. 0.1 to 3.0%
    b. 3.0 to 5.0%
    c. 4.0 to 7.0%
    d. 5.0 to 10.0%

64. Which has the USEPA chosen to suggest assaying water for as indicators for the inactivation of viruses?

    a. Total coliforms
    b. Fecal coliforms
    c. *E-coli*
    d. Coliphages

65. Which date does the climatic year for the U.S. Geological Survey use in its publication of stream flow records?

   a. October 1
   b. June 1
   c. July 1
   d. January 1

66. Coccidioidomycosis is caused by

   a. viruses.
   b. bacteria.
   c. fungi.
   d. protozoans.

67. Consistency is maintained by providing certified laboratories with a known concentration of a contaminant. Who oversees the analytical results of the samples provided to the certified laboratories?

   a. USEPA and state primacy programs
   b. Primacy agency for each state
   c. American Chemical Society
   d. US Department of Public Health

68. Which percentage equipment standby for chlorination facilities do most state regulations require?

   a. 50%
   b. 75%
   c. 90%
   d. 100%

69. Water treatment plants are increasingly using hypochlorination because it is

   a. relatively safe compared to using gaseous chlorine.
   b. very inexpensive compared to other disinfectants.
   c. a stronger oxidant than gaseous chlorine.
   d. not going to form trihalomethanes.

70. Which bacteria genus has the bacteria species that cause anthrax?

   a. *Clostridium*
   b. *Enterococcus*
   c. *Bacillus*
   d. *Yersinia*

Additional Practice Questions – Evaluate Characteristics of Source Water

Answers on page 221

1. Which adverse effects does the secondary contaminant iron have?

   a. Unappealing to drink, undesirable taste, and possible indication of corrosion
   b. Discolored laundry brown and changed taste of water, coffee, tea, and other beverages
   c. Undesirable metallic taste and possible indication of corrosion
   d. Added total dissolved solids and scale, indication of sewage contamination, and tastes

2. Which gas occurs mainly in groundwater, is heavier than air, and is odoriferous?

   a. Hydrogen sulfide
   b. Carbon dioxide
   c. Radon
   d. Methane

3. Which gas is radioactive, occurs in many groundwater supplies, and is colorless and odorless?

   a. Neon
   b. Argon
   c. Boron
   d. Radon

4. Groundwater pH is generally

   a. Exactly 7.0 units.
   b. 4.0 to 6.0 units.
   c. 6.0 to 8.5 units.
   d. 8.5 to 9.5 units.

5. Surface water pH is generally

   a. Exactly 7.0 units.
   b. 4.0 to 6.0 units.
   c. 6.5 to 8.5 units.
   d. 8.5 to 9.5 units.

6. The movement of water from leaves, grasses, and other plants to the atmosphere is called

   a. organic evaporation.
   b. condensation.
   c. transpiration.
   d. capillary action.

7. A term used to describe cloud formation is

   a. transpiration.
   b. evaporation.
   c. condensation.
   d. precipitation.

8. A diatom is a type of

   a. bacterium.
   b. algae.
   c. virus.
   d. protozoan.

9. Which radioactive nuclide is a gas?

   a. Radium
   b. Uranium
   c. Radon
   d. Thorium

10. Which is the term applied to all suspended matter in water?

    a. Colloids
    b. Nonsettleable particulates
    c. Stabilized particulates
    d. Turbidity

11. Which is the safe dosage for most species of fish when using copper sulfate to control algae in a body of water?

    a. 0.3 mg/L
    b. 0.5 mg/L
    c. 0.8 mg/L
    d. 1.0 mg/L

12. A lake would be considered polluted if it had a preponderance of

    a. green algae.
    b. blue-green algae.
    c. yellow-green algae.
    d. golden-brown algae.

13. Which of the following is easiest to control at the water treatment plant?

    a. *Giardia*
    b. *Cryptosporidium*
    c. Coliform bacteria
    d. Viruses

14. Enteric protozoans

    a. usually have four stages in their life cycles.
    b. are easy to culture.
    c. have cysts that can penetrate filters.
    d. can be found in very high densities in most lakes.

15. Which type of adverse effects does the secondary contaminant color have?

    a. Unappealing appearance and indication that dissolved organics may be present
    b. Added total dissolved solids and scale, indication of contamination, and tastes
    c. Taste, scale, corrosion, and hardness
    d. Undesirable taste and appearance

16. The most common forms of manganese found in nature are

    a. sulfides, inosilicates, and native metal.
    b. sulfides and phosphates.
    c. oxides, carbonates, and hydroxides.
    d. phyllosilicates, tectosilicates, and inosilicates.

17. Precursors to the formation of trihalomethanes (THMs) would most likely come from which source?

    a. Domestic and commercial activities
    b. Wastewater treatment plants and industrial waters
    c. Humic materials
    d. Reactions that occur within the treatment plant

18. Immediately below the water table is the

    a. piezometric surface.
    b. capillary zone.
    c. saturated zone.
    d. recharge zone.

19. Which type of wells are commonly used near the shore of a lake or near a river?

    a. Monitoring wells
    b. Bedrock wells
    c. Gravel wall wells
    d. Radial wells

20. Many streams would dry up after a rain if it were not for

    a. human construction influences.
    b. surface runoff.
    c. capillary action.
    d. groundwater flow.

21. Which range represents the usual levels of carbon dioxide in surface waters?

    a. 0 to 5 mg/L
    b. 5 to 10 mg/L
    c. 10 to 25 mg/L
    d. 25 to 40 mg/L

22. Microorganisms from smallest to largest are as follows:

    a. Viruses, protozoans, and bacteria
    b. Bacteria, viruses, and protozoans
    c. Viruses, bacteria, and protozoans
    d. Protozoans, bacteria, and viruses

23. Water channels will most likely occur in which type of rock?

    a. Sandstone cemented with calcite
    b. Limestone
    c. Volcanic breccia
    d. Mélange

24. An artesian aquifer could occur in a(n)

    a. confined aquifer.
    b. unconfined aquifer.
    c. water table aquifer.
    d. shale formation.

25. Immediately above the water table is the

    a. aquifer zone.
    b. unsaturated zone.
    c. unconfined aquifer.
    d. recharge zone.

26. Which would most likely be the source of contaminants that include pesticides, herbicides, and nitrates?

    a. Land application of wastewater
    b. Agricultural activities
    c. Municipal landfills
    d. Liquid waste leaching ponds

27. Which statement is true concerning dissolved oxygen (DO)?

    a. Warm water contains more DO than cold water
    b. Cold water contains more DO than warm water
    c. Saline water contains more DO than fresh water
    d. Temperature and salinity have no impact on DO

28. The copper sulfate dose for controlling algae in lakes is predominately based on a lake's

    a. temperature.
    b. pH.
    c. turbidity.
    d. alkalinity.

29. To obtain the best and most lasting control of algae, which is the optimum alkalinity range of the water being treated when using copper sulfate for algae control?

    a. <50 mg/L
    b. 50–75 mg/L
    c. 76–100 mg/L
    d. >101 mg/L

30. For operational corrosion control, the total alkalinity concentration should be measured every

    a. 8 hours.
    b. 12 hours.
    c. 4 hours.
    d. Week.

31. How often should the Langelier Index of the raw and treated water be calculated?

    a. Every 8 hours
    b. Every 12 hours
    c. Every day
    d. Every week

32. The process whereby water moves with the air currents in the atmosphere is called

    a. transpiration.
    b. evaporation.
    c. interception.
    d. advection.

33. The polarity of water causes the hydrogen part of one water molecule to be weakly bonded to the oxygen of another water molecule. This bonding is called

    a. ionic.
    b. covalent.
    c. van der Waals.
    d. hydrogen bonding.

34. Oligotrophic is defined as a(n)

    a. organism that requires very little light.
    b. organism that feeds on the bottom.
    c. low nutrient body of water with abundant oxygen.
    d. organism that is sessile in the immature stage, but later becomes free swimming in the adult stage.

35. Where will lake water first become anaerobic ?

    a. In the hypolimnion
    b. In the mesolimnion
    c. In the thermocline
    d. In the epilimnion

36. Which type of hardness is considered permanent hardness?

    a. Carbonate hardness
    b. Noncarbonated hardness
    c. Calcium hardness
    d. Magnesium hardness

37. Streams that flow only occasionally are called

    a. annual streams.
    b. perennial streams.
    c. ephemeral streams.
    d. intermittent streams.

38. Which is the main problem that dinoflagellates cause in water supplies?

    a. Oxygen depletion
    b. Clogging filters
    c. Taste and odors
    d. Toxicity

39. Which group of algae do *Euglena* species belong to?

    a. Green algae
    b. Blue-green algae
    c. Flagellates
    d. Diatoms

40. The purple or green sulfur bacteria

    a. are able to photosynthesize.
    b. require sulfur as well as a rich oxygen environment for their metabolism.
    c. oxidize elemental sulfur to form sulfides.
    d. produce organic acids, sulfuric acids, and hydrogen sulfide during their metabolism.

41. Which organisms are prokaryotes and release odorous compounds such as geosmin and 2-methylisoborneal?

    a. *Methylomonas*
    b. *Chlorobium*
    c. *Clostridium*
    d. Cyanobacteria

42. Which type of adverse effect does the secondary contaminant zinc have?

    a. Unappealing taste and possible indication of contamination
    b. Added total dissolved solids and stained laundry
    c. A laxative effect and undesirable metallic taste
    d. Undesirable taste and a milky appearance

43. Which statement is true concerning dissolved oxygen?

    a. Dissolved oxygen will increase the absorption of carbon dioxide
    b. High dissolved oxygen has adverse health effects
    c. Most consumers do not prefer high levels of dissolved oxygen in the water
    d. Dissolved oxygen may oxidize iron and manganese

44. Which is the most common radionuclide in water?

    a. Radium
    b. Uranium
    c. Radon
    d. Thorium

45. How much water a material will hold depends on its

    a. size and volume.
    b. permeability.
    c. hydraulic conductivity.
    d. porosity.

46. Which would most likely be the source of contaminants that include nitrates, sulfates, organic compounds (solvents), sodium, and microbiological contaminants?

    a. Septic tanks and leech fields
    b. Agricultural activities
    c. Infiltration of urban runoff
    d. Municipal landfills

47. Bored wells, constructed with an auger driven into the earth, are limited in depth to approximately

    a. 20 ft.
    b. 30 ft.
    c. 50 ft.
    d. 60 ft.

48. During a pumping test for a public water supply well, which must be held constant?

    a. The pumping water level
    b. The pump's amperage
    c. The drawdown
    d. The pumping rate

49. Destratification of a lake will often reduce

    a. turbidity.
    b. nutrients.
    c. fish.
    d. algae.

50. Zebra mussels multiply and spread so fast because

    a. the female produces over 250,000 eggs in a single season.
    b. the larvae are free swimming for 2 or 3 months before attaching themselves.
    c. they commonly adhere to boats so they move around and disperse eggs.
    d. they can live out of water for as long as 3 months.

51. Which raw water characteristic may require the addition of a weighing agent?

    a. High pH
    b. Low water temperature
    c. High organic content
    d. Low turbidity

52. Which is an unconsolidated material?

    a. Limestone
    b. Sand
    c. Fractured rock
    d. Volcanic rock

53. Beta radiation consists of a(n)

    a. electron.
    b. neutron.
    c. proton.
    d. muon.

54. Which is the unit of gamma radiation?

    a. Muon
    b. Neutrino
    c. X-ray
    d. Photon

55. Which of the following causes lung cancer?

    a. Uranium
    b. Radium
    c. Radon
    d. Thorium

56. Which genera of cyanobacteria primarily release neurotoxins?

    a. *Nodularia*
    b. *Microcystis*
    c. *Anabaena*
    d. *Cylindrospermopsis*

57. Which one of the following would be best to use for controlling algae in large water bodies?

    a. Hydrogen peroxide
    b. Granular activated carbon
    c. Powdered activated carbon
    d. Magnesium sulfate

58. Trout can be killed if the copper sulfate application exceeds what dosage level?

   a. 0.08 mg/L
   b. 0.10 mg/L
   c. 0.12 mg/L
   d. 0.14 mg/L

59. Which method would be the most effective treatment for zebra mussels at the inlet?

   a. Continuous dosing of chloramines at 0.5 to 1.0 mg/L
   b. Continuous dosing of chlorine at 0.5 to 1.0 mg/L
   c. Shock dosages of potassium permanganate at 1.0 to 2.0 mg/L for at least 15 minutes
   d. Shock treatment using chlorine at a dosage of 10.0 mg/L for 30 minutes

60. The bulk of synthetic organic compounds (SOCs) are

   a. solvents.
   b. PAHs, PCBs, and polynuclear aromatic hydrocarbons.
   c. pesticides.
   d. volatiles.

61. Which naturally occurring radionuclide is most prevalent in drinking water?

   a. Gamma
   b. Alpha
   c. Beta
   d. Muson

62. If artificial radionuclides are present in the environment, which element is most likely to be found?

   a. Potassium 20
   b. Carbon 14
   c. Argon 40
   d. Tritium

63. How much water an aquifer will yield depends

   a. on its size or volume.
   b. on its porosity and permeability.
   c. almost entirely on the type of rock formation.
   d. on the pore spaces.

64. When drilling a well it is important to identify (log) the various formations below the surface. Which type of logging measures how controlled radiation penetrates the different formations below the surface?

   a. Electric logging
   b. Magnetic-reversal logging
   c. Gamma-ray logging
   d. X-ray logging

65. Which would most likely be the source of contaminants that include inorganic compounds, petroleum products, and heavy metals?

   a. Land application of wastewater
   b. Deicing activities to control snow and ice on roads
   c. Liquid waste storage ponds
   d. Infiltration of urban runoff

66. Which is the general test period to confirm design capacity for a public water supply well that is in a confined aquifer?

    a. At least 24 hours
    b. At least 48 hours
    c. At least 72 hours
    d. At least 96 hours

67. Zebra mussels will breed only when the water temperature is above

    a. 8°C.
    b. 12°C.
    c. 16°C.
    d. 20°C.

68. A conventional water treatment plant in a rich farming area has to treat water high in nitrates. Which is its best course of action in the long term?

    a. Manage watershed to prevent nitrates from entering the water system
    b. Change its treatment to ion exchange
    c. Change its treatment to reverse osmosis
    d. Add a strong oxidation chemical before coagulation

69. Water primarily becomes capable of dissolving iron and manganese because

    a. of acid rain.
    b. oxygen reacts with iron and manganese.
    c. oxygen is removed by the decomposition of organic materials making the water capable of dissolving iron and manganese.
    d. erosional forces bring iron and manganese into contact with water capable of dissolving iron and manganese due to acid rain.

70. The outside boundary of a watershed would be defined by a

    a. saddle.
    b. ridge.
    c. river.
    d. county, state, or country boundary.

71. Which national law authorizes the government to clean up contaminants from hazardous waste sites or contaminants caused by chemical spills that could possibly threaten the environment?

    a. The Toxic Substances Control Act (TSCA)
    b. The Comprehensive Environmental Response, Compensation, and Liability Act (Superfund)
    c. The Resource Conservation and Recovery Act (RCRA)
    d. The Safe Drinking Water Act

72. Viruses are considered

    a. dissolved solids.
    b. colloidal solids.
    c. flocculated solids.
    d. suspended solids.

73. In general, higher removal is required as the TOC _____ and the alkalinity _____.

    a. increases; decreases
    b. decreases; increases
    c. increases; increases
    d. decreases; decreases

74. Algae can be very beneficial, but they cause a problem when they remove

    a. phosphorous.
    b. ammonium.
    c. nitrates.
    d. iron.

75. Which is the source of most radioactive contamination in water?

    a. It is naturally occurring
    b. It comes from medical waste deposited in landfills that works its way into the water cycle
    c. It comes from improper mining operations
    d. It comes from nuclear power plant waste, past nuclear testing, and nuclear plant accidents

Math for Water Treatment Operators Levels I & II

Sample Questions

Answers on Page 229

1. Which is the hardness in mg/L of a treatment plant's well water if the hardness is 18.44 grains per gallon (gpg)?

    a. 1.1 mg/L
    b. 315.7 mg/L
    c. 415.7 mg/L
    d. 535.2 mg/L

2. Find the specific yield in gpm/ft if a well produces 105 gpm and the drawdown for the well is 16.3 ft.

    a. 6.00 gpm/ft
    b. 6.44 gpm/ft
    c. 7.20 gpm/ft
    d. 7.28 gpm/ft

3. If the static level in the well was 138.6 ft below ground level and the drawdown was 21.1 ft, which must have been the pumping water level in the well?

    a. 117.5 ft
    b. 129.0 ft
    c. 150.0 ft
    d. 159.7 ft

4. Which is the pressure in lb/ft^2, 189 ft below a lake's surface if the lake is 386 ft in depth?

    a. 302 lb/ft^2
    b. 1,169 lb/ft^2
    c. 11,790 lb/ft^2
    d. 72,950 lb/ft^2

5. A circular clarifier has a weir length of 162 ft. Which is the weir overflow rate in gpd/ft, if the flow is 2,330,000 gallons per day (gpd)?

    a. 4,580 gpd/ft
    b. 14,380 gpd/ft
    c. 28,770 gpd/ft
    d. 65,530 gpd/ft

6. Convert 35.1 cfs to gpm.

    a. 14,200 gpm
    b. 15,800 gpm
    c. 17,600 gpm
    d. 18,300 gpm

7. Convert 7.7 million gallons a day (mgd) into cubic feet per second (cfs).

   a. 11 cfs
   b. 12 cfs
   c. 15 cfs
   d. 19 cfs

8. How many million gallons (mil gal) are there in 318 acre-ft?

   a. 104 mil gal
   b. 107 mil gal
   c. 110 mil gal
   d. 116 mil gal

9. Convert 68 degrees Fahrenheit to degrees Celsius.

   a. 20°C
   b. 37°C
   c. 45°C
   d. 65°C

10. Calculate 81.5% of 316.

    a. 219
    b. 232
    c. 258
    d. 267

11. If 8.25 pounds of soda ash are mixed into 45 gallons of water, which is the percent of soda ash in the slurry?

    a. 2.0% soda ash slurry
    b. 2.1% soda ash slurry
    c. 2.2% soda ash slurry
    d. 2.3% soda ash slurry

12. Calculate the area of a circular reservoir in ft^2, with a diameter of 411 ft.

    a. 108,000 ft^2
    b. 112,000 ft^2
    c. 125,000 ft^2
    d. 133,000 ft^2

13. Determine the circumference of a clarifier, if the radius is 95 ft.

    a. 300 ft
    b. 400 ft
    c. 500 ft
    d. 600 ft

14. Determine the volume in gallons for a pipe completely full of water given the following data:

    - Diameter = 1.5 feet
    - Length = 1.09 miles

    a. 65,000 gal
    b. 68,000 gal
    c. 74,000 gal
    d. 76,000 gal

15. Which is the concentration of alum in mg/L, if 5.0 mL of a 0.30 grams/liter alum solution is added to 1,000 mL of deionized water?

    a. 1.2 mg/L alum
    b. 1.5 mg/L alum
    c. 1.8 mg/L alum
    d. 1.9 mg/L alum

16. Which is the phenolphthalein alkalinity as mg/L $CaCO_3$ of a water sample given the following parameters?

    - Sample size = 100 mL
    - Normality of the sulfuric acid = 0.02 N
    - Titrant used to pH of 8.3 = 1.8 mL (designated by convention as A)

    a. 1.8 mg/L as $CaCO_3$
    b. 3.6 mg/L as $CaCO_3$
    c. 18 mg/L as $CaCO_3$
    d. 36 mg/L as $CaCO_3$

17. How many gallons are there in 28.65 acre-ft?

    a. 9,354,282 gal
    b. 9,322,137 gal
    c. 9,335,000 gal
    d. 9,763,599 gal

18. If 7.3 lb of polymer (assume 100%) are mixed into 35 gal of water, determine the percentage of polymer in the slurry.

    a. 2.1% slurry
    b. 2.4% slurry
    c. 2.5% slurry
    d. 2.8% slurry

19. Which is the percentage of removal across a settling basin, if the influent is 17.1 ntu and the effluent is 1.13 ntu?

    a. 90.5% ntu removed
    b. 92.5% ntu removed
    c. 93.0% ntu removed
    d. 93.4% ntu removed

20. Find the detention time in hours for a clarifier that has an inner diameter of 112.2 ft and a water depth of 10.33 ft if the flow rate is 7.26 mgd.

    a. 2.10 hr
    b. 2.14 hr
    c. 2.52 hr
    d. 2.96 hr

142   WATER OPERATOR CERTIFICATION STUDY GUIDE

21. Calculate the lime dosage in mg/L that is required given the following parameters:
    - Jar test determines the alum dosage = 8.5 mg/L
    - Raw alkalinity = 9.0 mg/L
    - Residual alkalinity needed for precipitation = 14 mg/L

    Know: 1 mg/L of alum reacts with 0.45 mg/L alkalinity
           1 mg/L of alum reacts with 0.35 mg/L lime

    a. 6.9 mg/L lime
    b. 11.3 mg/L lime
    c. 11.34 mg/L lime
    d. 20.9 mg/L lime

22. A dosage of 0.35 mg/L of copper sulfate pentahydrate is desired to control algae in an 8,850 acre-ft capacity reservoir. If the available copper is 25%, how many pounds of copper sulfate pentahydrate are required?
    Know: 1 ac-ft = 43,560 ft^3

    a. 2,884 lb copper sulfate
    b. 4,500 lb copper sulfate
    c. 8,418 lb copper sulfate
    d. 33,672 lb copper sulfate

23. Determine the specific gravity (SG) for a solution that weighs 11.87 lb/gal.

    a. 1.38 SG
    b. 1.40 SG
    c. 1.42 SG
    d. 2.58 SG

24. A filter is 24 ft by 28 ft. Calculate the filtration rate in gpm, if it receives a flow of 3,250 gpm.

    a. 4.4 gpm/ft^2
    b. 4.8 gpm/ft^2
    c. 5.0 gpm/ft^2
    d. 5.1 gpm/ft^2

25. Determine the backwash rate in gpm/ft^2 given the following:
    - Backwash flow of 13 cfs (ft^3/sec)
    - Filter is 25 ft by 18.2 ft

    a. 12.8 gpm/ft^2
    b. 12.9 gpm/ft^2
    c. 13 gpm/ft^2
    d. 14 gpm/ft^2

26. Calculate the backwash pumping rate if a filter requires a backwash rate of 18 gpm/ft^2 and the filter is 20.0 ft by 24.0 ft.

    a. 8,600 gpm
    b. 8,640 gpm
    c. 8,700 gpm
    d. 8,780 gpm

27. Which is the raw water alkalinity in mg/L as $CaCO_3$ of a water sample with a beginning pH of 7.18, given the following parameters?

    - Sample size = 100 mL
    - Normality of the sulfuric acid = 0.02 N
    - Titrant used to pH of 4.6 = 11.4 mL

    a. 100 mg/L as $CaCO_3$
    b. 110 mg/L as $CaCO_3$
    c. 114 mg/L as $CaCO_3$
    d. 120 mg/L as $CaCO_3$

28. A chemical metering pump is pumping 26.3 gallons per day (gal/day). How many mL/min is this?

    a. 47.6 mL/min
    b. 60.8 mL/min
    c. 69.1 mL/min
    d. 75.4 mL/min

29. Find the density of a solution, if it weighs 112.7 lb/ft^3.

    a. 1.81
    b. 2.02
    c. 2.28
    d. 2.40

30. Find the drawdown of a well that has a specific yield of 28.4, if the well yields 325 gpm.

    a. 9.8 ft
    b. 11.4 ft
    c. 12.9 ft
    d. 14.1 ft

31. A water treatment plant has an emergency shutdown. How many water supply hours are left in a 119.8-ft diameter tank given the following data?

    - Tank's water level = 27.6 ft
    - Water cannot go below 16.0 ft at any time to comply with fire control
    - Water usage averages 483 gpm

    a. 33.7 hr
    b. 35.0 hr
    c. 35.4 hr
    d. 36.2 hr

32. A lime tank is conical at the bottom and cylindrical at the top. If the diameter of the cylinder is 14 ft with a depth of 24 ft and the cone depth is 12.5 ft, calculate the volume of the tank in cubic feet.

    a. 3,700 ft^3
    b. 4,000 ft^3
    c. 4,200 ft^3
    d. 4,300 ft^3

33. Flow through a channel 5.8 ft wide is 20.3 cfs. If the velocity is 1.4 ft/sec, how deep is the water in the channel?

   a. 2.3 ft
   b. 2.5 ft
   c. 2.6 ft
   d. 2.7 ft

34. A lake is 107 ft deep. Which is the psi on the bottom?

   a. 44.8 psi
   b. 45.2 psi
   c. 45.6 psi
   d. 46.3 psi

35. How many gallons per day flow through a 2.18-mil gal capacity sedimentation basin, if the detention time is 2.79 hr?

   a. 17,100,000 gal/day
   b. 18,800,000 gal/day
   c. 19,700,000 gal/day
   d. 20,300,000 gal/day

36. An ion exchange softener is treating a flow rate of 245 gpm. Which is the operating time in hours, if the softener unit treats 434,000 gal before it requires regeneration?

   a. 27.7 hr
   b. 28.6 hr
   c. 29.5 hr
   d. 30.1 hr

37. Zinc orthophosphate (ZOP) is used at a treatment plant for corrosion control. The plant is treating 12.1 mgd with a dosage of 0.15 mg/L. Determine the feeder setting for ZOP in mL/min, if the specific gravity of the ZOP is 1.63.

   a. 2.9 mL/min of ZOP
   b. 3.2 mL/min of ZOP
   c. 3.5 mL/min of ZOP
   d. 3.9 mL/min of ZOP

38. Determine the feed rate for alum in mL/min with the following conditions:

   - Plant flow = 30.9 mgd
   - Alum dosage rate = 10.4 mg/L
   - Alum percentage = 48.4%
   - Alum specific gravity = 1.31

   a. 1,250 mL/min
   b. 1,270 mL/min
   c. 1,330 mL/min
   d. 1,410 mL/min

39. A 1.50-ft diameter pipe that is 1.62 miles long was disinfected with chlorine. If 47.2 lb of chlorine were used, which was the dosage in mg/L?

    a. 25.0 mg/L
    b. 30.0 mg/L
    c. 50.0 mg/L
    d. 60.0 mg/L

40. Calculate the specific gravity (SG) for an unknown liquid with a density of 87.6 lb/ft^3.

    a. 1.40 SG
    b. 1.43 SG
    c. 1.51 SG
    d. 1.62 SG

Math for Water Treatment Operators Levels III & IV

Sample Questions

Answers on Page 237

1. The alum dosage for a plant with a flow of 26.5 cfs is 655 mL/min. If the raw water flow rate is adjusted to 18.5 cfs, which should be the theoretical alum dosage in mL/min, if all water parameters remain the same?

    a. 410 mL/min
    b. 418 mL/min
    c. 436 mL/min
    d. 457 mL/min

2. Which is the percentage recovery for a reverse osmosis unit with a 4-2-1 arrangement given the following data?

    - Product flow is 570 gpm
    - Feed flow is 1.03 mgd

    a. 79.7%
    b. 80.0%
    c. 80.2%
    d. 80.5%

3. How many lb/day of sodium fluorosilicate (Na_2SiF_6) are required given the following parameters?

    - Flow rate is 1,750 gpm
    - Fluoride desired is 1.20 mg/L
    - Fluoride in raw water is 0.15 mg/L
    - Sodium fluorosilicate is 98.1% pure
    - Fluoride (F) ion percent is 60.6%

    a. 34 lb/day, F
    b. 37 lb/day, F
    c. 42 lb/day, F
    d. 48 lb/day, F

4. An ion exchange softener is treating a flow rate of 125 gpm. Which is the operating time in hours if the softener unit treats 297,000 gallons before it requires regeneration?

    a. 37 hrs
    b. 39.6 hrs
    c. 40 hrs
    d. 41.1 hrs

148   WATER OPERATOR CERTIFICATION STUDY GUIDE

5. A water treatment plant has 8 filters with an average flow rate of 4.89 gpm/ft². If the plant flow is 32.7 cfs, what is the filtration area of each filter?

   a. 375 ft²/filter
   b. 398 ft²/filter
   c. 400 ft²/filter
   d. 410 ft²/filter

6. Calculate the amount of iron removed in pounds per year from a water plant that treats an average of 20.2 mgd if the average iron concentration is 0.52 mg/L and the removal efficiency is 84%.

   a. 26,000 lb/yr of Fe removed
   b. 27,000 lb/yr of Fe removed
   c. 31,975 lb/yr of Fe removed
   d. 32,000 lb/yr of Fe removed

7. Determine the percent mineral rejection from a reverse osmosis plant if the feedwater contains 1,230 mg/L TDS and the product water contains 135 mg/L TDS.

   a. 88%
   b. 89%
   c. 90%
   d. 91%

8. Calculate the log removal for a water treatment plant if the samples show a raw water coliform count of 295/100 mL (through extrapolation) and the finished water shows 2.0/100 mL.

   a. 1.8 log removal
   b. 1.9 log removal
   c. 2.0 log removal
   d. 2.2 log removal

9. A 0.25 Normal solution of H_3PO_4 (phosphoric acid) is to be prepared. If 4.5 liters of solution is desired, how many grams of H_3PO_4 are required? The gram formula for H_3PO_4 is 98.00. Give answer to nearest 100th of a gram.

   a. 8.17 grams
   b. 24.50 grams
   c. 36.75 grams
   d. 42.31 grams

10. A 2.00% stock polymer solution (20,000 ppm or 20,000 mg/L) is desired for performing a jar test. If the polymer has a specific gravity of 1.27 and is 84.5% polymer, how many milliliters are required to make exactly 1,000 mL stock solution?

    a. 12.3 mL polymer
    b. 15.3 mL polymer
    c. 18.6 mL polymer
    d. 21.8 mL polymer

11. Calculate the theoretical detention time in hours for the following water treatment plant:

    - Flow rate of 12.2 mgd
    - Four flocculation basins measuring: 45.0 ft by 10.0 ft by 11.0 ft in average depth each
    - Sedimentation basin measuring: 285 ft by 65.0 ft by 11.4 ft in average depth
    - Eight filters measuring: 35.0 ft by 28.0 ft by 12.3 ft in depth each
    - Clear well averages 2.05 million gallons (mil gal)

    a. 7.61 hr
    b. 8.60 hr
    c. 8.78 hr
    d. 8.85 hr

12. Calculate the CT and inactivation ratio for a water treatment plant that has the following parameters; and does this treatment facility meet the CT? (see appendix B)

    - Daily Parameters:
    - Detention time = 83 min
    - pH = 7.6
    - Lowest Temperature = 12°C
    - Lowest chlorine residual = 0.60 mg/L
    - A 1.5 log removal is required for this system

    a. 0.86 inactivation ratio
    b. 1.2 inactivation ratio
    c. 1.3 inactivation ratio
    d. 1.35 inactivation ratio

13. Calculate the feed rate for fluorosilicic acid in mL/min given the following data:

    - Flow rate is 11.8 mgd
    - Fluoride desired is 1.20 mg/L
    - Fluoride in raw water is 0.20 mg/L
    - Treated with 20.5% solution of H_2SiF_6
    - Fluoride ion percent is 79.0%
    - H_2SiF_6 weighs 9.8 lb/gal

    a. 33 mL/min, H_2SiF_6
    b. 62 mL/min, H_2SiF_6
    c. 145 mL/min, H_2SiF_6
    d. 163 mL/min, H_2SiF_6

MOLECULAR WEIGHTS OF CHEMICAL COMPOUNDS	
COMPOUND	MOLECULAR WEIGHT
Alkalinity, as $CaCO_3$	100.1
Carbon Dioxide, CO_2	44.0
Hardness, as $CaCO_3$	100.1
Hydrated Lime, $Ca(OH)_2$	74.1
Magnesium, Mg^{2+}	24.3
Magnesium Hydroxide, $Mg(OH)_2$	58.3
Quicklime, CaO	56.1
Soda Ash, Na_2CO_3	106.0

Use this table to solve problem 14.

14. Determine the hydrated lime dose required in mg/L for water with the following characteristics:

	Source Water	Softened Water
Total Alkalinity, mg/L	165 mg/L as $CaCO_3$	39 mg/L
Total Hardness, mg/L	248 mg/L as $CaCO_3$	76 mg/L
CO_2, mg/L	13 mg/L	0 mg/L
Mg^{2+}	21 mg/L	7.8 mg/L
pH	7.0	7.8
Lime Purity	92%	

Use an excess lime dosage of 15% (115% or 1.15 in decimal form)

    a.   190 mg/L, $Ca(OH)_2$
    b.   194 mg/L, $Ca(OH)_2$
    c.   195 mg/L, $Ca(OH)_2$
    d.   200 mg/L, $Ca(OH)_2$

15. What is the flow through a membrane unit in gpd/ft^2, if the water flux of the unit is 4.75×10^{-4} $gm/cm^2/s$?

    a.   10 gpd/ft^2
    b.   10.1 gpd/ft^2
    c.   100 gpd/ft^2
    d.   101 gpd/ft^2

16. A conventional water treatment plant had to discontinue pre-chlorination, that is, no addition of chlorine to the flocculation basins and the sedimentation basin due to elevated trihalomethane levels. Consequently, the chlorine dose was increased before the filters and the clear well and a lithium chloride tracer study was performed. The plant requires a 1.5 log removal for *Giardia* cysts. Given the following parameters on the first day of this process change and referring to the CT values table in Appendix B, determine if this plant is in CT compliance:

Unit Process or Piping	T^{10} Value, Min	Lowest Chlorine Residual
Filtration	12	0.45 mg/L
Piping (filter to clear well)	4.5	0.40 mg/L
Clearwell	51	1.20 mg/L

Unit Process or Piping	Temperature	pH	CT Value, Tables
Filtration	11	6.9	47.5
Piping (filter to clear well)	11	6.9	47.1
Clear well	12	7.6	45.5

   a. 0.9 inactivation ratio, plant is out of compliance
   b. 1.1 inactivation ratio, plant is in compliance
   c. 1.13 inactivation ratio, plant is in compliance
   d. 1.27 inactivation ratio, plant is in compliance

17. A 5-min drawdown test result showed that 106 mL of a polymer aid was being used to help treat the raw water. The specific gravity (SG) of the polymer aid is 1.26. If the plant is treating 3,225 gpm, what is the polymer dosage in mg/L? Give answer to three significant figures.

   a. 1.74 mg/L polymer aid
   b. 2.19 mg/L polymer aid
   c. 10.94 mg/L polymer aid
   d. 15.12 mg/L polymer aid

18. A lime tank is conical at the bottom and cylindrical at the top. If the diameter of the cylinder is 15 ft with a depth of 28 ft and the cone depth is 12 ft, what is the volume of the tank in cubic feet? Give answer to three significant figures.

   a. 1,040 ft³
   b. 5,510 ft³
   c. 5,650 ft³
   d. 7,060 ft³

19. A watershed, 158 square miles, receives an average of 22.6 inches of rain each year. The amount of rain collected for treatment is 6.75%. How many million gallons (mil gal) of water are available per year for the small community that resides there?

   a. 4,060 mil gal/year
   b. 4,190 mil gal/year
   c. 25,400 mil gal/year
   d. 50,300 mil gal/year

20. A softener unit has 118 ft³ of resin with a capacity of 25.5 kilograins/ft³. How many gallons of water will the unit treat, if the water contains 14.2 gpg?

    a. 195,000 gal
    b. 210,000 gal
    c. 211,900 gal
    d. 212,500 gal

21. A solution of lime needs to be prepared for a jar test. How many grams of quicklime, CaO, would you mix with 1 L of water to make a 1.0% (Wt-volume) solution?

    a. 0.1 g of CaO
    b. 1 g of CaO
    c. 10 g of CaO
    d. 100 g of CaO

22. Calculate the percent removal across a settling basin and filter complex, if the raw water influent is 5.45 ntu and the effluent (post filters) is 0.018 ntu. Give answer to three significant figures.

    a. 98.4%
    b. 99.0%
    c. 99.3%
    d. 99.7%

23. Determine the psi at the bottom of an alum storage tank if the level of the alum in the tank is 8.95 ft and the density of the alum is 11.32 lb/gal.

    a. 5.13 psi
    b. 5.26 psi
    c. 5.37 psi
    d. 5.41 psi

24. Find the detention time in minutes for a clarifier that has a diameter of 152 ft and a water depth of 14.8 ft, if the flow rate is 4.25 mgd.

    a. 650 min
    b. 680 min
    c. 700 min
    d. 710 min

25. A 3-minute drawdown test used 191 mL of polymer for treating the raw water. The specific gravity of the polymer is 1.34. Which is the polymer dosage in mg/L, if the plant is treating 3,280 gpm?

    a. 6.63 mg/L
    b. 6.72 mg/L
    c. 6.87 mg/L
    d. 6.99 mg/L

26. A polymer solution has a specific gravity of 1.35 and a concentration of 80%. How many microliters are required to do a jar test, if the test uses 2-liter jars and the dosage needed is 6 mg/L? Give result to nearest tenth of a microliter.

    a. 11.1 microliters
    b. 11.3 microliters
    c. 11.4 microliters
    d. 11.7 microliters

27. Which should be the chemical feeder setting in lb/min if 12.5 mgd is treated with 7.25 mg/L of soda ash?

    a. 0.486 lb/min
    b. 0.525 lb/min
    c. 0.548 lb/min
    d. 0.561 lb/min

28. How many pounds per day of 65% calcium hypochlorite are required for maintaining a 2.5 mg/L dosage for a 2,575 gpm treatment plant?

    a. 100 lb/day
    b. 110 lb/day
    c. 120 lb/day
    d. 130 lb/day

29. Determine the chemical feeder setting in mL/min for a polymer solution, if the desired dosage is 3.95 mg/L and the treatment plant is treating 10.0 mgd. The specific gravity of the polymer is 1.33.

    a. 70.1 mL/min
    b. 73.9 mL/min
    c. 76.2 mL/min
    d. 78.1 mL/min

30. Four filters have a surface area of 450 ft each, measured to the nearest foot. Calculate the filtration rate in gpm, if the flow received is 21.5 ft^3/s. Give answer to three significant figures.

    a. 5.11 gpm/ft^2
    b. 5.21 gpm/ft^2
    c. 5.36 gpm/ft^2
    d. 5.58 gpm/ft^2

31. Determine the hardness of a particular body of raw water in mg/L, if the hardness of a water sample is 19.4 grains per gallon (gpg).

    a. 332 mg/L
    b. 339 mg/L
    c. 342 mg/L
    d. 350 mg/L

32. The exchange capacity of a softener is 8,850,000 grains. The softener treats water with an average hardness of 347 mg/L. Which is the capacity of the softener in gallons?

    a. 429,000 gal
    b. 437,000 gal
    c. 444,000 gal
    d. 449,000 gal

33. Find the backwash rate in gpm per ft^2, if a filter has an area of 620 ft^2 with a backwash rate of 13.5 cfs.

    a. 9.1 gpm/ft^2
    b. 9.4 gpm/ft^2
    c. 9.6 gpm/ft^2
    d. 9.8 gpm/ft^2

154   WATER OPERATOR CERTIFICATION STUDY GUIDE

34. The level in a storage tank drops 4.25 ft in exactly 12 hours. If the tank has a diameter of 50.0 ft and the plant is producing 2.95 mgd, which is the average discharge rate of the treated water discharge pumps in gallons per minute?

   a.   2,090 gpm
   b.   2,100 gpm
   c.   2,120 gpm
   d.   2,140 gpm

35. Ten filters have a surface area of 480 ft each. Calculate the filtration rate in gpm, if the total flow through the filters is 16.5 cubic feet per second.

   a.   1.2 gpm/ft^2
   b.   1.3 gpm/ft^2
   c.   1.5 gpm/ft^2
   d.   1.9 gpm/ft^2

36. Determine the amount of iron removed per year, if the iron concentration is 0.21 mg/L, the plant treats an average of 14.1 mgd, and the removal efficiency is 95.7% (0.957).

   a.   8,000 lb/yr
   b.   8,200 lb/yr
   c.   8,600 lb/yr
   d.   9,000 lb/yr

Answers to Water Treatment Operator Questions

Monitor, Evaluate, & Adjust Treatment Processes–Chemical Addition

Sample Questions for Level I—Answers

1. Answer: **c.** concentration of chlorine and contact time.

   Reference: *Water Treatment,* 4th edition. 2010. Nicholas G. Pizzi, Editor. American Water Works Association. Chapter 7.

2. Answer: **b.** stabilization.

   Reference: *Water Treatment,* 4th edition. 2010. Nicholas G. Pizzi, Editor. American Water Works Association. Chapter 9.

3. Answer: **b.** pH.

   Reference: *Water Treatment Operator Handbook,* Revised edition. 2005. Nicholas G. Pizzi. American Water Works Association. Chapter 12.

4. Answer: **c.** Cryptosporidiosis

   Reference: *Basic Chemistry for Water & Wastewater Operators,* Revised edition. 2005. Darshan Singh Sarai. American Water Works Association. Chapter 15.

5. Answer: **c.** 12 to 24 hours; 50 mg/L

   Reference: M20, *Water Chlorination/Chlorination Practices and Principles,* 2nd edition. 2006. American Water Works Association. Appendix C.

Sample Questions for Level II—Answers

1. Answer: **b.** Langelier saturation index

   Reference: *Basic Chemistry for Water & Wastewater Operators,* Revised edition. 2005. Darshan Singh Sarai. American Water Works Association. Chapter 14.

2. Answer: **d.** has little pH effect.

   Reference: *Water Treatment,* 4th edition. 2010. Nicholas G. Pizzi, Editor. American Water Works Association. Chapter 7.

3. Answer: **c.** sometimes produce disinfection by-products known to be carcinogenic.

   Reference: *Water Treatment Operator Handbook,* Revised edition. 2005. Nicholas G. Pizzi. American Water Works Association. Chapter 8.

4. Answer: **b.** concentration of chlorine, contact time, pH, and temperature.

   Reference: M20, *Water Chlorination/Chlorination Practices and Principles,* 2nd edition. 2006. American Water Works Association. Chapter 3.

5. Answer: **c.** every day.

   Reference: *Water Treatment,* 4th edition. 2010. Nicholas G. Pizzi, Editor. American Water Works Association. Chapter 8.

Sample Questions for Level III—Answers

1. Answer: **d.** There is potential for the light bulbs to be coated with light-obscuring material, preventing the UV light from killing microorganisms

   Reference: *Basic Microbiology for Drinking Water Personnel*, 2nd edition. 2006. Dennis R. Hill. American Water Works Association. Chapter 5.

2. Answer: **d.** removing iron.

   Reference: *Water Treatment*, 4th edition. 2010. Nicholas G. Pizzi, Editor. American Water Works Association. Chapter 7.

3. Answer: **a.** is a much stronger oxidant.

   Reference: *Water Treatment*, 4th edition. 2010. Nicholas G. Pizzi, Editor. American Water Works Association. Chapter 7.

4. Answer: **a.** Chlorine dioxide

   Reference: *Water Treatment*, 4th edition. 2010. Nicholas G. Pizzi, Editor. American Water Works Association. Chapter 7.

5. Answer: **a.** 6 inches.

   Reference: *Water Treatment*, 4th edition. 2010. Nicholas G. Pizzi, Editor. American Water Works Association. Chapter 8.

Sample Questions for Level IV—Answers

1. Answer: **a.** Ozone

   Reference: *Basic Microbiology for Drinking Water Personnel*, 2nd edition. 2006. Dennis R. Hill. American Water Works Association. Chapter 5.

2. Answer: **d.** potassium permanganate

   Reference: *Water Treatment*, 4th edition. 2010. Nicholas G. Pizzi, Editor. American Water Works Association. Chapter 7.

3. Answer: **a.** must be supplied with extremely dry air.

   Reference: *Water Treatment*, 4th edition. 2010. Nicholas G. Pizzi, Editor. American Water Works Association. Chapter 7.

4. Answer: **c.** (1) potassium permanganate; (2) chlorine; (3) chlorine; (4) chloramines

   Reference: M20, *Water Chlorination/Chlorination Practices and Principles*, 2nd edition. 2006. American Water Works Association. Chapter 6.

5. Answer: **a.** 1 to 3%

   Reference: *Basic Chemistry for Water & Wastewater Operators*, Revised edition. 2005. Darshan Singh Sarai. American Water Works Association. Chapter 15.

Monitor, Evaluate, & Adjust Treatment Processes–Coagulation and Flocculation

Sample Questions for Level I—Answers

1. Answer: **b.** 15 to 45 minutes.

   Reference: *Water Treatment Operator Handbook*, Revised edition. 2005. Nicholas G. Pizzi. American Water Works Association. Chapter 5.

2. Answer: **c.** 5.8 to 7.5.

   Reference: *Water Treatment,* 4th edition. 2010. Nicholas G. Pizzi, Editor. American Water Works Association. Chapter 4.

3. Answer: **a.** Colloidal particles are so small that gravity has little effect on them

   Reference: *Basic Chemistry for Water & Wastewater Operators,* Revised edition. 2005. Darshan Singh Sarai. American Water Works Association. Chapter 8.

4. Answer: **c.** Zeta potential

   Reference: *Water Treatment,* 4th edition. 2010. Nicholas G. Pizzi, Editor. American Water Works Association. Chapter 4.

5. Answer: **a.** electrons

   Reference: *Water Treatment,* 4th edition. 2010. Nicholas G. Pizzi, Editor. American Water Works Association. Chapter 4.

Sample Questions for Level II—Answers

1. Answer: **c.** increasing the alum dosage.

   Reference: *Basic Chemistry for Water & Wastewater Operators,* Revised edition. 2005. Darshan Singh Sarai. American Water Works Association. Chapter 8.

2. Answer: **c.** 6.5 to 7.2

   Reference: M37, *Operational Control of Coagulation and Filtration Processes,* 3rd edition. 2011. American Water Works Association. Introduction.

3. Answer: **a.** van der Waals forces

   Reference: *Water Treatment,* 4th edition. 2010. Nicholas G. Pizzi, Editor. American Water Works Association. Chapter 4.

4. Answer: **a.** Excessive flocculation time

   Reference: *Water Treatment Operator Handbook,* Revised edition. 2005. Nicholas G. Pizzi. American Water Works Association. Chapter 5.

5. Answer: **c.** More dense floc

   Reference: *Basic Chemistry for Water & Wastewater Operators,* Revised edition. 2005. Darshan Singh Sarai. American Water Works Association. Chapter 8.

Sample Questions for Level III—Answers

1. Answer: **a.** 0.5 mg/L

   Reference: *Water Treatment,* 4th edition. 2010. Nicholas G. Pizzi, Editor. American Water Works Association. Chapter 4.

2. Answer: **a.** NaCl

   Reference: *Basic Chemistry for Water & Wastewater Operators,* Revised edition. 2005. Darshan Singh Sarai. American Water Works Association. Chapter 13.

3. Answer: **c.** excellent.

   Reference: *Water Treatment,* 4th edition. 2010. Nicholas G. Pizzi, Editor. American Water Works Association. Chapter 4.

4. Answer: **c.** They increase head loss

   Reference: *Water Treatment Operator Handbook*, Revised edition. 2005. Nicholas G. Pizzi. American Water Works Association. Chapter 5.

5. Answer: **a.** 3.5 to 9.0

   Reference: *Basic Chemistry for Water & Wastewater Operators,* Revised edition. 2005. Darshan Singh Sarai. American Water Works Association. Chapter 8.

Sample Questions for Level IV—Answers

1. Answer: **b.** 3

   Reference: *Water Treatment*, 4th edition. 2010. Nicholas G. Pizzi, Editor. American Water Works Association. Chapter 4.

2. Answer: **d.** Nonionic and anionic polymers

   Reference: *Water Treatment Operator Handbook*, Revised edition. 2005. Nicholas G. Pizzi. American Water Works Association. Chapter 5.

3. Answer: **b.** Optimize coagulation, flocculation, and filtration

   Reference: M7, *Problem Organisms in Water: Identification and Treatment,* 3rd edition. 2003. American Water Works Association. Chapter 11.

4. Answer: **d.** the addition of an acid to lower pH before coagulation.

   Reference: *Water Treatment Operator Handbook,* Revised edition. 2005. Nicholas G. Pizzi. American Water Works Association. Chapter 5.

5. Answer: **b.** it could inhibit floc formation.

   Reference: *Water Treatment Operator Handbook,* Revised edition. 2005. Nicholas G. Pizzi. American Water Works Association. Chapter 5.

Monitor, Evaluate, & Adjust Treatment Processes–Clarification and Sedimentation

Sample Questions for Level I—Answers

1. Answer: **d.** Effluent launder

   Reference: *Water Treatment*, 4th edition. 2010. Nicholas G. Pizzi, Editor. American Water Works Association. Chapter 5.

2. Answer: **d.** At least twice per day

   Reference: *Water Treatment*, 4th edition. 2010. Nicholas G. Pizzi, Editor. American Water Works Association. Chapter 5.

3. Answer: **d.** to draw off the liquid from a vessel of any size without stirring up bottom sediment.

   Reference: *The Water Dictionary: A Comprehensive Reference of Water Terminology,* 2nd edition. 2010. Nancy McTigue, Editor & James M. Symons, Editor Emeritus. American Water Works Association.

4. Answer: **b.** Once a year

   Reference: *Water Treatment*, 4th edition. 2010. Nicholas G. Pizzi, Editor. American Water Works Association. Chapter 5.

5. Answer: **b.** To remove pathogens

   Reference: *Water Treatment Operator Handbook,* Revised edition. 2005. Nicholas G. Pizzi. American Water Works Association. Chapter 1.

Sample Questions for Level II—Answers

1. Answer: **c.** Use multibarrier approach – coagulation, flocculation, sedimentation, and filtration

   Reference: M7, *Problem Organisms in Water: Identification and Treatment,* 3rd edition. 2003. American Water Works Association. Chapter 11.

2. Answer: **c.** Poor inlet baffling

   Reference: *Water Treatment,* 4th edition. 2010. Nicholas G. Pizzi, Editor. American Water Works Association. Chapter 5.

3. Answer: **a.** less than 0.5-log

   Reference: *Water Treatment Operator Handbook,* Revised edition. 2005. Nicholas G. Pizzi. American Water Works Association. Chapter 6.

4. Answer: **d.** 10 gpm/ft

   Reference: *Water Treatment,* 4th edition. 2010. Nicholas G. Pizzi, Editor. American Water Works Association. Chapter 5.

5. Answer: **d.** algae.

   Reference: *Water Treatment,* 4th edition. 2010. Nicholas G. Pizzi, Editor. American Water Works Association. Chapter 5.

Sample Questions for Level III—Answers

1. Answer: **d.** Magnitude of the charges

   Reference: *Water Treatment,* 4th edition. 2010. Nicholas G. Pizzi, Editor. American Water Works Association. Chapter 4.

2. Answer: **c.** Solids can become resuspended or taste and odors can develop

   Reference: *Water Treatment,* 4th edition. 2010. Nicholas G. Pizzi, Editor. American Water Works Association. Chapter 5.

3. Answer: **d.** Floc must have good settling characteristics

   Reference: *Water Treatment,* 4th edition. 2010. Nicholas G. Pizzi, Editor. American Water Works Association. Chapter 5.

4. Answer: **d.** Spiral-flow basins

   Reference: *Water Treatment,* 4th edition. 2010. Nicholas G. Pizzi, Editor. American Water Works Association. Chapter 5.

5. Answer: **a.** 50°

   Reference: *Water Treatment,* 4th edition. 2010. Nicholas G. Pizzi, Editor. American Water Works Association. Chapter 5.

Sample Questions for Level IV—Answers

1. Answer: **a.** Optimize the settling process

    Reference: M7, *Problem Organisms in Water: Identification and Treatment,* 3rd edition. 2003. American Water Works Association. Chapter 11.

2. Answer: **b.** 45°

    Reference: *Water Treatment,* 4th edition. 2010. Nicholas G. Pizzi, Editor. American Water Works Association. Chapter 5.

3. Answer: **d.** Because of the recycled materials from the sludge blanket, the chemical reactions occur more quickly and completely in the mixing area

    Reference: *Water Treatment,* 4th edition. 2010. Nicholas G. Pizzi, Editor. American Water Works Association. Chapter 5.

4. Answer: **a.** maintain a uniform sludge blanket layer.

    Reference: *Water Treatment,* 4th edition. 2010. Nicholas G. Pizzi, Editor. American Water Works Association. Chapter 5.

5. Answer: **b.** high in color and low in turbidity.

    Reference: *Water Treatment,* 4th edition. 2010. Nicholas G. Pizzi, Editor. American Water Works Association. Chapter 5.

Monitor, Evaluate, & Adjust Treatment Processes–Filtration

Sample Questions for Level I—Answers

1. Answer: **b.** 2 to 3 gpm/ft^2

    Reference: *Iron and Manganese Removal Handbook.* 1999. Elmer O. Sommerfield. American Water Works Association. Chapter 3.

2. Answer: **c.** it is becoming more efficient in particle removal.

    Reference: *Water Treatment Operator Handbook,* Revised edition. 2005. Nicholas G. Pizzi. American Water Works Association. Chapter 7.

3. Answer: **c.** 100 mg/L Cl_2.

    Reference: *Iron and Manganese Removal Handbook.* 1999. Elmer O. Sommerfield. American Water Works Association. Chapter 6.

4. Answer: **a.** Minor

    Reference: *Water Treatment,* 4th edition. 2010. Nicholas G. Pizzi, Editor. American Water Works Association. Chapter 6.

5. Answer: **a.** 1 to 2 ntu.

    Reference: *Water Treatment,* 4th edition. 2010. Nicholas G. Pizzi, Editor. American Water Works Association. Chapter 5.

Sample Questions for Level II—Answers

1. Answer: **b.** Coagulation, flocculation, sedimentation, and filtration

    Reference: *Chemistry of Water Treatment,* 2nd edition. 1998. Samuel D. Faust & Osman M. Aly. CRC Press.

2. Answer: **d.** Floc breakthrough

   Reference: *Water Treatment,* 4th edition. 2010. Nicholas G. Pizzi, Editor. American Water Works Association. Chapter 6.

3. Answer: **c.** a sand boil.

   Reference: *Water Treatment,* 4th edition. 2010. Nicholas G. Pizzi, Editor. American Water Works Association. Chapter 6.

4. Answer: **b.** requires regeneration with potassium permanganate (1 hour soak with 60 grams $KMnO_4/ft^3$).

   Reference: *Iron and Manganese Removal Handbook.* 1999. Elmer O. Sommerfield. American Water Works Association. Chapter 6.

5. Answer: **d.** Sedimentation

   Reference: *Water Treatment,* 4th edition. 2010. Nicholas G. Pizzi, Editor. American Water Works Association. Chapter 6.

Sample Questions for Level III—Answers

1. Answer: **d.** Use a disinfectant that targets the specific organisms in question

   Reference: M7, *Problem Organisms in Water: Identification and Treatment,* 3rd edition. 2003. American Water Works Association. Chapter 11.

2. Answer: **c.** *Cryptosporidium*

   Reference: *Basic Microbiology for Drinking Water Personnel,* 2nd edition. 2006. Dennis R. Hill. American Water Works Association. Chapter 3.

3. Answer: **c.** higher pressure.

   Reference: *Water Treatment Operator Handbook,* Revised edition. 2005. Nicholas G. Pizzi. American Water Works Association. Chapter 11.

4. Answer: **d.** Shorter filter runs

   Reference: *Iron and Manganese Removal Handbook.* 1999. Elmer O. Sommerfield. American Water Works Association. Chapter 6.

5. Answer: **c.** media depth and media size.

   Reference: *Water Treatment Operator Handbook,* Revised edition. 2005. Nicholas G. Pizzi. American Water Works Association. Chapter 7.

Sample Questions for Level IV—Answers

1. Answer: **c.** Superchlorinate

   Reference: M7, *Problem Organisms in Water: Identification and Treatment,* 3rd edition. 2003. American Water Works Association. Chapter 11.

2. Answer: **b.** Sustain the same bed expansion without media loss by reducing or increasing backwash flow rate

   Reference: *Water Treatment Operator Handbook,* Revised edition. 2005. Nicholas G. Pizzi. American Water Works Association. Chapter 7.

3. Answer: **b.** They are cleaned with an acid wash

   Reference: *Water Treatment Operator Handbook,* Revised edition. 2005. Nicholas G. Pizzi. American Water Works Association. Chapter 11.

4. Answer: **c.** Reverse osmosis

   Reference: *Water Treatment Operator Handbook,* Revised edition. 2005. Nicholas G. Pizzi. American Water Works Association. Chapter 11.

5. Answer: **a.** feed pressure.

   Reference: *Water Treatment,* 4th edition. 2010. Nicholas G. Pizzi, Editor. American Water Works Association. Chapter 15.

Monitor, Evaluate, & Adjust Treatment Processes–Residuals Disposal

Sample Questions, General, for Level II—Answers

1. Answer: **b.** 5%

   Reference: *Water Treatment Operator Handbook,* Revised edition. 2005. Nicholas G. Pizzi. American Water Works Association. Chapter 9.

2. Answer: **c.** Lagoons

   Reference: *Water Treatment Plant Operation,* Volume II, 5th edition. 2006. Ken Kerri. California State University – Sacramento. Chapter 17 – Residuals, Page 233.

3. Answer: **a.** daily.

   Reference: *Water Treatment Plant Operation,* Volume II, 5th edition. 2006. Ken Kerri. California State University – Sacramento. Chapter 17 – Residuals, Page 233.

4. Answer: **c.** Thickener

   Reference: *Water Treatment Plant Operation,* Volume II, 5th edition. 2006. Ken Kerri. California State University – Sacramento. Chapter 17 – Residuals, Page 236.

5. Answer: **b.** Sand bed

   Reference: *Water Treatment Plant Operation,* Volume II, 5th edition. 2006. Ken Kerri. California State University – Sacramento. Chapter 17 – Residuals, Page 236

Sample Questions for Level III—Answers

1. Answer: **c.** 35,000 to 45,000 mg/L

   Reference: *Water Treatment,* 4th edition. 2010. Nicholas G. Pizzi, Editor. American Water Works Association. Chapter 12.

2. Answer: **a.** 2–4%

   Reference: *Water Treatment Plant Operation*, Volume II, 5th edition. 2006. Ken Kerri. California State University – Sacramento. Chapter 17 – Residuals, Page 235.

3. Answer: **d.** Water table level

   Reference: *Water Treatment Plant Operation*, Volume II, 5th edition. 2006. Ken Kerri. California State University – Sacramento. Chapter 17 – Residuals, Page 237.

4. Answer: **c.** Filter Press

   Reference: *Water Treatment Plant Operation*, Volume II, 5th edition. 2006. Ken Kerri. California State University – Sacramento. Chapter 17 – Residuals, Page 245.

5. Answer: **b.** Vacuum Filters

   Reference: *Water Treatment Plant Operation*, Volume II, 5th edition. 2006. Ken Kerri. California State University – Sacramento. Chapter 17 – Residuals, Page 245.

Sample Questions for Level IV—Answers

1. Answer: **d.** The Safe Drinking Water Act

   Reference: *Water Sources*, 4th edition. 2010. Paul Koch, Editor. American Water Works Association. Chapter 7.

2. Answer: **a.** 0.1 to 2.0%

   Reference: *Water Treatment*, 4th edition. 2010. Nicholas G. Pizzi, Editor. American Water Works Association. Chapter 5.

3. Answer: **d.** Sludge collection and removal

   Reference: *Water Treatment*, 4th edition. 2010. Nicholas G. Pizzi, Editor. American Water Works Association. Chapter 5.

4. Answer: **a.** One truckload: 0.2% × 100,000 gal/5% = 4,000 gal

   Reference: *Water Treatment Operator Handbook*, Revised edition. 2005. Nicholas G. Pizzi. American Water Works Association. Chapter 6.

5. Answer: **c.** can be very gelatinous and difficult to handle.

   Reference: *Water Treatment Operator Handbook*, Revised edition. 2005. Nicholas G. Pizzi. American Water Works Association. Chapter 6.

Monitor, Evaluate, & Adjust Treatment Processes–Additional Treatment Tasks

Sample Questions for Level I—Answers

1. Answer: **d.** Iron and manganese

   Reference: *Iron and Manganese Removal Handbook*. 1999. Elmer O. Sommerfield. American Water Works Association. Chapter 2.

2. Answer: **d.** Polyphosphates and chlorine

   Reference: *Iron and Manganese Removal Handbook*. 1999. Elmer O. Sommerfield. American Water Works Association. Chapter 7.

3. Answer: **d.** Right after the water leaves the well

   Reference: *Water Treatment*, 4th edition. 2010. Nicholas G. Pizzi, Editor. American Water Works Association. Chapter 10.

4. Answer: **a.** adding CO_2 to the water.

   Reference: *Water Treatment Operator Handbook*, Revised edition. 2005. Nicholas G. Pizzi. American Water Works Association. Chapter 9.

5. Answer: **c.** regenerated.

   Reference: *Water Treatment Operator Handbook,* Revised edition. 2005. Nicholas G. Pizzi. American Water Works Association. Chapter 9.

Sample Questions for Level II—Answers

1. Answer: **b.** iron

   Reference: *Water Treatment,* 4th edition. 2010. Nicholas G. Pizzi, Editor. American Water Works Association. Chapter 11.

2. Answer: **d.** on the filters.

   Reference: *Water Treatment,* 4th edition. 2010. Nicholas G. Pizzi, Editor. American Water Works Association. Chapter 11.

3. Answer: **d.** Rock salt or pellet-type salt

   Reference: *Water Treatment,* 4th edition. 2010. Nicholas G. Pizzi, Editor. American Water Works Association. Chapter 12.

4. Answer: **b.** organic compounds responsible for tastes and odors.

   Reference: *Water Treatment,* 4th edition. 2010. Nicholas G. Pizzi, Editor. American Water Works Association. Chapter 13.

5. Answer: **a.** all hardness.

   Reference: *Basic Science Concepts and Applications,* 4th edition. 2010. Nicholas G. Pizzi, Editor. American Water Works Association. Chemistry 6.

Sample Questions for Level III—Answers

1. Answer: **c.** 25 mg/L

   Reference: *Basic Chemistry for Water & Wastewater Operators,* Revised edition. 2005. Darshan Singh Sarai. American Water Works Association. Chapter 13.

2. Answer: **b.** Raise the pH of the water

   Reference: *Water Treatment Operator Training Handbook*, 3rd edition. William C. Lauer and Nicolas Pizzi. American Water Works Association. Chapter 9.

3. Answer: **a.** total organic carbon.

   Reference: *Water Treatment,* 4th edition. 2010. Nicholas G. Pizzi, Editor. American Water Works Association. Chapter 12.

4. Answer: **a.** 5%

   Reference: *Water Treatment,* 4th edition. 2010. Nicholas G. Pizzi, Editor. American Water Works Association. Chapter 13.

5. Answer: **a.** Raw water intake

   Reference: *Water Treatment,* 4th edition. 2010. Nicholas G. Pizzi, Editor. American Water Works Association. Chapter 13.

Sample Questions for Level IV—Answers

1. Answer: **b.** Granular activated carbon

   Reference: *Chemistry of Water Treatment,* 2nd edition. 1998. Samuel D. Faust & Osman M. Aly. CRC Press.

2. Answer: **a.** 2 inches.

   Reference: *Water Treatment,* 4th edition. 2010. Nicholas G. Pizzi, Editor. American Water Works Association. Chapter 13.

3. Answer: **c.** Anion exchange

   Reference: *Chemistry of Water Treatment,* 2nd edition. 1998. Samuel D. Faust & Osman M. Aly. CRC Press.

4. Answer: **d.** Coal layer is too coarse

   Reference: *Iron and Manganese Removal Handbook.* 1999. Elmer O. Sommerfield. American Water Works Association. Chapter 8.

5. Answer: **d.** Granular activated carbon

   Reference: *Chemistry of Water Treatment,* 2nd edition. 1998. Samuel D. Faust & Osman M. Aly. CRC Press.

Laboratory Analyses

Sample Questions for Level I—Answers

1. Answer: **a.** Bacteriological

   Reference: *Water Quality*, 4th edition. 2010. Joseph A. Ritter, Editor. American Water Works Association. Chapter 2.

2. Answer: **c.** based on population.

   Reference: *Water Treatment Operator Handbook*, Revised edition. 2005. Nicholas G. Pizzi. American Water Works Association. Chapter 12.

3. Answer: **b.** Coliform group bacteria

   Reference: *Water Quality*, 4th edition. 2010. Joseph A. Ritter, Editor. American Water Works Association. Chapter 4.

4. Answer: **d.** A solution

   Reference: *Basic Science Concepts and Applications*, 4th edition. 2010. Nicholas G. Pizzi, Editor. American Water Works Association. Chemistry 4.

5. Answer: **d.** inorganic compounds.

   Reference: *Basic Science Concepts and Applications*, 4th edition. 2010. Nicholas G. Pizzi, Editor. American Water Works Association. Chemistry 5.

Sample Questions for Level II—Answers

1. Answer: **b.** Grab sample

   Reference: *Water Treatment Operator Handbook*, Revised edition. 2005. Nicholas G. Pizzi. American Water Works Association. Chapter 12.

2. Answer: **c.** 8

    Reference: *Water Treatment Operator Handbook*, Revised edition. 2005. Nicholas G. Pizzi. American Water Works Association. Chapter 12.

3. Answer: **a.** <2.0.

    Reference: *The Water Dictionary: A Comprehensive Reference of Water Terminology*, 2nd edition. 2010. Nancy McTigue, Editor & James M. Symons, Editor Emeritus. American Water Works Association.

4. Answer: **a.** $Na_2S_2O_3$

    Reference: *The Water Dictionary: A Comprehensive Reference of Water Terminology*, 2nd edition. 2010. Nancy McTigue, Editor & James M. Symons, Editor Emeritus. American Water Works Association.

5. Answer: **c.** Biological activity

    Reference: *Water Quality*, 4th edition. 2010. Joseph A. Ritter, Editor. American Water Works Association. Chapter 2.

Sample Questions for Level III—Answers

1. Answer: **d.** concentrated nitric acid.

    Reference: *Water Quality*, 4th edition. 2010. Joseph A. Ritter, Editor. American Water Works Association. Chapter 6.

2. Answer: **b.** standardized solution.

    Reference: *Basic Chemistry for Water & Wastewater Operators*, Revised edition. 2005. Darshan Singh Sarai. American Water Works Association. Chapter 7.

3. Answer: **c.** Membrane filtration

    Reference: *Basic Microbiology for Drinking Water Personnel*, 2nd edition. 2006. Dennis R. Hill. American Water Works Association. Chapter 4.

4. Answer: **b.** Pink to dark red with green metallic surface sheen

    Reference: *Water Quality*, 4th edition. 2010. Joseph A. Ritter, Editor. American Water Works Association. Chapter 4.

5. Answer: **b.** 24 to 48 hours

    Reference: *Water Treatment Operator Handbook*, Revised edition. 2005. Nicholas G. Pizzi. American Water Works Association. Chapter 12.

Sample Questions for Level IV—Answers

1. Answer: **c.** polypropylene.

    Reference: *Water Quality*, 4th edition. 2010. Joseph A. Ritter, Editor. American Water Works Association. Chapter 3.

2. Answer: **c.** 0.5 mL

    Reference: *Water Quality*, 4th edition. 2010. Joseph A. Ritter, Editor. American Water Works Association. Chapter 5.

3. Answer: **d.** determining the degree of mineralization of the water.

   Reference: *Water Quality*, 4th edition. 2010. Joseph A. Ritter, Editor. American Water Works Association. Chapter 5.

4. Answer: **c.** dissolved metals; preserving

   Reference: *Water Quality*, 4th edition. 2010. Joseph A. Ritter, Editor. American Water Works Association. Chapter 6.

5. Answer: **c.** Dramatic decline in dissolved oxygen

   Reference: M7, *Problem Organisms in Water: Identification and Treatment*, 3rd edition. 2003. American Water Works Association. Chapter 3.

Comply with Drinking Water Regulations

Sample Questions for Level I—Answers

1. Answer: **a.** Tier I

   Reference: *Water Quality*, 4th edition. 2010. Joseph A. Ritter, Editor. American Water Works Association. Chapter 1.

2. Answer: **a.** 1

   Reference: *Water Quality*, 4th edition. 2010. Joseph A. Ritter, Editor. American Water Works Association. Chapter 1.

3. Answer: **b.** 24 hours.

   Reference: *Water Quality*, 4th edition. 2010. Joseph A. Ritter, Editor. American Water Works Association. Chapter 1.

4. Answer: **c.** 15; 25; 60

   Reference: *Water Quality*, 4th edition. 2010. Joseph A. Ritter, Editor. American Water Works Association. Chapter 1.

5. Answer: **a.** At consumers' faucets

   Reference: *Water Quality*, 4th edition. 2010. Joseph A. Ritter, Editor. American Water Works Association. Chapter 1.

Sample Questions for Level II—Answers

1. Answer: **b.** 0.2 mg/L

   Reference: *Water Quality*, 4th edition. 2010. Joseph A. Ritter, Editor. American Water Works Association. Chapter 1.

2. Answer: **c.** 1.3 mg/L

   Reference: *Water Quality*, 4th edition. 2010. Joseph A. Ritter, Editor. American Water Works Association. Chapter 1.

3. Answer: **c.** Point(s) where water enters the distribution system, including all filter effluents if a surface water treatment plant

   Reference: *Water Quality*, 4th edition. 2010. Joseph A. Ritter, Editor. American Water Works Association. Chapter 1.

4. Answer: **a.** coliform samples.

   Reference: *Water Quality*, 4th edition. 2010. Joseph A. Ritter, Editor. American Water Works Association. Chapter 2.

5. Answer: **b.** 0.060 mg/L

   Reference: *Water Quality*, 4th edition. 2010. Joseph A. Ritter, Editor. American Water Works Association. Chapter 1.

Sample Questions for Level III—Answers

1. Answer: **d.** No more than 5%

   Reference: National Primary Drinking Water Regulations, Subpart G – Maximum Contaminant Levels and Maximum Residual Disinfectant Levels, §141.63 – Maximum contaminant levels (MCLs) for microbiological contaminants (a)(1)

2. Answer: **d.** 4 log

   Reference: National Primary Drinking Water Regulations, Subpart H – Filtration and Disinfection, §141.70 – General requirements (a)(2)

3. Answer: **c.** Entry points to the distribution system

   Reference: *Water Quality*, 4th edition. 2010. Joseph A. Ritter, Editor. American Water Works Association. Chapter 1.

4. Answer: **c.** Each entry point to the system

   Reference: *Water Quality*, 4th edition. 2010. Joseph A. Ritter, Editor. American Water Works Association. Chapter 1.

5. Answer: **a.** population served is > 3,300 people.

   Reference: *Water Quality*, 4th edition. 2010. Joseph A. Ritter, Editor. American Water Works Association. Chapter 6.

Sample Questions for Level IV—Answers

1. Answer: **d.** 100,000 people.

   Reference: *Water Quality*, 4th edition. 2010. Joseph A. Ritter, Editor. American Water Works Association. Chapter 1.

2. Answer: **b.** < 2.0 mg/L

   Reference: *Water Treatment*, 4th edition. 2010. Nicholas G. Pizzi, Editor. American Water Works Association. Chapter 4.

3. Answer: **c.** 20/100 mL; 90%

   Reference: National Primary Drinking Water Regulations, Subpart H – Filtration and Disinfection, §141.71 – Criteria for avoiding filtration (a)(1)

4. Answer: **d.** During the quarter that previously yielded the highest analytical result

   Reference: National Primary Drinking Water Regulations, Subpart C – Monitoring and Analytical Requirements, §141.24 – Organic chemicals, sampling and analytical requirements, (f)(11)(iii)

5. Answer: **d.** Lead and Copper Rule

   Reference: *Water Treatment Operator Handbook*, Revised edition. 2005. Nicholas G. Pizzi. American Water Works Association. Chapter 1.

Operate and Maintain Equipment

Sample Questions for Level I—Answers

1. Answer: **b.** control water leakage from the stuffing box.

   Reference: *Pumps & Pumping*, 9th edition. 2010. E.E. "Skeet" Arasmith, Mitch Scheele, & Kimon Zentz. ACR Publications.

2. Answer: **b.** control water leakage along the pump's shaft.

   Reference: *Pumps & Pumping*, 9th edition. 2010. E.E. "Skeet" Arasmith, Mitch Scheele, & Kimon Zentz. ACR Publications.

3. Answer: **b.** The stuffing box

   Reference: *Pumps & Pumping*, 9th edition. 2010. E.E. "Skeet" Arasmith, Mitch Scheele, & Kimon Zentz. ACR Publications.

4. Answer: **b.** Centrifugal pump

   Reference: *Pumps & Pumping*, 9th edition. 2010. E.E. "Skeet" Arasmith, Mitch Scheele, & Kimon Zentz. ACR Publications.

5. Answer: **c.** Clogging

   Reference: *Water Treatment*, 4th edition. 2010. Nicholas G. Pizzi, Editor. American Water Works Association. Chapter 3.

Sample Questions for Level II—Answers

1. Answer: **c.** Packing gland

   Reference: *Pumping: Fundamentals for the Water & Wastewater Maintenance Operator Series*. 2001. Frank R. Spellman and Joanne Drinan. Technomic Publishing Company. Chapter 3.

2. Answer: **b.** 90 degrees.

   Reference: *Pumps & Pumping*, 9th edition. 2010. E.E. "Skeet" Arasmith, Mitch Scheele, & Kimon Zentz. ACR Publications.

3. Answer: **c.** $V^2/2g$

   Reference: *Pumps & Pumping*, 9th edition. 2010. E.E. "Skeet" Arasmith, Mitch Scheele, & Kimon Zentz. ACR Publications.

4. Answer: **c.** Semi-open

   Reference: *Pumping: Fundamentals for the Water & Wastewater Maintenance Operator Series*. 2001. Frank R. Spellman and Joanne Drinan. Technomic Publishing Company. Chapter 4.

5. Answer: **b.** torque

   Reference: *Pumps & Pumping*, 9th edition. 2010. E.E. "Skeet" Arasmith, Mitch Scheele, & Kimon Zentz. ACR Publications.

Sample Questions for Level III—Answers

1. Answer: **d.** Add another cylinder and feed from both

   Reference: *Water Treatment*, 4th edition. 2010. Nicholas G. Pizzi, Editor. American Water Works Association. Chapter 7.

2. Answer: **a.** loss of electrical power.

   Reference: M20, *Water Chlorination/Chlorination Practices and Principles*, 2nd edition. 2006. American Water Works Association. Chapter 4.

3. Answer: **d.** blockage.

   Reference: *Water Quality*, 4th edition. 2010. Joseph A. Ritter, Editor. American Water Works Association. Chapter 5.

4. Answer: **a.** 20 to 30 minutes

   Reference: *Water Treatment*, 4th edition. 2010. Nicholas G. Pizzi, Editor. American Water Works Association. Chapter 11.

5. Answer: **d.** 160°F or higher

   Reference: *Water Treatment*, 4th edition. 2010. Nicholas G. Pizzi, Editor. American Water Works Association. Chapter 11.

Sample Questions for Level IV—Answers

1. Answer: **b.** 10 to 15 psi higher

   Reference: *Pumps & Pumping*, 9th edition. 2010. E.E. "Skeet" Arasmith, Mitch Scheele, & Kimon Zentz. ACR Publications.

2. Answer: **a.** High head conditions

   Reference: *Pumps & Pumping*, 9th edition. 2010. E.E. "Skeet" Arasmith, Mitch Scheele, & Kimon Zentz. ACR Publications.

3. Answer: **b.** 20%

   Reference: *Pumps & Pumping*, 9th edition. 2010. E.E. "Skeet" Arasmith, Mitch Scheele, & Kimon Zentz. ACR Publications.

4. Answer: **b.** discharge head.

   Reference: *Pumping: Fundamentals for the Water & Wastewater Maintenance Operator Series.* 2001. Frank R. Spellman and Joanne Drinan. Technomic Publishing Company. Chapter 2.

5. Answer: **b.** 2.0 degrees

   Reference: *Pumping: Fundamentals for the Water & Wastewater Maintenance Operator Series.* 2001. Frank R. Spellman and Joanne Drinan. Technomic Publishing Company. Chapter 4.

Perform Security, Safety, and Administrative Procedures

Sample Questions for Level I—Answers

1. Answer: **a.** Operator safety

    Reference: *Water Treatment Operator Handbook*, Revised edition. 2005. Nicholas G. Pizzi. American Water Works Association. Chapter 13.

2. Answer: **c.** 157 to 162°F

    Reference: *Water Treatment Operator Handbook*, Revised edition. 2005. Nicholas G. Pizzi. American Water Works Association. Chapter 13.

3. Answer: **a.** every time.

    Reference: M20, *Water Chlorination/Chlorination Practices and Principles*, 2nd edition. 2006. American Water Works Association. Chapter 4.

4. Answer: **d.** Lead gasket

    Reference: M20, *Water Chlorination/Chlorination Practices and Principles*, 2nd edition. 2006. American Water Works Association. Chapter 4.

5. Answer: **c.** Eyes

    Reference: *Water Quality*, 4th edition. 2010. Joseph A. Ritter, Editor. American Water Works Association. Chapter 3.

Sample Questions for Level II—Answers

1. Answer: **a.** Highly explosive hydrogen gas will be released

    Reference: *Water Treatment*, 4th edition. 2010. Nicholas G. Pizzi, Editor. American Water Works Association. Chapter 4.

2. Answer: **c.** will support combustion.

    Reference: *Water Treatment*, 4th edition. 2010. Nicholas G. Pizzi, Editor. American Water Works Association. Chapter 7.

3. Answer: **d.** have a window in the door so an operator can look into the room to detect any abnormal conditions.

    Reference: *Water Treatment*, 4th edition. 2010. Nicholas G. Pizzi, Editor. American Water Works Association. Chapter 7.

4. Answer: **b.** the internal atmosphere of the space must be tested to measure oxygen content, presence of flammable gases and vapors, and potentially toxic air contaminants.

    Reference: *Water Treatment Operator Handbook*, Revised edition. 2005. Nicholas G. Pizzi. American Water Works Association. Chapter 14.

5. Answer: **c.** calcium.

    Reference: *Water Treatment Operator Handbook*, Revised edition. 2005. Nicholas G. Pizzi. American Water Works Association. Chapter 10.

Sample Questions for Level III—Answers

1. Answer: **c.** the National Institute of Occupational Safety and Health (NIOSH).

   Reference: *Water Treatment*, 4th edition. 2010. Nicholas G. Pizzi, Editor. American Water Works Association. Chapter 7.

2. Answer: **b.** The Chemical Transportation Emergency Center

   Reference: *Water Treatment*, 4th edition. 2010. Nicholas G. Pizzi, Editor. American Water Works Association. Chapter 7.

3. Answer: **b.** Oxygen

   Reference: *Water Treatment*, 4th edition. 2010. Nicholas G. Pizzi, Editor. American Water Works Association. Chapter 13.

4. Answer: **b.** Time-Weighted Average (TWA)

   Reference: M20, *Water Chlorination/Chlorination Practices and Principles*, 2nd edition. 2006. American Water Works Association. Chapter 5.

5. Answer: **b.** 1,500 lb

   Reference: M20, *Water Chlorination/Chlorination Practices and Principles*, 2nd edition. 2006. American Water Works Association. Chapter 5.

Sample Questions for Level IV—Answers

1. Answer: **c.** Henry's constant

   Reference: *Water Treatment*, 4th edition. 2010. Nicholas G. Pizzi, Editor. American Water Works Association. Chapter 14.

2. Answer: **d.** 2,000 ppm.

   Reference: *Water Treatment Operator Handbook*, Revised edition. 2005. Nicholas G. Pizzi. American Water Works Association. Chapter 8.

3. Answer: **c.** Kit C

   Reference: *Water Treatment*, 4th edition. 2010. Nicholas G. Pizzi, Editor. American Water Works Association. Chapter 7.

4. Answer: **c.** 67%

   Reference: *Water Treatment Operator Handbook*, Revised edition. 2005. Nicholas G. Pizzi. American Water Works Association. Chapter 13.

5. Answer: **d.** Draw water from an adjoining system or systems

   Reference: *Water Sources*, 4th edition. 2010. Paul Koch, Editor. American Water Works Association. Chapter 4.

Evaluate Characteristics of Source Water

Sample Questions for Level I—Answers

1. Answer: **c.** coliform bacteria.

   Reference: *Water Treatment*, 4th edition. 2010. Nicholas G. Pizzi, Editor. American Water Works Association. Chapter 7.

2. Answer: **b.** Discolored laundry and changed taste of water, coffee, tea, and other beverages

   Reference: *Water Quality*, 4th edition. 2010. Joseph A. Ritter, Editor. American Water Works Association. Chapter 1.

3. Answer: **c.** Anaerobic decomposition

   Reference: *Water Quality*, 4th edition. 2010. Joseph A. Ritter, Editor. American Water Works Association. Chapter 6.

4. Answer: **a.** turbidity.

   Reference: *Water Sources*, 4th edition. 2010. Paul Koch, Editor. American Water Works Association. Chapter 6.

5. Answer: **d.** Calcium and magnesium

   Reference: *Water Treatment*, 4th edition. 2010. Nicholas G. Pizzi, Editor. American Water Works Association. Chapter 11.

Sample Questions for Level II—Answers

1. Answer **d.** calcium bicarbonate.

   Reference: *Basic Chemistry for Water & Wastewater Operators*, Revised edition. 2005. Darshan Singh Sarai. American Water Works Association. Chapter 13.

2. Answer **a.** Blue-green algae

   Reference: *Water Sources*, 4th edition. 2010. Paul Koch, Editor. American Water Works Association. Chapter 6.

3. Answer **d.** temperature, pressure, and salinity.

   Reference: *Water Quality*, 4th edition. 2010. Joseph A. Ritter, Editor. American Water Works Association. Chapter 6.

4. Answer **b.** biological oxidation of organic matter.

   Reference: *Water Quality*, 4th edition. 2010. Joseph A. Ritter, Editor. American Water Works Association. Chapter 6.

5. Answer **d.** 4.0°C

   Reference: *Basic Chemistry for Water & Wastewater Operators*, Revised edition. 2005. Darshan Singh Sarai. American Water Works Association. Chapter 9.

Sample Questions for Level III—Answers

1. Answer: **c.** one-ninth

   Reference: *Basic Chemistry for Water & Wastewater Operators*, Revised edition. 2005. Darshan Singh Sarai. American Water Works Association. Chapter 9.

2. Answer: **b.** Covalent

   Reference: *Basic Chemistry for Water & Wastewater Operators*, Revised edition. 2005. Darshan Singh Sarai. American Water Works Association. Chapter 9.

3. Answer: **d.** Control nutrients

   Reference: M7, *Problem Organisms in Water: Identification and Treatment*, 3rd edition. 2003. American Water Works Association. Chapter 11.

4. Answer: **c.** 10.00 mg/L.

   Reference: *Water Treatment Operator Handbook*, Revised edition. 2005. Nicholas G. Pizzi. American Water Works Association. Chapter 4.

5. Answers: **a.** 1.0 mg/L.

   Reference: *Water Quality*, 4th edition. 2010. Joseph A. Ritter, Editor. American Water Works Association. Chapter 6.

Sample Questions for Level IV—Answers

1. Answer: **d.** Where aerobic and anaerobic conditions meet

   Reference: *Iron and Manganese Removal Handbook*. 1999. Elmer O. Sommerfield. American Water Works Association. Chapter 4.

2. Answer: **b.** Aerobic activities

   Reference: *Iron and Manganese Removal Handbook*. 1999. Elmer O. Sommerfield. American Water Works Association. Chapter 4.

3. Answer: **c.** radium 226.

   Reference: *Water Quality*, 4th edition. 2010. Joseph A. Ritter, Editor. American Water Works Association. Chapter 7.

4. Answer: **a.** *Volvox*

   Reference: M7, *Problem Organisms in Water: Identification and Treatment*, 3rd edition. 2003. American Water Works Association. Chapter 10.

5. Answer: **c.** infectious hepatitis.

   Reference: *Water Quality*, 4th edition. 2010. Joseph A. Ritter, Editor. American Water Works Association. Chapter 4.

Answers to Additional Water Treatment Operator Practice Questions

Monitor, Evaluate, and Adjust Treatment Processes

Chemical Addition

1. Answer: **d.** Calcium carbonate

    Reference: *Water Quality,* 4th edition. 2010. Joseph A. Ritter, Editor. American Water Works Association. Chapter 5.

2. Answer: **d.** Potassium permanganate

    Reference: *Water Treatment*, 4th edition. 2010. Nicholas G. Pizzi, Editor. American Water Works Association. Chapter 7.

3. Answer: **d.** 500 psi.

    Reference: *Water Treatment*, 4th edition. 2010. Nicholas G. Pizzi, Editor. American Water Works Association. Chapter 7.

4. Answer: **d.** tuberculation.

    Reference: *Water Treatment*, 4th edition. 2010. Nicholas G. Pizzi, Editor. American Water Works Association. Chapter 9.

5. Answer: **c.** should be fed before chlorine.

    Reference: *Water Treatment*, 4th edition. 2010. Nicholas G. Pizzi, Editor. American Water Works Association. Chapter 7.

6. Answer: **d.** Alkalinity

    Reference: *Water Treatment*, 4th edition. 2010. Nicholas G. Pizzi, Editor. American Water Works Association. Chapter 4.

7. Answer: **b.** Potassium permanganate

    Reference: *Water Treatment*, 4th edition. 2010. Nicholas G. Pizzi, Editor. American Water Works Association. Chapter 7.

8. Answer: **c.** oxidant to reduce iron bacteria.

    Reference: *Water Treatment*, 4th edition. 2010. Nicholas G. Pizzi, Editor. American Water Works Association. Chapter 2.

9. Answer: **a.** $Ca(OH)_2$

    Reference: *Basic Microbiology for Drinking Water Personnel,* 2nd edition. 2006. Dennis R. Hill. American Water Works Association. Chapter 5.

10. Answer: **a.** gypsum.

    Reference: *Water Treatment Operator Handbook,* Revised edition. 2005. Nicholas G. Pizzi. American Water Works Association. Chapter 9.

11. Answer: **d.** Yellow

    Reference: *Water Treatment Operator Handbook*, Revised edition, 2005. Nicholas G. Pizzi. American Water Works Association. Chapter 12.

12. Answer: **c.** sonic flow.

    Reference: M20, *Water Chlorination/Chlorination Practices and Principles*, 2nd edition. 2006. American Water Works Association. Chapter 4.

13. Answer: **b.** Powder of calcium hypochlorite tablet

    Reference: *Water Sources*, 4th edition. 2010. Paul Koch, Editor. American Water Works Association. Chapter 2.

14. Answer: **d.** One-ton containers are equipped with two valves, one for liquid and one for gas

    Reference: *Water Treatment*, 4th edition. 2010. Nicholas G. Pizzi, Editor. American Water Works Association. Chapter 7.

15. Answer: **b.** 4 mg/L

    Reference: *Water Treatment*, 4th edition. 2010. Nicholas G. Pizzi, Editor. American Water Works Association. Chapter 8.

16. Answer: **a.** 800

    Reference: *Water Treatment Operator Handbook*, Revised edition. 2005. Nicholas G. Pizzi. American Water Works Association. Chapter 8.

17. Answer: **b.** total trihalomethanes.

    Reference: *Water Treatment Operator Handbook*, Revised edition. 2005. Nicholas G. Pizzi. American Water Works Association. Chapter 8.

18. Answer: **d.** Every hour

    Reference: *Water Treatment Operator Handbook*, Revised edition. 2005. Nicholas G. Pizzi. American Water Works Association. Chapter 10.

19. Answer: **a.** haloacetic acids.

    Reference: *Water Treatment Operator Handbook*, Revised edition. 2005. Nicholas G. Pizzi. American Water Works Association. Chapter 8.

20. Answer: **b.** mottling of the teeth.

    Reference: *Water Treatment Operator Handbook*, Revised edition. 2005. Nicholas G. Pizzi. American Water Works Association. Chapter 10.

21. Answer: **b.** 460

    Reference: M20, *Water Chlorination/Chlorination Practices and Principles*, 2nd edition. 2006. American Water Works Association, Chapter 4.

22. Answer: **a.** A few seconds to a few minutes

    Reference: *Water Treatment*, 4th edition. 2010. Nicholas G. Pizzi, Editor. American Water Works Association. Chapter 7.

23. Answer: **d.** is odorless.

    Reference: *Water Treatment*, 4th edition. 2010. Nicholas G. Pizzi, Editor. American Water Works Association. Chapter 8.

ANSWERS TO ADDITIONAL WATER TREATMENT OPERATOR PRACTICE QUESTIONS 177

24. Answer: **a.** pH

    Reference: *Water Treatment*, 4th edition. 2010. Nicholas G. Pizzi, Editor. American Water Works Association. Chapter 9.

25. Answer: **c.** Lime

    Reference: *Water Treatment*, 4th edition. 2010. Nicholas G. Pizzi, Editor. American Water Works Association. Chapter 9.

26. Answer: **c.** Every day

    Reference: *Water Treatment*, 4th edition. 2010. Nicholas G. Pizzi, Editor. American Water Works Association. Chapter 8.

27. Answer: **b.** 7.0

    Reference: *Water Treatment*, 4th edition. 2010. Nicholas G. Pizzi, Editor. American Water Works Association. Chapter 8.

28. Answer: **b.** 2,650 angstroms

    Reference: *Basic Chemistry for Water & Wastewater Operators*, Revised edition. 2005. Darshan Singh Sarai. American Water Works Association. Chapter 15.

29. Answer: **c.** Vacuum

    Reference: *Water Treatment Operator Handbook*, Revised edition. 2005. Nicholas G. Pizzi. American Water Works Association. Chapter 8.

30. Answer: **b.** Nitrates

    Reference: *Basic Chemistry for Water & Wastewater Operators*, Revised edition. 2005. Darshan Singh Sarai. American Water Works Association. Chapter 14.

31. Answer: **b.** $CaCO_3$

    Reference: *Water Treatment*, 4th edition. 2010. Nicholas G. Pizzi, Editor. American Water Works Association. Chapter 9.

32. Answer: **b.** air.

    Reference: *Basic Chemistry for Water & Wastewater Operators*, Revised edition. 2005. Darshan Singh Sarai. American Water Works Association. Chapter 1.

33. Answer: **b.** Peroxone

    Reference: M7, *Problem Organisms in Water: Identification and Treatment,* 3rd edition. 2003. American Water Works Association. Chapter 1.

34. Answer: **c.** Ozone

    Reference: *Water Treatment*, 4th edition. 2010. Nicholas G. Pizzi, Editor. American Water Works Association. Chapter 7.

35. Answer: **b.** 85 to 90%

    Reference: *Water Treatment*, 4th edition. 2010. Nicholas G. Pizzi, Editor. American Water Works Association. Chapter 7.

36. Answer: **d.** 9.0 to 11.0 pH units.

    Reference: *Water Treatment*, 4th edition. 2010. Nicholas G. Pizzi, Editor. American Water Works Association. Chapter 7.

37. Answer: **d.** After water has received complete treatment

    Reference: *Water Treatment*, 4th edition. 2010. Nicholas G. Pizzi, Editor. American Water Works Association. Chapter 8.

38. Answer: **b.** CO_2

    Reference: *Water Treatment*, 4th edition. 2010. Nicholas G. Pizzi, Editor. American Water Works Association. Chapter 9.

39. Answer: **d.** Polyphosphates

    Reference: *Water Treatment*, 4th edition. 2010. Nicholas G. Pizzi, Editor. American Water Works Association. Chapter 9.

40. Answer: **c.** 54°F

    Reference: *Water Treatment*, 4th edition. 2010. Nicholas G. Pizzi, Editor. American Water Works Association. Chapter 9.

41. Answer: **c.** ammonia and form chloramines.

    Reference: *Water Treatment Operator Handbook*, Revised edition. 2005. Nicholas G. Pizzi. American Water Works Association. Chapter 8.

42. Answer: **a.** Water temperature

    Reference: *Water Treatment*, 4th edition. 2010. Nicholas G. Pizzi, Editor. American Water Works Association. Chapter 4.

43. Answer: **c.** low pH values.

    Reference: *Iron and Manganese Removal Handbook*. 1999. Elmer O. Sommerfield. American Water Works Association. Chapter 7.

44. Answer: **b.** Hypochlorous acid

    Reference: *Chemistry of Water Treatment,* 2nd edition. 1998. Samuel D. Faust & Osman M. Aly. CRC Press.

45. Answer: **c.** Chloramines

    Reference: *Water Treatment*, 4th edition. 2010. Nicholas G. Pizzi, Editor. American Water Works Association. Chapter 7.

46. Answer: **b.** 1.50

    Reference: *Water Treatment*, 4th edition. 2010. Nicholas G. Pizzi, Editor. American Water Works Association. Chapter 7.

47. Answer: **c.** Diffusers

    Reference: *Water Treatment*, 4th edition. 2010. Nicholas G. Pizzi, Editor. American Water Works Association. Chapter 7.

48. Answer: **b.** temperature of the chlorine liquid.

    Reference: *Water Treatment*, 4th edition. 2010. Nicholas G. Pizzi, Editor. American Water Works Association. Chapter 7.

49. Answer: **b.** 100°F.

    Reference: *Water Treatment*, 4th edition. 2010. Nicholas G. Pizzi, Editor. American Water Works Association. Chapter 9.

ANSWERS TO ADDITIONAL WATER TREATMENT OPERATOR PRACTICE QUESTIONS 179

50. Answer: **a.** Zone 1

    Reference: M20, *Water Chlorination/Chlorination Practices and Principles,* 2nd edition. 2006. American Water Works Association. Chapter 3.

51. Answer: **a.** 6

    Reference: M20, *Water Chlorination/Chlorination Practices and Principles*, 2nd edition. 2006. American Water Works Association. Chapter 3.

52. Answer: **c.** Amber

    Reference: M20, *Water Chlorination/Chlorination Practices and Principles,* 2nd edition. 2006. American Water Works Association. Chapter 4.

53. Answer: **b.** –30°F.

    Reference: M20, *Water Chlorination/Chlorination Practices and Principles,* 2nd edition. 2006. American Water Works Association. Chapter 4.

54. Answer: **b.** pH and alkalinity adjustment with lime

    Reference: *Basic Science Concepts and Applications,* 4th edition. 2010. Nicholas G. Pizzi, Editor. Chemistry 6.

55. Answer: **c.** $Ca(OH)_2$

    Reference: *Water Treatment*, 4th edition. 2010. Nicholas G. Pizzi, Editor. American Water Works Association. Chapter 4.

56. Answer: **a.** at increasingly shorter intervals.

    Reference: M7, *Problem Organisms in Water: Identification and Treatment,* 3rd edition. 2003. American Water Works Association. Chapter 2.

57. Answer: **b.** Air surging, valved surge blocks, or jetting

    Reference: M7, *Problem Organisms in Water: Identification and Treatment,* 3rd edition. 2003. American Water Works Association. Chapter 2.

58. Answer: **c.** 50 mg/L

    Reference: *Water Sources,* 4th edition. 2010. Paul Koch, Editor. American Water Works Association. Chapter 2.

59. Answer: **a.** Slaked lime

    Reference: *Basic Chemistry for Water & Wastewater Operators*, Revised edition. 2005. Darshan Singh Sarai. American Water Works Association. Chapter 7.

60. Answer: **b.** Superchlorinate reservoirs and storage tanks

    Reference: M7, *Problem Organisms in Water: Identification and Treatment,* 3rd edition. 2003. American Water Works Association. Chapter 11.

61. Answer: **a.** nitrous oxides.

    Reference: *The Water Dictionary: A Comprehensive Reference of Water Terminology*, 2nd edition. 2010. Nancy McTigue, Editor, & James M. Symons, Editor Emeritus. American Water Works Association.

62. Answer: **a.** $HOCl > OCl > NH_2Cl > HCl$

    Reference: *Water Treatment Operator Handbook*, Revised edition. 2005. Nicholas G. Pizzi. American Water Works Association. Chapter 8.

63. Answer: **b.** 250 to 265 nm

    Reference: *Water Treatment*, 4th edition. 2010. Nicholas G. Pizzi, Editor. American Water Works Association. Chapter 7.

64. Answer: **c.** Point 4

    Reference: *Water Treatment*, 4th edition. 2010. Nicholas G. Pizzi, Editor. American Water Works Association. Chapter 7.

65. Answer: **b.** Carbon dioxide

    Reference: *Water Treatment*, 4th edition. 2010. Nicholas G. Pizzi, Editor. American Water Works Association. Chapter 9.

66. Answer: **d.** Oxygen

    Reference: *Water Treatment*, 4th edition. 2010. Nicholas G. Pizzi, Editor. American Water Works Association. Chapter 7.

67. Answer: **b.** Zinc

    Reference: *Water Treatment*, 4th edition. 2010. Nicholas G. Pizzi, Editor. American Water Works Association. Chapter 9.

68. Answer: **c.** oxygen.

    Reference: *Water Treatment*, 4th edition. 2010. Nicholas G. Pizzi, Editor. American Water Works Association. Chapter 10.

69. Answer: **d.** Chloroorganics and chloramines

    Reference: *Water Treatment Operator Handbook*, Revised edition. 2005. Nicholas G. Pizzi. American Water Works Association. Chapter 8.

70. Answer: **d.** Chlorine dioxide

    Reference: *Chemistry of Water Treatment,* 2nd edition. 1998. Samuel D. Faust & Osman M. Aly. CRC Press.

71. Answer: **c.** Ozone

    Reference: *Chemistry of Water Treatment,* 2nd edition. 1998. Samuel D. Faust & Osman M. Aly. CRC Press.

72. Answer: **a.** Ozone

    Reference: *Water Treatment*, 4th edition. 2010. Nicholas G. Pizzi, Editor. American Water Works Association. Chapter 7.

73. Answer: **a.** Oxygen

    Reference: *Water Treatment*, 4th edition. 2010. Nicholas G. Pizzi, Editor. American Water Works Association. Chapter 7.

74. Answer: **d.** Because it is neutral

    Reference: M20, *Water Chlorination/Chlorination Practices and Principles*, 2nd edition. 2006. American Water Works Association. Chapter 3.

75. Answer: **b.** Monochloramine

    Reference: M20, *Water Chlorination/Chlorination Practices and Principles*, 2nd edition. 2006. American Water Works Association. Chapter 3.

ANSWERS TO ADDITIONAL WATER TREATMENT OPERATOR PRACTICE QUESTIONS 181

76. Answer: **c.** Centrifugal pumps

    Reference: M20, *Water Chlorination/Chlorination Practices and Principles*, 2nd edition. 2006. American Water Works Association. Chapter 4.

77. Answer: **a.** their vulnerability to nitrification.

    Reference: M20, *Water Chlorination/Chlorination Practices and Principles*, 2nd edition. 2006. American Water Works Association. Chapter 6.

78. Answer: **d.** Post clearwell

    Reference: M20, *Water Chlorination/Chlorination Practices and Principles*, 2nd edition. 2006. American Water Works Association. Chapter 6.

79. Answer: **a.** Chelation

    Reference: *Basic Science Concepts and Applications*, 4th edition. 2010. Nicholas G. Pizzi, Editor. Chemistry 6.

80. Answer: **d.** 20.0 times

    Reference: *Basic Chemistry for Water & Wastewater Operators*, Revised edition. 2005. Darshan Singh Sarai. American Water Works Association. Chapter 15.

81. Answer: **a.** 1:2

    Reference: *Basic Chemistry for Water & Wastewater Operators*, Revised edition. 2005. Darshan Singh Sarai. American Water Works Association. Chapter 15.

82. Answer: **d.** 5.0 parts chlorine to 1 part nitrite

    Reference: M7, *Problem Organisms in Water: Identification and Treatment*, 3rd edition. 2003. American Water Works Association. Chapter 4.

83. Answer: **b.** 2 to 4%

    Reference: *Water Treatment*, 4th edition. 2010. Nicholas G. Pizzi, Editor. American Water Works Association. Chapter 7.

84. Answer: **a.** 2% ozone.

    Reference: *Water Treatment*, 4th edition. 2010. Nicholas G. Pizzi, Editor. American Water Works Association. Chapter 7.

85. Answer: **b.** Add chlorine to water, then this water to hydrochloric acid and sodium chlorate

    Reference: *Water Treatment*, 4th edition. 2010. Nicholas G. Pizzi, Editor. American Water Works Association. Chapter 7.

86. Answer: **a.** Polyphosphate

    Reference: *Water Treatment*, 4th edition. 2010. Nicholas G. Pizzi, Editor. American Water Works Association. Chapter 9.

87. Answer: **d.** It is completely reversible

    Reference: M20, *Water Chlorination/Chlorination Practices and Principles*, 2nd edition. 2006. American Water Works Association. Chapter 3.

88. Answer: **b.** antimony.

    Reference: *Water Treatment Operator Handbook*, Revised edition. 2005. Nicholas G. Pizzi. American Water Works Association. Chapter 8.

89. Answer: **b.** ozone.

    Reference: M7, *Problem Organisms in Water: Identification and Treatment,* 3rd edition. 2003. American Water Works Association. Chapter 10.

90. Answer: **d.** Chlorine dioxide

    Reference: *Water Treatment,* 4th edition. 2010. Nicholas G. Pizzi, Editor. American Water Works Association. Chapter 7.

91. Answer: **b.** has no by-products.

    Reference: *Water Treatment,* 4th edition. 2010. Nicholas G. Pizzi, Editor. American Water Works Association. Chapter 7.

92. Answer: **c.** Inject chlorine gas under vacuum into a stream of chlorite solution

    Reference: *Water Treatment,* 4th edition. 2010. Nicholas G. Pizzi, Editor. American Water Works Association. Chapter 7.

93. Answer: **a.** 20 to 30%

    Reference: *Water Treatment,* 4th edition. 2010. Nicholas G. Pizzi, Editor. American Water Works Association. Chapter 8.

94. Answer: **c.** Crystalline sodium fluoride

    Reference: *Water Treatment,* 4th edition. 2010. Nicholas G. Pizzi, Editor. American Water Works Association. Chapter 8.

95. Answer: **c.** UV light

    Reference: M20, *Water Chlorination/Chlorination Practices and Principles,* 2nd edition. 2006. American Water Works Association. Chapter 2.

96. Answer: **a.** Titanium and tantalum

    Reference: M20, *Water Chlorination/Chlorination Practices and Principles,* 2nd edition. 2006. American Water Works Association. Chapter 2.

97. Answer: **d.** 2,000 lb/d.

    Reference: M20, *Water Chlorination/Chlorination Practices and Principles,* 2nd edition. 2006. American Water Works Association. Chapter 4.

98. Answer: **c.** higher temperatures.

    Reference: M20, *Water Chlorination/Chlorination Practices and Principles,* 2nd edition. 2006. American Water Works Association. Chapter 6.

99. Answer: **b.** maintaining a chlorine to ammonia-nitrogen ratio between 4.5:1 and 5.0:1.

    Reference: M20, *Water Chlorination/Chlorination Practices and Principles,* 2nd edition. 2006. American Water Works Association. Chapter 6.

100. Answer: **b.** Sodium hypochlorite

    Reference: *Water Treatment,* 4th edition. 2010. Nicholas G. Pizzi, Editor. American Water Works Association. Chapter 7.

Coagulation and Flocculation

1. Answer: **a.** 3

    Reference: *Basic Chemistry for Water & Wastewater Operators*, Revised edition. 2005. Darshan Singh Sarai. American Water Works Association. Chapter 8.

2. Answer: **d.** water-soluble organic polyelectrolytes.

    Reference: *The Water Dictionary: A Comprehensive Reference of Water Terminology*, 2nd edition. 2010. Nancy McTigue, Editor, & James M. Symons, Editor Emeritus. American Water Works Association.

3. Answer: **a.** Electrical charge on a suspended particle

    Reference: *Water Quality,* 4th edition. 2010. Joseph A. Ritter, Editor. American Water Works Association. Chapter 5.

4. Answer: **c.** electrically charged.

    Reference: *Basic Chemistry for Water & Wastewater Operators*, Revised edition. 2005. Darshan Singh Sarai. American Water Works Association. Chapter 8.

5. Answer: **c.** Brownian movement.

    Reference: M37, *Operational Control of Coagulation and Filtration Processes,* 3rd edition. 2011. American Water Works Association. Chapter 4.

6. Answer: **c.** Sodium aluminate

    Reference: *Water Treatment*, 4th edition. 2010. Nicholas G. Pizzi, Editor. American Water Works Association. Chapter 4.

7. Answer: **a.** aluminum hydroxide.

    Reference: *Water Treatment Operator Handbook*, Revised edition. 2005. Nicholas G. Pizzi. American Water Works Association. Chapter 5.

8. Answer: **a.** ferric hydroxide.

    Reference: *Water Treatment Operator Handbook*, Revised edition. 2005. Nicholas G. Pizzi. American Water Works Association. Chapter 5.

9. Answer: **b.** 15 to 45 seconds.

    Reference: *Water Treatment Operator Handbook*, Revised edition. 2005. Nicholas G. Pizzi. American Water Works Association. Chapter 5.

10. Answer: **a.** 1 to 2 seconds.

    Reference: *Water Treatment*, 4th edition. 2010. Nicholas G. Pizzi, Editor. American Water Works Association. Chapter 4.

11. Answer: **b.** negatively charged.

    Reference: *Water Treatment*, 4th edition. 2010. Nicholas G. Pizzi, Editor. American Water Works Association. Chapter 4.

12. Answer: **c.** Coagulation and filtration

    Reference: *Basic Science Concepts and Applications,* 4th edition. 2010. Nicholas G. Pizzi, Editor. Chemistry 6.

13. Answer: **b.** zeta potential is stronger than the van der Waals forces.

    Reference: *Water Treatment*, 4th edition. 2010. Nicholas G. Pizzi, Editor. American Water Works Association. Chapter 4.

14. Answer: **b.** 3:1 aluminum sulfate to hydrated lime.

    Reference: *Water Treatment*, 4th edition. 2010. Nicholas G. Pizzi, Editor. American Water Works Association. Chapter 4.

15. Answer: **c.** 1.0 ft/sec

    Reference: *Water Treatment Operator Handbook*, Revised edition. 2005. Nicholas G. Pizzi. American Water Works Association. Chapter 5.

16. Answer: **a.** 50 to 60 times

    Reference: *Water Treatment*, 4th edition. 2010. Nicholas G. Pizzi, Editor. American Water Works Association. Chapter 4.

17. Answer: **b.** 8.0 fps

    Reference: M37, *Operational Control of Coagulation and Filtration Processes*, 3rd edition. 2011. American Water Works Association. Introduction.

18. Answer: **b.** Static mixers

    Reference: *Water Treatment*, 4th edition. 2010. Nicholas G. Pizzi, Editor. American Water Works Association. Chapter 4.

19. Answer: **d.** Polymers

    Reference: *Water Treatment*, 4th edition. 2010. Nicholas G. Pizzi, Editor. American Water Works Association. Chapter 4.

20. Answer: **c.** low in turbidity and mineral content and high in color.

    Reference: *Water Treatment*, 4th edition. 2010. Nicholas G. Pizzi, Editor. American Water Works Association. Chapter 4.

21. Answer: **a.** mechanical mixers.

    Reference: *Water Treatment*, 4th edition. 2010. Nicholas G. Pizzi, Editor. American Water Works Association. Chapter 4.

22. Answer: **d.** maximum.

    Reference: *Water Treatment*, 4th edition. 2010. Nicholas G. Pizzi, Editor. American Water Works Association. Chapter 4.

23. Answer: **d.** 8:1 ferric sulfate to chlorine.

    Reference: *Water Treatment*, 4th edition. 2010. Nicholas G. Pizzi, Editor. American Water Works Association. Chapter 4.

24. Answer: **c.** a pump (adding coagulant on suction side).

    Reference: *Water Treatment*, 4th edition. 2010. Nicholas G. Pizzi, Editor. American Water Works Association. Chapter 4.

25. Answer: **a.** Initial application at point of highest mixing intensity

    Reference: *Water Treatment Operator Handbook*, Revised edition. 2005. Nicholas G. Pizzi. American Water Works Association. Chapter 5.

ANSWERS TO ADDITIONAL WATER TREATMENT OPERATOR PRACTICE QUESTIONS 185

Clarification and Sedimentation

1. Answer: **b.** gravity.

   Reference: *Water Treatment Operator Handbook*, Revised edition. 2005. Nicholas G. Pizzi. American Water Works Association. Chapter 6.

2. Answer: **d.** 4 hours

   Reference: *Water Treatment Operator Handbook*, Revised edition. 2005. Nicholas G. Pizzi. American Water Works Association. Chapter 6.

3. Answer: **b.** 90% or more

   Reference: *Water Treatment Operator Handbook*, Revised edition. 2005. Nicholas G. Pizzi. American Water Works Association. Chapter 6.

4. Answer: **b.** 4

   Reference: *Water Treatment*, 4th edition. 2010. Nicholas G. Pizzi, Editor. American Water Works Association. Chapter 5.

5. Answer: **b.** Settling zone

   Reference: *Water Treatment*, 4th edition. 2010. Nicholas G. Pizzi, Editor. American Water Works Association. Chapter 5.

6. Answer: **c.** float.

   Reference: *Water Treatment*, 4th edition. 2010. Nicholas G. Pizzi, Editor. American Water Works Association. Chapter 5.

7. Answer: **b.** 20,000 gpd/ft

   Reference: *Water Treatment*, 4th edition. 2010. Nicholas G. Pizzi, Editor. American Water Works Association. Chapter 5.

8. Answer: **c.** Mixing and settling zones

   Reference: *Water Treatment*, 4th edition. 2010. Nicholas G. Pizzi, Editor. American Water Works Association. Chapter 11.

9. Answer: **c.** 55.0 degrees.

   Reference: *Water Treatment Operator Handbook*, Revised edition. 2005. Nicholas G. Pizzi. American Water Works Association. Chapter 6.

10. Answer: **d.** Dissolved air flotation

    Reference: *Water Treatment Operator Handbook*, Revised edition. 2005. Nicholas G. Pizzi. American Water Works Association. Chapter 6.

11. Answer: **a.** copper sulfate and lime.

    Reference: *Water Treatment Operator Handbook*, Revised edition. 2005. Nicholas G. Pizzi. American Water Works Association. Chapter 6.

12. Answer: **b.** Short-circuiting

    Reference: *Water Treatment*, 4th edition. 2010. Nicholas G. Pizzi, Editor. American Water Works Association. Chapter 5.

13. Answer: **b.** Every 40 to 50 seconds

    Reference: *Water Treatment*, 4th edition. 2010. Nicholas G. Pizzi, Editor. American Water Works Association. Chapter 5.

14. Answer: **a.** 4 to 8 hours.

    Reference: *Water Treatment*, 4th edition. 2010. Nicholas G. Pizzi, Editor. American Water Works Association. Chapter 5.

15. Answer: **d.** Density current

    Reference: *Water Treatment*, 4th edition. 2010. Nicholas G. Pizzi, Editor. American Water Works Association. Chapter 5.

16. Answer: **a.** are actually inclined about 5 degrees.

    Reference: *Water Treatment Operator Handbook*, Revised edition. 2005. Nicholas G. Pizzi. American Water Works Association. Chapter 6.

17. Answer: **c.** Actiflo process

    Reference: *Water Treatment Operator Handbook*, Revised edition. 2005. Nicholas G. Pizzi. American Water Works Association. Chapter 6.

18. Answer: **b.** Superpulsators

    Reference: *Water Treatment Operator Handbook*, Revised edition. 2005. Nicholas G. Pizzi. American Water Works Association. Chapter 6.

19. Answer: **d.** 2.0-log

    Reference: *Water Treatment Operator Handbook*, Revised edition. 2005. Nicholas G. Pizzi. American Water Works Association. Chapter 6.

20. Answer: **b.** beach.

    Reference: *Water Treatment*, 4th edition. 2010. Nicholas G. Pizzi, Editor. American Water Works Association. Chapter 5.

21. Answer: **d.** Settling

    Reference: M7, *Problem Organisms in Water: Identification and Treatment,* 3rd edition. 2003. American Water Works Association. Chapter 5.

22. Answer: **d.** 60°

    Reference: *Water Treatment*, 4th edition. 2010. Nicholas G. Pizzi, Editor. American Water Works Association. Chapter 5.

23. Answer: **c.** Actiflo process

    Reference: *Water Treatment Operator Handbook*, Revised edition. 2005. Nicholas G. Pizzi. American Water Works Association. Chapter 6.

24. Answer: **c.** 16 gpm/ft^2

    Reference: *Water Treatment Operator Handbook*, Revised edition. 2005. Nicholas G. Pizzi. American Water Works Association. Chapter 6.

25. Answer: **b.** Coagulation, sedimentation, and filtration

    Reference: *Chemistry of Water Treatment,* 2nd edition. 1998. Samuel D. Faust & Osman M. Aly. CRC Press.

Filtration

1. Answer: **c.** potassium permanganate solution during backwashing.

    Reference: *Water Treatment*, 4th edition. 2010. Nicholas G. Pizzi, Editor. American Water Works Association. Chapter 10.

ANSWERS TO ADDITIONAL WATER TREATMENT OPERATOR PRACTICE QUESTIONS   187

2. Answer: **d.** Glauconite sand coated with manganese dioxide

   Reference: *Iron and Manganese Removal Handbook.* 1999. Elmer O. Sommerfield. American Water Works Association. Chapter 3.

3. Answer: **c.** Schmutzdecke

   Reference: *The Water Dictionary: A Comprehensive Reference of Water Terminology,* 2nd edition. 2010. Nancy McTigue, Editor, & James M. Symons, Editor Emeritus. American Water Works Association.

4. Answer: **a.** Diatomaceous earth filter

   Reference: *Water Treatment,* 4th edition. 2010. Nicholas G. Pizzi, Editor. American Water Works Association. Chapter 6.

5. Answer: **c.** 6 months.

   Reference: *Water Treatment Operator Handbook,* Revised edition. 2005. Nicholas G. Pizzi. American Water Works Association. Chapter 7.

6. Answer: **b.** 8 ft.

   Reference: *Water Treatment Operator Handbook,* Revised edition. 2005. Nicholas G. Pizzi. American Water Works Association. Chapter 7.

7. Answer: **d.** Diatoms

   Reference: *Water Treatment,* 4th edition. 2010. Nicholas G. Pizzi, Editor. American Water Works Association. Chapter 2.

8. Answer: **c.** 10 to 12 gpm/ft^2

   Reference: *Iron and Manganese Removal Handbook.* 1999. Elmer O. Sommerfield. American Water Works Association. Chapter 3.

9. Answer: **b.** 24 hours.

   Reference: *Water Treatment,* 4th edition. 2010. Nicholas G. Pizzi, Editor. American Water Works Association. Chapter 6.

10. Answer: **b.** Microfiltration, ultrafiltration, nanofiltration, and reverse osmosis

    Reference: *Basic Microbiology for Drinking Water Personnel,* Dennis R. Hill, Second edition, Chapter 5.

11. Answer: **b.** 0.05 mg/L.

    Reference: *Iron and Manganese Removal Handbook.* 1999. Elmer O. Sommerfield. American Water Works Association. Chapter 6.

12. Answer: **a.** 10 μm.

    Reference: *Iron and Manganese Removal Handbook.* 1999. Elmer O. Sommerfield. American Water Works Association. Chapter 6.

13. Answer: **b.** Activated carbon

    Reference: *The Water Dictionary: A Comprehensive Reference of Water Terminology,* 2nd edition. 2010. Nancy McTigue, Editor, & James M. Symons, Editor Emeritus. American Water Works Association.

14. Answer: **b.** 3.0 to 12.0 gpm/ft^2

    Reference: *Water Treatment,* 4th edition. 2010. Nicholas G. Pizzi, Editor. American Water Works Association. Chapter 6.

15. Answer: **c.** 8 to 9%

    Reference: *Iron and Manganese Removal Handbook.* 1999. Elmer O. Sommerfield. American Water Works Association. Chapter 6.

16. Answer: **d.** unstratified.

    Reference: *Water Treatment*, 4th edition. 2010. Nicholas G. Pizzi, Editor. American Water Works Association. Chapter 6.

17. Answer: **a.** electrostatic charges.

    Reference: *The Water Dictionary: A Comprehensive Reference of Water Terminology*, 2nd edition. 2010. Nancy McTigue, Editor, & James M. Symons, Editor Emeritus. American Water Works Association.

18. Answer: **b.** 2 to 4%

    Reference: *Water Treatment*, 4th edition. 2010. Nicholas G. Pizzi, Editor. American Water Works Association. Chapter 6.

19. Answer: **b.** Because it is lighter and less dense

    Reference: *Water Treatment*, 4th edition. 2010. Nicholas G. Pizzi, Editor. American Water Works Association. Chapter 6.

20. Answer: **d.** adding a layer of anthracite above the greensand.

    Reference: *Water Treatment*, 4th edition. 2010. Nicholas G. Pizzi, Editor. American Water Works Association. Chapter 10.

21. Answer: **c.** Butterfly

    Reference: *Water Treatment Operator Handbook*, Revised edition. 2005. Nicholas G. Pizzi. American Water Works Association. Chapter 7.

22. Answer: **a.** adsorption and oxidation.

    Reference: *Water Treatment Operator Handbook*, Revised edition. 2005. Nicholas G. Pizzi. American Water Works Association. Chapter 10.

23. Answer: **a.** Coagulation, flocculation, and filtration

    Reference: *Chemistry of Water Treatment,* 2nd edition. 1998. Samuel D. Faust & Osman M. Aly. CRC Press.

24. Answer: **a.** Reverse osmosis

    Reference: *Chemistry of Water Treatment,* 2nd edition. 1998. Samuel D. Faust & Osman M. Aly. CRC Press.

25. Answer: **b.** with higher temperature.

    Reference: *Water Treatment Operator Handbook*, Revised edition. 2005. Nicholas G. Pizzi. American Water Works Association. Chapter 11.

26. Answer: **a.** Reverse osmosis

    Reference: *Chemistry of Water Treatment,* 2nd edition. 1998. Samuel D. Faust & Osman M. Aly. CRC Press.

27. Answer: **b.** 2.0 gpm/ft^2

    Reference: *Water Treatment*, 4th edition. 2010. Nicholas G. Pizzi, Editor. American Water Works Association. Chapter 13.

ANSWERS TO ADDITIONAL WATER TREATMENT OPERATOR PRACTICE QUESTIONS

28. Answer: **a.** 0.50 mg/L.

    Reference: *Iron and Manganese Removal Handbook*. 1999. Elmer O. Sommerfield. American Water Works Association. Chapter 6.

29. Answer: **a.** increasing solution pH.

    Reference: *Iron and Manganese Removal Handbook*. 1999. Elmer O. Sommerfield. American Water Works Association. Chapter 6.

30. Answer: **b.** up to 25 ntu.

    Reference: *Water Treatment*, 4th edition. 2010. Nicholas G. Pizzi, Editor. American Water Works Association. Chapter 6.

31. Answer: **c.** are used only for water with low turbidity.

    Reference: *Water Treatment*, 4th edition. 2010. Nicholas G. Pizzi, Editor. American Water Works Association. Chapter 6.

32. Answer: **c.** pyrolusite filter beds.

    Reference: *Iron and Manganese Removal Handbook*. 1999. Elmer O. Sommerfield. American Water Works Association. Chapter 5.

33. Answer: **d.** The water being filtered has limited contact time with the GAC

    Reference: *Water Treatment*, 4th edition. 2010. Nicholas G. Pizzi, Editor. American Water Works Association. Chapter 6.

34. Answer: **b.** it does not require a rate-of-flow controller.

    Reference: *Water Treatment*, 4th edition. 2010. Nicholas G. Pizzi, Editor. American Water Works Association. Chapter 6.

35. Answer: **a.** Increase chlorination and backwash

    Reference: M7, *Problem Organisms in Water: Identification and Treatment,* 3rd edition. 2003. American Water Works Association. Chapter 11.

Residuals Disposal

1. Answer: **d.** Water table level

   Reference: *Water Treatment Plant Operation*, Volume II, 5th edition. 2006. Ken Kerri, California State University – Sacramento. Chapter 17 – Residuals, Page 237.

2. Answer: **b.** 3

   Reference: *Water Treatment Plant Operation,* Volume II, 5th edition. 2006. Ken Kerri, California State University – Sacramento. Chapter 17 – Residuals, Page 240.

3. Answer: **d.** Belt filter press

   Reference: *Water Treatment Plant Operation,* Volume II, 5th edition. 2006. Ken Kerri, California State University – Sacramento. Chapter 17 – Residuals, Page 241.

4. Answer: **a.** Centrifuge

   Reference: *Water Treatment Plant Operation*, Volume II, 5th edition. 2006. Ken Kerri, California State University – Sacramento. Chapter 17 – Residuals, Page 244.

5. Answer: **c.** Filter press

   Reference: *Water Treatment Plant Operation,* Volume II, 5th edition. 2006. Ken Kerri, California State University – Sacramento. Chapter 17 – Residuals, Page 245.

6. Answer: **b.** Vacuum filters

   Reference: *Water Treatment Plant Operation,* Volume II, 5th edition. 2006. Ken Kerri, California State University – Sacramento. Chapter 17 – Residuals, Page 245.

Additional Treatment Tasks

1. Answer: **c.** Brine

   Reference: *Water Treatment Operator Handbook*, Revised edition. 2005. Nicholas G. Pizzi. American Water Works Association. Chapter 9.

2. Answer: **b.** kept in solution by certain chemicals.

   Reference: *Water Treatment Operator Handbook*, Revised edition. 2005. Nicholas G. Pizzi. American Water Works Association. Chapter 10.

3. Answer: **b.** Ferric sulfate

   Reference: *Water Treatment*, 4th edition. 2010. Nicholas G. Pizzi, Editor. American Water Works Association. Chapter 4.

4. Answer: **a.** 0.2 mg/L.

   Reference: *Water Treatment*, 4th edition. 2010. Nicholas G. Pizzi, Editor. American Water Works Association. Chapter 10.

5. Answer: **b.** Two times as much lime

   Reference: *Water Treatment Operator Handbook*, Revised edition. 2005. Nicholas G. Pizzi. American Water Works Association. Chapter 9.

6. Answer: **d.** Lime and soda ash

   Reference: *Water Treatment Operator Handbook*, Revised edition. 2005. Nicholas G. Pizzi. American Water Works Association. Chapter 11

7. Answer: **d.** exhausted.

   Reference: *Water Treatment Operator Handbook*, Revised edition. 2005. Nicholas G. Pizzi. American Water Works Association. Chapter 9.

8. Answer: **b.** Lime softening

   Reference: *Chemistry of Water Treatment,* 2nd edition. 1998. Samuel D. Faust & Osman M. Aly. CRC Press.

9. Answer: **b.** A few minutes before chlorine application

   Reference: *Iron and Manganese Removal Handbook*. 1999. Elmer O. Sommerfield. American Water Works Association. Chapter 7.

10. Answer: **d.** strengthen the floc.

    Reference: *Water Treatment*, 4th edition. 2010. Nicholas G. Pizzi, Editor. American Water Works Association. Chapter 4.

11. Answer: **b.** Carbon dioxide

    Reference: *Water Treatment*, 4th edition. 2010. Nicholas G. Pizzi, Editor. American Water Works Association. Chapter 9.

ANSWERS TO ADDITIONAL WATER TREATMENT OPERATOR PRACTICE QUESTIONS   191

12. Answer: **c.** is a sequestering agent.

    Reference: *Water Treatment*, 4th edition. 2010. Nicholas G. Pizzi, Editor. American Water Works Association. Chapter 9.

13. Answer: **d.** little or no noncarbonated hardness.

    Reference: *Water Treatment*, 4th edition. 2010. Nicholas G. Pizzi, Editor. American Water Works Association. Chapter 11.

14. Answer: **d.** only on the amount of noncarbonated hardness to be removed.

    Reference: *Water Treatment*, 4th edition. 2010. Nicholas G. Pizzi, Editor. American Water Works Association. Chapter 11.

15. Answer: **a.** before the normal coagulation-flocculation step.

    Reference: *Water Treatment*, 4th edition. 2010. Nicholas G. Pizzi, Editor. American Water Works Association. Chapter 13.

16. Answer: **a.** Sodium

    Reference: *Water Treatment Operator Handbook*, Revised edition. 2005. Nicholas G. Pizzi. American Water Works Association. Chapter 9.

17. Answer: **a.** oxidized and made insoluble.

    Reference: *Water Treatment Operator Handbook*, Revised edition. 2005. Nicholas G. Pizzi. American Water Works Association. Chapter 10.

18. Answer: **b.** iron

    Reference: *Water Treatment*, 4th edition. 2010. Nicholas G. Pizzi, Editor. American Water Works Association. Chapter 11.

19. Answer: **c.** Red

    Reference: M20, *Water Chlorination/Chlorination Practices and Principles*, 2nd edition. 2006. American Water Works Association. Chapter 3.

20. Answer: **d.** cementation of filter media.

    Reference: *Basic Science Concepts and Applications*, 4th edition. 2010. Nicholas G. Pizzi, Editor. Chemistry 6.

21. Answer: **b.** Carbon dioxide

    Reference: *Water Treatment*, 4th edition. 2010. Nicholas G. Pizzi, Editor. American Water Works Association. Chapter 9.

22. Answer: **a.** Activated carbon adsorption

    Reference: M7, *Problem Organisms in Water: Identification and Treatment*, 3rd edition. 2003. American Water Works Association. Chapter 1.

23. Answer: **c.** electrolysis.

    Reference: *Basic Chemistry for Water & Wastewater Operators*, Revised edition. 2005. Darshan Singh Sarai. American Water Works Association. Chapter 10

24. Answer: **a.** no dissolved oxygen.

    Reference: *Water Treatment*, 4th edition. 2010. Nicholas G. Pizzi, Editor. American Water Works Association. Chapter 10.

25. Answer: **c.** A dissolution-dispersion step using acidification and surfactants

    Reference: M7, *Problem Organisms in Water: Identification and Treatment,* 3rd edition. 2003. American Water Works Association. Chapter 2.

26. Answer: **c.** Noncarbonated hardness

    Reference: *Iron and Manganese Removal Handbook.* 1999. Elmer O. Sommerfield. American Water Works Association. Chapter 7.

27. Answer: **b.** pH.

    Reference: *Water Treatment,* 4th edition. 2010. Nicholas G. Pizzi, Editor. American Water Works Association. Chapter 12.

28. Answer: **d.** 1.28 mg/L

    Reference: *Basic Science Concepts and Applications,* 4th edition. 2010. Nicholas G. Pizzi, Editor. Chemistry 6.

29. Answer: **b.** At the time of installation and every six months thereafter

    Reference: *Water Treatment,* 4th edition. 2010. Nicholas G. Pizzi, Editor. American Water Works Association. Chapter 13.

30. Answer: **d.** Lime softening

    Reference: *Chemistry of Water Treatment,* 2nd edition. 1998. Samuel D. Faust & Osman M. Aly. CRC Press.

31. Answer: **b.** pH and alkalinity

    Reference: *Water Treatment,* 4th edition. 2010. Nicholas G. Pizzi, Editor. American Water Works Association. Chapter 12.

32. Answer: **a.** 5 μm

    Reference: *Iron and Manganese Removal Handbook.* 1999. Elmer O. Sommerfield. American Water Works Association. Chapter 6.

33. Answer: **b.** Use activated carbon

    Reference: M7, *Problem Organisms in Water: Identification and Treatment,* 3rd edition. 2003. American Water Works Association. Chapter 11.

34. Answer: **a.** Na_2SiO_3

    Reference: *Water Treatment,* 4th edition. 2010. Nicholas G. Pizzi, Editor. American Water Works Association. Chapter 4.

35. Answer: **a.** Magnesium hardness

    Reference: *Water Treatment,* 4th edition. 2010. Nicholas G. Pizzi, Editor. American Water Works Association. Chapter 11.

36. Answer: **b.** accomplishes greater removal of magnesium.

    Reference: *Water Treatment,* 4th edition. 2010. Nicholas G. Pizzi, Editor. American Water Works Association. Chapter 11.

37. Answer: **c.** 8.3.

    Reference: *Water Treatment,* 4th edition. 2010. Nicholas G. Pizzi, Editor. American Water Works Association. Chapter 11.

ANSWERS TO ADDITIONAL WATER TREATMENT OPERATOR PRACTICE QUESTIONS 193

38. Answer: **b.** high pH that is required to operate the process.

    Reference: *Water Treatment*, 4th edition. 2010. Nicholas G. Pizzi, Editor. American Water Works Association. Chapter 11.

39. Answer: **c.** 50 to 75%

    Reference: *Water Treatment*, 4th edition. 2010. Nicholas G. Pizzi, Editor. American Water Works Association. Chapter 12.

40. Answer: **d.** 50%

    Reference: *Water Treatment*, 4th edition. 2010. Nicholas G. Pizzi, Editor. American Water Works Association. Chapter 13.

41. Answer: **c.** Cascade aerator

    Reference: *Water Treatment*, 4th edition. 2010. Nicholas G. Pizzi, Editor. American Water Works Association. Chapter 14.

42. Answer: **d.** Draft aerator

    Reference: *Water Treatment*, 4th edition. 2010. Nicholas G. Pizzi, Editor. American Water Works Association. Chapter 14.

43. Answer: **c.** Solids-contact unit

    Reference: *Water Treatment Operator Handbook*, Revised edition. 2005. Nicholas G. Pizzi. American Water Works Association. Chapter 6.

44. Answer: **b.** Calcium carbonate crystals

    Reference: *Water Treatment Operator Handbook*, Revised edition. 2005. Nicholas G. Pizzi. American Water Works Association. Chapter 9.

45. Answer: **b.** 3 to 5 kilograins/ft^3

    Reference: *Water Treatment Operator Handbook*, Revised edition. 2005. Nicholas G. Pizzi. American Water Works Association. Chapter 9.

46. Answer: **a.** Particulate removal and particulate adsorption

    Reference: *Water Treatment Operator Handbook*, Revised edition. 2005. Nicholas G. Pizzi. American Water Works Association. Chapter 10.

47. Answer: **d.** Aeration and stripping

    Reference: *Chemistry of Water Treatment,* 2nd edition. 1998. Samuel D. Faust & Osman M. Aly. CRC Press.

48. Answer: **d.** strong-acid resins.

    Reference: *Water Treatment*, 4th edition. 2010. Nicholas G. Pizzi, Editor. American Water Works Association. Chapter 12.

49. Answer: **a.** 1.0 mg/L.

    Reference: *Water Treatment*, 4th edition. 2010. Nicholas G. Pizzi, Editor. American Water Works Association. Chapter 10.

50. Answer: **a.** Polyelectrolytes

    Reference: *Water Treatment*, 4th edition. 2010. Nicholas G. Pizzi, Editor. American Water Works Association. Chapter 4.

51. Answer: **a.** empty-bed contact time.

    Reference: *Water Treatment Operator Handbook,* Revised edition. 2005. Nicholas G. Pizzi. Editor. Chapter 10.

52. Answer: **b.** Granular activated carbon

    Reference: *Chemistry of Water Treatment,* 2nd edition. 1998. Samuel D. Faust & Osman M. Aly. CRC Press.

53. Answer: **b.** Aeration and stripping

    Reference: *Chemistry of Water Treatment,* 2nd edition. 1998. Samuel D. Faust & Osman M. Aly. CRC Press.

54. Answer: **c.** $Mg(OH)_2$

    Reference: *Water Treatment Operator Handbook,* Revised edition. 2005. Nicholas G. Pizzi. American Water Works Association. Chapter 9.

55. Answer: **b.** 15.

    Reference: *Water Treatment,* 4th edition. 2010. Nicholas G. Pizzi, Editor. American Water Works Association. Chapter 15.

56. Answer: **a.** Chlorine is being fed too far ahead of the sequestering agent

    Reference: *Iron and Manganese Removal Handbook*. 1999. Elmer O. Sommerfield. American Water Works Association. Chapter 7.

57. Answer: **d.** increasing levels of calcium and magnesium.

    Reference: *Iron and Manganese Removal Handbook*. 1999. Elmer O. Sommerfield. American Water Works Association. Chapter 7.

58. Answer: **d.** Activated carbon and ozone

    Reference: M7, *Problem Organisms in Water: Identification and Treatment,* 3rd edition. 2003. American Water Works Association. Chapter 10.

59. Answer: **c.** Risk of introduction of pathogenic microorganisms into the finished water if not properly designed

    Reference: *Water Treatment,* 4th edition. 2010. Nicholas G. Pizzi, Editor. American Water Works Association. Chapter 6.

60. Answer: **b.** 5 to 15%.

    Reference: *Water Treatment,* 4th edition. 2010. Nicholas G. Pizzi, Editor. American Water Works Association. Chapter 11.

61. Answer: **d.** pH is lower than 8.2.

    Reference: *Water Treatment,* 4th edition. 2010. Nicholas G. Pizzi, Editor. American Water Works Association. Chapter 12.

62. Answer: **a.** 0.15 mg/L

    Reference: *Water Treatment,* 4th edition. 2010. Nicholas G. Pizzi, Editor. American Water Works Association. Chapter 12.

63. Answer: **a.** Aluminum

    Reference: *Water Treatment,* 4th edition. 2010. Nicholas G. Pizzi, Editor. American Water Works Association. Chapter 12.

64. Answer: **c.** PAC and chlorine are added too close to each other

    Reference: *Water Treatment*, 4th edition. 2010. Nicholas G. Pizzi, Editor. American Water Works Association. Chapter 13.

65. Answer: **c.** 25:1

    Reference: *Water Treatment*, 4th edition. 2010. Nicholas G. Pizzi, Editor. American Water Works Association. Chapter 14.

66. Answer: **a.** monitor filters for carbonate deposition.

    Reference: *Water Treatment Operator Handbook,* Revised edition. 2005. Nicholas G. Pizzi. American Water Works Association. Chapter 9.

67. Answer: **d.** Replace or reactivate the GAC

    Reference: *Water Treatment Operator Handbook,* Revised edition. 2005. Nicholas G. Pizzi. American Water Works Association. Chapter 10.

68. Answer: **d.** Granular activated carbon

    Reference: *Chemistry of Water Treatment,* 2nd edition. 1998. Samuel D. Faust & Osman M. Aly. CRC Press.

69. Answer: **a.** Strong base anion exchange

    Reference: *Chemistry of Water Treatment,* 2nd edition. 1998. Samuel D. Faust & Osman M. Aly. CRC Press.

70. Answer: **c.** before chlorine addition.

    Reference: M20, *Water Chlorination/Chlorination Practices and Principles*, 2nd edition. 2006. American Water Works Association. Chapter 6.

71. Answer: **d.** Cation permeable membrane, plastic spacer, anion permeable membrane

    Reference: *Water Treatment Plant Operation*, Volume II, 5th edition. 2006. Ken Kerri. California State University–Sacramento. Page 213.

72. Answer: **d.** 110°F.

    Reference: *Water Treatment Plant Operation*, Volume II, 5th edition. 2006. Ken Kerri. California State University–Sacramento. Page 213.

73. Answer: **c.** scaling or membrane fouling.

    Reference: *Water Treatment Plant Operation*, Volume II, 5th edition. 2006. Ken Kerri. California State University–Sacramento. Page 213.

74. Answer: **d.** Oxidants

    Reference: US Environmental Protection Agency. EPA's recommendations for enhanced monitoring for Hexavalent Chromium (Chromium-6) in Drinking Water. 2011. http://water.epa.gov/drink/info/chromium/guidance.cfm

75. Answer: **a.** Ion chromatography

    Reference: US Environmental Protection Agency. EPA's recommendations for enhanced monitoring for Hexavalent Chromium (Chromium-6) in Drinking Water. 2011. http://water.epa.gov/drink/info/chromium/guidance.cfm

Answers to Additional Practice Questions – Laboratory Analyses

1. Answer: **a.** as soon as possible.

   Reference: *Water Quality*, 4th edition. 2010. Joseph A. Ritter, Editor. American Water Works Association. Chapter 5.

2. Answer: **a.** 100 mL

   Reference: *Water Quality*, 4th edition. 2010. Joseph A. Ritter, Editor. American Water Works Association. Chapter 5.

3. Answer: **c.** 100 mL

   Reference: *Water Quality*, 4th edition. 2010. Joseph A. Ritter, Editor. American Water Works Association. Chapter 5.

4. Answer: **b.** Fluoride

   Reference: *Water Treatment Operator Handbook*, Revised edition. 2005. Nicholas G. Pizzi. American Water Works Association. Chapter 12.

5. Answer: **c.** 1 to 5°C.

   Reference: *Water Quality*, 4th edition. 2010. Joseph A. Ritter, Editor. American Water Works Association. Chapter 3.

6. Answer: **d.** 6 months

   Reference: *Water Quality*, 4th edition. 2010. Joseph A. Ritter, Editor. American Water Works Association. Chapter 6.

7. Answer: **a.** nitric acid.

   Reference: *Water Quality*, 4th edition. 2010. Joseph A. Ritter, Editor. American Water Works Association. Chapter 6.

8. Answer: **b.** red to purple.

   Reference: *Basic Microbiology for Drinking Water Personnel*, 2nd edition. 2006. Dennis R. Hill. American Water Works Association. Chapter 4.

9. Answer: **c.** NH_3

   Reference: *Basic Microbiology for Drinking Water Personnel*, 2nd edition. 2006. Dennis R. Hill. American Water Works Association. Chapter 5.

10. Answer: **b.** ions.

    Reference: *Iron and Manganese Removal Handbook*. 1999. Elmer O. Sommerfield. American Water Works Association. Chapter 2.

11. Answer: **d.** Daily

    Reference: *Water Quality*, 4th edition. 2010. Joseph A. Ritter, Editor. American Water Works Association. Chapter 5.

12. Answer: **c.** Methyl orange test

    Reference: *Water Treatment Operator Handbook*, Revised edition. 2005. Nicholas G. Pizzi. American Water Works Association. Chapter 9.

13. Answer: **b.** Langelier Index

    Reference: *Water Quality*, 4th edition. 2010. Joseph A. Ritter, Editor. American Water Works Association. Chapter 5.

ANSWERS TO ADDITIONAL WATER TREATMENT OPERATOR PRACTICE QUESTIONS 197

14. Answer: **a.** The sampler

    Reference: *Water Quality*, 4th edition. 2010. Joseph A. Ritter, Editor. American Water Works Association. Chapter 2.

15. Answer: **b.** Nitric acid

    Reference: *Water Quality*, 4th edition. 2010. Joseph A. Ritter, Editor. American Water Works Association. Chapter 5.

16. Answer: **c.** 7 days

    Reference: *Water Quality*, 4th edition. 2010. Joseph A. Ritter, Editor. American Water Works Association. Chapter 6.

17. Answer: **a.** Zinc

    Reference: *Water Treatment Operator Handbook*, Revised edition. 2005. Nicholas G. Pizzi. American Water Works Association. Chapter 12.

18. Answer: **a.** <2.0 pH units.

    Reference: *Water Quality*, 4th edition. 2010. Joseph A. Ritter, Editor. American Water Works Association. Chapter 6.

19. Answer: **c.** roundworms.

    Reference: *Basic Microbiology for Drinking Water Personnel*, 2nd edition. 2006. Dennis R. Hill. American Water Works Association. Chapter 3.

20. Answer: **c.** dull powdery appearance with fuzzy border.

    Reference: M7, *Problem Organisms in Water: Identification and Treatment,* 3rd edition. 2003. American Water Works Association. Chapter 1.

21. Answer: **a.** 5 TON

    Reference: *Water Quality*, 4th edition. 2010. Joseph A. Ritter, Editor. American Water Works Association. Chapter 5.

22. Answer: **b.** 4 hours.

    Reference: *Water Treatment,* 4th edition. 2010. Nicholas G. Pizzi, Editor. American Water Works Association. Chapter 4.

23. Answer: **d.** > 300 mg/L

    Reference: *Water Treatment,* 4th edition. 2010. Nicholas G. Pizzi, Editor. American Water Works Association. Chapter 11.

24. Answer: **c.** soap scum.

    Reference: *Water Treatment,* 4th edition. 2010. Nicholas G. Pizzi, Editor. American Water Works Association. Chapter 11.

25. Answer: **a.** TOC correlates with production of disinfection by-products

    Reference: *Water Treatment Operator Handbook*, Revised edition. 2005. Nicholas G. Pizzi. American Water Works Association. Chapter 1.

26. Answer: **a.** Hydrogen

    Reference: *Basic Chemistry for Water & Wastewater Operators*, Revised edition. 2005. Darshan Singh Sarai. American Water Works Association. Chapter 6.

27. Answer: **a.** a salt.

Reference: *Basic Chemistry for Water & Wastewater Operators*, Revised edition. 2005. Darshan Singh Sarai. American Water Works Association. Chapter 6.

28. Answer: **b.** acids and hydroxides.

Reference: *Basic Chemistry for Water & Wastewater Operators*, Revised edition. 2005. Darshan Singh Sarai. American Water Works Association. Chapter 6.

29. Answer: **a.** cause acidity.

Reference: *Basic Chemistry for Water & Wastewater Operators*, Revised edition. 2005. Darshan Singh Sarai. American Water Works Association. Chapter 6.

30. Answer: **b.** LI = pH − pHs

Reference: *Water Quality*, 4th edition. 2010. Joseph A. Ritter, Editor. American Water Works Association. Chapter 5.

31. Answer: **c.** protozoans.

Reference: *Water Treatment,* 4th edition. 2010. Nicholas G. Pizzi, Editor. American Water Works Association. Chapter 7.

32. Answer: **b.** 4°C.

Reference: *Basic Science Concepts and Applications,* 4th edition. 2010. Nicholas G. Pizzi, Editor. American Water Works Association. Hydraulics 1.

33. Answer: **a.** conventional treatment.

Reference: M37, *Operational Control of Coagulation and Filtration Processes,* 3rd edition. 2011. American Water Works Association. Chapter 1.

34. Answer: **b.** 24 hours.

Reference: *Water Quality*, 4th edition. 2010. Joseph A. Ritter, Editor. American Water Works Association. Chapter 4.

35. Answer: **b.** Burette

Reference: *Water Quality*, 4th edition. 2010. Joseph A. Ritter, Editor. American Water Works Association. Chapter 3.

36. Answer: **c.** *Escherichia coli.*

Reference: *Water Quality*, 4th edition. 2010. Joseph A. Ritter, Editor. American Water Works Association. Chapter 4.

37. Answer: **d.** <500 colonies per mL

Reference: *Water Quality*, 4th edition. 2010. Joseph A. Ritter, Editor. American Water Works Association. Chapter 4.

38. Answer: **c.** Changes in temperature will significantly affect results

Reference: *Water Quality*, 4th edition. 2010. Joseph A. Ritter, Editor. American Water Works Association. Chapter 5.

39. Answer: **c.** Titration with ethylenediaminetetraacetic acid

Reference: *Water Quality*, 4th edition. 2010. Joseph A. Ritter, Editor. American Water Works Association. Chapter 5.

40. Answer: **c.** SPADNS

    Reference: *Water Quality*, 4th edition. 2010. Joseph A. Ritter, Editor. American Water Works Association. Chapter 6.

41. Answer: **c.** blue.

    Reference: *Water Treatment Operator Handbook*, Revised edition. 2005. Nicholas G. Pizzi. American Water Works Association. Chapter 9.

42. Answer: **b.** Transfer pipette

    Reference: *Water Quality*, 4th edition. 2010. Joseph A. Ritter, Editor. American Water Works Association. Chapter 3.

43. Answer: **a.** pH meter

    Reference: *Water Quality*, 4th edition. 2010. Joseph A. Ritter, Editor. American Water Works Association. Chapter 5.

44. Answer: **a.** multiplying by 17.12.

    Reference: *Water Treatment Operator Handbook*, Revised edition. 2005. Nicholas G. Pizzi. American Water Works Association. Chapter 9.

45. Answer: **d.** 28 days.

    Reference: *Water Quality*, 4th edition. 2010. Joseph A. Ritter, Editor. American Water Works Association. Chapter 5.

46. Answer: **c.** Glass bottles, cleaned with detergent and rinsed with distilled water

    Reference: *Water Quality*, 4th edition. 2010. Joseph A. Ritter, Editor. American Water Works Association. Chapter 5.

47. Answer: **a.** <2 pH units.

    Reference: *Water Quality*, 4th edition. 2010. Joseph A. Ritter, Editor. American Water Works Association. Chapter 6.

48. Answer: **c.** the measure of a solution's ionic strength and an indirect measure of the total dissolved solids.

    Reference: *Water Quality*, 4th edition. 2010. Joseph A. Ritter, Editor. American Water Works Association. Chapter 5.

49. Answer: **b.** 3 TON

    Reference: *Water Quality*, 4th edition. 2010. Joseph A. Ritter, Editor. American Water Works Association. Chapter 5.

50. Answer: **c.** unacidified and pass through 0.45 μm membrane.

    Reference: *Water Quality*, 4th edition. 2010. Joseph A. Ritter, Editor. American Water Works Association. Chapter 6.

51. Answer: **a.** bacteria.

    Reference: *Water Treatment,* 4th edition. 2010. Nicholas G. Pizzi, Editor. American Water Works Association. Chapter 7.

52. Answer: **a.** producing data that is precise and unbiased.

    Reference: *Water Treatment Operator Handbook*, Revised edition. 2005. Nicholas G. Pizzi. American Water Works Association. Chapter 12.

53. Answer: **a.** isotopes.

    Reference: *Basic Chemistry for Water & Wastewater Operators*, Revised edition. 2005. Darshan Singh Sarai. American Water Works Association. Chapter 3.

54. Answer: **b.** valence

    Reference: *Basic Chemistry for Water & Wastewater Operators*, Revised edition. 2005. Darshan Singh Sarai. American Water Works Association. Chapter 4.

55. Answer: **a.** electrons are lost by the species being oxidized.

    Reference: *Basic Chemistry for Water & Wastewater Operators*, Revised edition. 2005. Darshan Singh Sarai. American Water Works Association. Chapter 4.

56. Answer: **b.** electrons are gained by the species being reduced.

    Reference: *Basic Chemistry for Water & Wastewater Operators*, Revised edition. 2005. Darshan Singh Sarai. American Water Works Association. Chapter 4.

57. Answer: **c.** an ionic reaction.

    Reference: *Basic Chemistry for Water & Wastewater Operators*, Revised edition. 2005. Darshan Singh Sarai. American Water Works Association. Chapter 5.

58. Answer: **a.** $Cl_2 + 2e^-$

    Reference: *Basic Chemistry for Water & Wastewater Operators*, Revised edition. 2005. Darshan Singh Sarai. American Water Works Association. Chapter 5.

59. Answer: **b.** Colloids

    Reference: *Basic Chemistry for Water & Wastewater Operators*, Revised edition. 2005. Darshan Singh Sarai. American Water Works Association. Chapter 7.

60. Answer: **d.** moles of solute per liter of solution.

    Reference: *Basic Chemistry for Water & Wastewater Operators*, Revised edition. 2005. Darshan Singh Sarai. American Water Works Association. Chapter 7.

61. Answer: **b.** Chlorine gas

    Reference: *Basic Chemistry for Water & Wastewater Operators*, Revised edition. 2005. Darshan Singh Sarai. American Water Works Association. Chapter 7.

62. Answer: **d.** hydrates.

    Reference: *Basic Chemistry for Water & Wastewater Operators*, Revised edition. 2005. Darshan Singh Sarai. American Water Works Association. Chapter 9.

63. Answer: **c.** Hydration of ions

    Reference: *Basic Chemistry for Water & Wastewater Operators*, Revised edition. 2005. Darshan Singh Sarai. American Water Works Association. Chapter 10.

64. Answer: **d.** an alcohol.

    Reference: *Basic Chemistry for Water & Wastewater Operators*, Revised edition. 2005. Darshan Singh Sarai. American Water Works Association. Chapter 12.

65. Answer: **a.** $pH - pH_s = 0$

    Reference: *Water Quality*, 4th edition. 2010. Joseph A. Ritter, Editor. American Water Works Association. Chapter 5.

ANSWERS TO ADDITIONAL WATER TREATMENT OPERATOR PRACTICE QUESTIONS

66. Answer: **c.** In mg/L as $CaCO_3$

    Reference: *Water Quality,* 4th edition. 2010. Joseph A. Ritter, Editor. American Water Works Association. Chapter 5.

67. Answer: **c.** Increase chlorination and backwash

    Reference: M7, *Problem Organisms in Water: Identification and Treatment,* 3rd edition. 2003. American Water Works Association. Chapter 11.

68. Answer: **c.** Dissolved organic matter

    Reference: *Water Quality,* 4th edition. 2010. Joseph A. Ritter, Editor. American Water Works Association. Chapter 5.

69. Answer: **b.** *Rhizoclonium*

    Reference: M7, *Problem Organisms in Water: Identification and Treatment,* 3rd edition. 2003. American Water Works Association. Chapter 10.

70. Answer: **a.** OH^-

    Reference: *Water Treatment Operator Handbook,* Revised edition. 2005. Nicholas G. Pizzi. American Water Works Association. Chapter 9.

71. Answer: **c.** HCO_3^-

    Reference: *Water Treatment Operator Handbook,* Revised edition. 2005. Nicholas G. Pizzi. American Water Works Association. Chapter 9.

72. Answer: **b.** 6 to 7 mg/L

    Reference: *Water Sources,* 4th edition. 2010. Paul Koch, Editor. American Water Works Association. Chapter 6.

73. Answer: **c.** a transfer of electrons occurs.

    Reference: *Basic Chemistry for Water & Wastewater Operators,* Revised edition. 2005. Darshan Singh Sarai. American Water Works Association. Chapter 4.

74. Answer: **d.** electronegativity.

    Reference: *Basic Chemistry for Water & Wastewater Operators,* Revised edition. 2005. Darshan Singh Sarai. American Water Works Association. Chapter 4.

75. Answer: **a.** *Microcystis*

    Reference: M7, *Problem Organisms in Water: Identification and Treatment,* 3rd edition. 2003. American Water Works Association. Chapter 10.

76. Answer: **a.** ionic bonding will occur.

    Reference: *Basic Chemistry for Water & Wastewater Operators,* Revised edition. 2005. Darshan Singh Sarai. American Water Works Association. Chapter 4.

77. Answer: **a.** *Chromatium*

    Reference: M7, *Problem Organisms in Water: Identification and Treatment,* 3rd edition. 2003. American Water Works Association. Chapter 3.

78. Answer: **c.** Stoichiometry

    Reference: *Basic Chemistry for Water & Wastewater Operators,* Revised edition. 2005. Darshan Singh Sarai. American Water Works Association. Chapter 5.

79. Answer: **b.** an ester.

    Reference: *Basic Chemistry for Water & Wastewater Operators*, Revised edition. 2005. Darshan Singh Sarai. American Water Works Association. Chapter 12.

80. Answer: **c.** 6.0 to 7.0 mg/L

    Reference: *Water Quality*, 4th edition. 2010. Joseph A. Ritter, Editor. American Water Works Association. Chapter 3.

81. Answer: **c.** 4,000 mg/L

    Reference: *Water Quality*, 4th edition. 2010. Joseph A. Ritter, Editor. American Water Works Association. Chapter 5.

82. Answer: **c.** *Chlorobium*

    Reference: M7, *Problem Organisms in Water: Identification and Treatment*, 3rd edition. 2003. American Water Works Association. Chapter 3.

83. Answer: **c.** Calcium sulfate

    Reference: *Water Quality*, 4th edition. 2010. Joseph A. Ritter, Editor. American Water Works Association. Chapter 3.

84. Answer: **b.** Dry-heat incubator

    Reference: *Water Quality*, 4th edition. 2010. Joseph A. Ritter, Editor. American Water Works Association. Chapter 3.

85. Answer: **c.** Water-bath

    Reference: *Water Quality*, 4th edition. 2010. Joseph A. Ritter, Editor. American Water Works Association. Chapter 3.

86. Answer: **d.** Atomic absorption spectrophotometer

    Reference: *Water Quality*, 4th edition. 2010. Joseph A. Ritter, Editor. American Water Works Association. Chapter 3.

87. Answer: **b.** Green lactose bile broth

    Reference: *Water Quality*, 4th edition. 2010. Joseph A. Ritter, Editor. American Water Works Association. Chapter 4.

88. Answer: **a.** brilliant green bile tubes

    Reference: *Water Quality*, 4th edition. 2010. Joseph A. Ritter, Editor. American Water Works Association. Chapter 4.

89. Answer: **d.** 8.3 or less.

    Reference: *Water Quality*, 4th edition. 2010. Joseph A. Ritter, Editor. American Water Works Association. Chapter 5.

90. Answer: **d.** pH_s = Temperature + TDS − log(Ca^{+2}) − log(alkalinity)

    Reference: *Water Quality*, 4th edition. 2010. Joseph A. Ritter, Editor. American Water Works Association. Chapter 5.

Comply with Drinking Water Regulations

Answers to Additional Practice Questions

1. Answer: **b.** Containing not more than 0.02% lead

   Reference: National Primary Drinking Water Regulations, Subpart F – Maximum Contaminant Level Goals and Maximum Residual Disinfectant Goals, §141.52 – Maximum contaminant level goals for microbiological contaminants

2. Answer: **a.** Zero

   Reference: National Primary Drinking Water Regulations, Subpart F – Maximum Contaminant Level Goals and Maximum Residual Disinfectant Goals, §141.52 – Maximum contaminant level goals for microbiological contaminants

3. Answer: **b.** 4 mg/L

   Reference: National Primary Drinking Water Regulations, Subpart F – Maximum Contaminant Level Goals and Maximum Residual Disinfectant Goals, §141.54 – Maximum residual disinfectant level goals for disinfectants

4. Answer: **d.** No more than 5%

   Reference: National Primary Drinking Water Regulations, Subpart G – Maximum Contaminant Levels and Maximum Residual Disinfectant Levels, §141.63 – Maximum contaminant levels (MCLs) for microbiological contaminants (a)(1)

5. Answer: **b.** 1

   Reference: National Primary Drinking Water Regulations, Subpart G – Maximum Contaminant Levels and Maximum Residual Disinfectant Levels, §141.63 – Maximum contaminant levels (MCLs) for microbiological contaminants (a)(2)

6. Answer: **a.** 0.010 mg/L

   Reference: National Primary Drinking Water Regulations, Subpart G – Maximum Contaminant Levels and Maximum Residual Disinfectant Levels, §141.64 – Maximum contaminant levels for disinfection byproducts (a)

7. Answer: **c.** Lime softening

   Reference: National Primary Drinking Water Regulations, Subpart G – Maximum Contaminant Levels and Maximum Residual Disinfectant Levels, §141.66 – Maximum contaminant levels for radionuclides (g) Table B

8. Answer: **c.** At least 3-log removal

   Reference: National Primary Drinking Water Regulations, Subpart H – Filtration and Disinfection, §141.70 – General requirements (a)(1)

9. Answer: **d.** 4 log

   Reference: National Primary Drinking Water Regulations, Subpart H – Filtration and Disinfection, §141.70 – General requirements (a)(2)

10. Answer: **d.** 5.0 ntu

    Reference: National Primary Drinking Water Regulations, Subpart H – Filtration and Disinfection, §141.71 – Criteria for avoiding filtration (a)(1)

11. Answer: **d.** 5%

    Reference: National Primary Drinking Water Regulations, Subpart H – Filtration and Disinfection, §141.72 – Disinfection (a)(4)(i)

12. Answer: **b.** 0.3 ntu; 95%

    Reference: National Primary Drinking Water Regulations, Subpart H – Filtration and Disinfection, §141.73 – Filtration (a)(1)

13. Answer: **d.** ≤ 1.0 ntu in 95% of the measurements collected each month

    Reference: National Primary Drinking Water Regulations, Subpart H – Filtration and Disinfection, §141.73 – Filtration (b)(1)

14. Answer: **d.** ≤ 1.0 ntu in 95% of the measurements collected each month

    Reference: National Primary Drinking Water Regulations, Subpart H – Filtration and Disinfection, §141.73 – Filtration (c)(1)

15. Answer: **c.** State.

    Reference: National Primary Drinking Water Regulations, Subpart H – Filtration and Disinfection, §141.74 – Analytical and monitoring requirements (a)

16. Answer: **d.** USEPA or the State

    Reference: National Primary Drinking Water Regulations, Subpart H – Filtration and Disinfection, §141.74 – Analytical and monitoring requirements (a)

17. Answer: **a.** 8 hours

    Reference: National Primary Drinking Water Regulations, Subpart H – Filtration and Disinfection, §141.74 – Analytical and monitoring requirements (a)(1) Table explanation 2

18. Answer: **c.** 10°C.

    Reference: National Primary Drinking Water Regulations, Subpart H – Filtration and Disinfection, §141.74 – Analytical and monitoring requirements (a)(1) Table explanation 2

19. Answer: **a.** a running annual average at any sampling point.

    Reference: National Primary Drinking Water Regulations, Subpart C – Monitoring and Analytical Requirements, §141.23 – Inorganic Chemical Sampling and Analytical Requirements (i)(1)

20. Answer: **d.** Preserve by keeping it at 4°C

    Reference: National Primary Drinking Water Regulations, Subpart C – Monitoring and Analytical Requirements, §141.23 – Inorganic Chemical Sampling and Analytical Requirements (k)(2)

21. Answer: **c.** Preserve with H_2SO_4

    Reference: National Primary Drinking Water Regulations, Subpart C – Monitoring and Analytical Requirements, §141.23 – Inorganic Chemical Sampling and Analytical Requirements (k)(2)

22. Answer: **d.** 14 days

    Reference: National Primary Drinking Water Regulations, Subpart C – Monitoring and Analytical Requirements, §141.23 – Inorganic Chemical Sampling and Analytical Requirements (k)(2)

23. Answer: **d.** During the quarter which previously yielded the highest analytical result

    Reference: National Primary Drinking Water Regulations, Subpart C – Monitoring and Analytical Requirements, §141.24 – Organic chemicals, sampling and analytical requirements (f)(11)(iii)

24. Answer: **a.** 3

    Reference: National Primary Drinking Water Regulations, Subpart C – Monitoring and Analytical Requirements, §141.24 – Organic chemicals, sampling and analytical requirements (f)(11)(iv)

25. Answer: **c.** Vinyl chloride

    Reference: National Primary Drinking Water Regulations, Subpart C – Monitoring and Analytical Requirements, §141.24 – Organic chemicals, sampling and analytical requirements (f)(11)(v)

26. Answer: **b.** Laboratory, 14 days

    Reference: National Primary Drinking Water Regulations, Subpart C – Monitoring and Analytical Requirements, §141.24 – Organic chemicals, sampling and analytical requirements (f)(14)

27. Answer: **b.** 0.0005 mg/L

    Reference: National Primary Drinking Water Regulations, Subpart C – Monitoring and Analytical Requirements, §141.24 – Organic chemicals, sampling and analytical requirements (f)(14)(i)

28. Answer: **b.** Alkalinity

    Reference: National Primary Drinking Water Regulations, Subpart C – Monitoring and Analytical Requirements, §141.28 – Certified laboratories (a)

29. Answer: **b.** 48 hours.

    Reference: National Primary Drinking Water Regulations, Subpart D – Reporting and Recordkeeping, §141.31 – Reporting requirements (b)

30. Answer: **c.** When a state laboratory performs the analyses

    Reference: National Primary Drinking Water Regulations, Subpart D – Reporting and Recordkeeping, §141.31 – Reporting requirements (c)

31. Answer: **b.** 10 days

    Reference: National Primary Drinking Water Regulations, Subpart D – Reporting and Recordkeeping, §141.31 – Reporting requirements (d)

32. Answer: **b.** When the water supplier serves more than 10,000 people

    Reference: National Primary Drinking Water Regulations, Subpart E, – Monitoring requirements for unregulated contaminants, §141.40 – (a)(2)(i)(A)

33. Answer: **a.** 12-month period.

    Reference: National Primary Drinking Water Regulations, Subpart E – Special Regulations, Including Monitoring Regulations and Prohibition on Lead Use, §141.40 – Monitoring requirements for unregulated contaminants (a)(4)(A)

34. Answer: **a.** 7 days.

    Reference: National Primary Drinking Water Regulations, Subpart E – Special Regulations, Including Monitoring Regulations and Prohibition on Lead Use, §141.40 – Monitoring requirements for unregulated contaminants (a)(4)(I)(ii)(E)

35. Answer: **c.** Annually

    Reference: National Primary Drinking Water Regulations, Subpart E – Special Regulations, Including Monitoring Regulations and Prohibition on Lead Use, §141.41 – Special monitoring of sodium (a)

36. Answer: **d.** At least every three years

    Reference: National Primary Drinking Water Regulations, Subpart E – Special Regulations, Including Monitoring Regulations and Prohibition on Lead Use, §141.41 – Special monitoring of sodium (a)

37. Answer: **a.** repair of cast iron pipes using leaded joints.

    Reference: National Primary Drinking Water Regulations, Subpart E – Special Regulations, Including Monitoring Regulations and Prohibition on Lead Use, §141.43 – Prohibition on use of lead pipes, solder and flux (a)(ii)

38. Answer: **d.** 1,000 mL

    Reference: National Primary Drinking Water Regulations, Subpart I – Control of Lead and Copper, §141.86 – Monitoring requirements for lead and copper in tap water (b)(2)

39. Answer: **c.** 6 hours.

    Reference: National Primary Drinking Water Regulations, Subpart I – Control of Lead and Copper, §141.86 – Monitoring requirements for lead and copper in tap water (b)(2)

40. Answer: **d.** 14 days.

    Reference: National Primary Drinking Water Regulations, Subpart I – Control of Lead and Copper, §141.86 – Monitoring requirements for lead and copper in tap water (b)(2)

41. Answer: **c.** At the tap after the water has remained motionless for at least 4 hours

    Reference: National Primary Drinking Water Regulations, Subpart I – Control of Lead and Copper, §141.86 – Monitoring requirements for lead and copper in tap water (b)(3)(i, ii, iii)

42. Answer: **a.** When the system meets the lead and copper action level for 2 consecutive six-month monitoring periods

    Reference: National Primary Drinking Water Regulations, Subpart I – Control of Lead and Copper, §141.86 – Monitoring requirements for lead and copper in tap water (d)(i)(B)

43. Answer: **c.** When the 90th percentile for lead is ≤0.005 mg/L and for copper is ≤0.65 mg/L

    Reference: National Primary Drinking Water Regulations, Subpart I – Control of Lead and Copper, §141.86 – Monitoring requirements for lead and copper in tap water (d)(4)(iv)

44. Answer: **c.** At least once every 9 years

    Reference: National Primary Drinking Water Regulations, Subpart I – Control of Lead and Copper, §141.86 – Monitoring requirements for lead and copper in tap water (g)

45. Answer: **b.** 2 samples for each water quality parameter, per entry point to the distribution system during each monitoring period

    Reference: National Primary Drinking Water Regulations, Subpart I – Control of Lead and Copper, §141.87 – Monitoring requirements for water quality parameters (a)(2)(ii)

46. Answer: **c.** pH

    Reference: National Primary Drinking Water Regulations, Subpart I – Control of Lead and Copper, §141.87 – Monitoring requirements for water quality parameters (c)(2)(i)

47. Answer: **d.** Dosage rate of the chemical corrosion inhibitor

    Reference: National Primary Drinking Water Regulations, Subpart I – Control of Lead and Copper, §141.87 – Monitoring requirements for water quality parameters (c)(2)(iii)

48. Answer: **a.** reduce the number of sites sampled.

    Reference: National Primary Drinking Water Regulations, Subpart I – Control of Lead and Copper, §141.87 – Monitoring requirements for water quality parameters (e)(1)

49. Answer: **b.** reduce the frequency of sampling from 6-month periods to annually.

    Reference: National Primary Drinking Water Regulations, Subpart I – Control of Lead and Copper, §141.87 – Monitoring requirements for water quality parameters (e)(2)(i)

50. Answer: **c.** one sample at every entry point to the distribution system that is representative of each well after treatment.

    Reference: National Primary Drinking Water Regulations, Subpart I – Control of Lead and Copper, §141.88 – Monitoring requirements for lead and copper in source water (a)(i)

51. Answer: **c.** one sample after any application of treatment or in the distribution system at a point which is representative of each source after treatment.

    Reference: National Primary Drinking Water Regulations, Subpart I – Control of Lead and Copper, §141.88 – Monitoring requirements for lead and copper in source water (a)(ii)

52. Answer: **b.** 10 days

    Reference: National Primary Drinking Water Regulations, Subpart I – Control of Lead and Copper, §141.90 – Reporting requirements (a)(1)

53. Answer: **b.** 12 years.

    Reference: National Primary Drinking Water Regulations, Subpart I – Control of Lead and Copper, §141.91 – Record keeping requirements

54. Answer: **c.** Gross alpha particle activity scan

    Reference: *Water Quality*, 4th edition. 2010. Joseph A. Ritter, Editor. American Water Works Association. Chapter 7.

55. Answer: **a.** determine alternative total organic carbon removal requirements for a particular plant.

    Reference: M37, *Operational Control of Coagulation and Filtration Processes*, 3rd edition. 2011. American Water Works Association. Chapter 1.

56. Answer: **b.** quarterly basis.

    Reference: US Environmental Protection Agency. EPA's recommendations for enhanced monitoring for Hexavalent Chromium (Chromium-6) in Drinking Water. 2011. http://water.epa.gov/drink/info/chromium/guidance.cfm

57. Answer: **c.** semi-annual basis.

    Reference: US Environmental Protection Agency. EPA's recommendations for enhanced monitoring for Hexavalent Chromium (Chromium-6) in Drinking Water. 2011. http://water.epa.gov/drink/info/chromium/guidance.cfm

58. Answer: **c.** 0.1 mg/L

    Reference: US Environmental Protection Agency. EPA's recommendations for enhanced monitoring for Hexavalent Chromium (Chromium-6) in Drinking Water. 2011. http://water.epa.gov/drink/info/chromium/guidance.cfm

59. Answer: **c.** ≤0.040 mg/L and ≤0.030 mg/L for TTHM and HAA5, respectively.

    Reference: National Primary Drinking Water Regulations Subpart L – Disinfectant Residuals, Disinfectant By-products and Disinfectant By-product Precursors, §141.135 – Treatment technique for control of disinfection byproduct (DBP) precursors (a)(2)(iv)

60. Answer: **a.** <60 mg/L

    Reference: National Primary Drinking Water Regulations Subpart L – Disinfectant Residuals, Disinfectant By-products and Disinfectant By-product Precursors, §141.135 – Treatment technique for control of disinfection byproduct (DBP) precursors (a)(3)(i and ii)

61. Answer: **d.** 5.0 mg/L

    Reference: National Primary Drinking Water Regulations, Subpart O – Consumer Confidence Reports §141.154 – Required additional health information (c)

62. Answer: **b.** 12 months

    Reference: National Primary Drinking Water Regulations Subpart P—Enhanced Filtration and Disinfection—Systems Serving 10,000 or More People, §141.172 – Disinfection profiling and benchmarking (b)(2)

63. Answer: **d.** each day during peak hourly flows.

    Reference: National Primary Drinking Water Regulations Subpart P—Enhanced Filtration and Disinfection—Systems Serving 10,000 or More People, §141.172 – Disinfection profiling and benchmarking (b)(2)(i through iv)

64. Answer: **d.** Chloramines

    Reference: National Primary Drinking Water Regulations Subpart P—Enhanced Filtration and Disinfection—Systems Serving 10,000 or More People, §141.172 – Disinfection profiling and benchmarking (b)(5)

65. Answer: **b.** develop a disinfection benchmark.

    Reference: National Primary Drinking Water Regulations Subpart P—Enhanced Filtration and Disinfection—Systems Serving 10,000 or More People, §141.172 – Disinfection profiling and benchmarking (c)(1)

66. Answer: **d.** 5.5-log

    Reference: National Primary Drinking Water Regulations, Subpart W – Enhanced Treatment for *Cryptosporidium* Requirements for Sanitary Surveys Performed by EPA, §141.701 – Source water sampling (d)(1)

67. Answer: **c.** 6 years

    Reference: National Primary Drinking Water Regulations, Subpart W – Enhanced Treatment for *Cryptosporidium* Requirements for Sanitary Surveys Performed by EPA, §141.701 – Source water sampling (f)(3)

68. Answer: **a.** 0.01 oocysts/L

    Reference: National Primary Drinking Water Regulations, Subpart W – Enhanced Treatment for *Cryptosporidium* Requirements for Sanitary Surveys Performed by EPA, §141.712 – Unfiltered system *Cryptosporidium* treatment requirements (b)(1)

69. Answer: **b.** 48 hours.

    Reference: National Primary Drinking Water Regulations, Subpart D – Reporting and Recordkeeping, §141.31 Reporting requirements (b)

70. Answer: **b.** Quarterly for a minimum of 2 quarters

    Reference: Title 40: Protection of Environment Part 141 – National Primary Drinking Water regulations Supart C – Monitoring and Analytical Requirements §141.24 (f)(11) and (11)(i)

71. Answer: **a.** 3

    Reference: National Primary Drinking Water Regulations, Subpart C – Monitoring and Analytical Requirements, §141.24 Organic chemicals, sampling and analytical requirements (f)(11)(iv)

72. Answer: **d.** Within 60 days

    Reference: Title 40: Protection of Environment Part 141 – National Primary Drinking Water regulations Supart I – Control of Lead and Copper §141.90 (a)(4)(iii)

73. Answer: **d.** 5.0 ntu.

    Reference: National Primary Drinking Water Regulations, Subpart H – Filtration and Disinfection, §141.75 – Reporting and record keeping requirements (b)(1)(iii)

74. Answer: **a.** The hydraulic conveyance used to transport the recycled flows

    Reference: National Primary Drinking Water Regulations, Subpart H – Filtration and Disinfection, §141.76 – Recycle provisions (b)(1)

Answers to Additional Practice Questions – Operate and Maintain Equipment

1. Answer: **a.** metering pump.

   Reference: *Water Treatment*, 4th edition. 2010. Nicholas G. Pizzi, Editor. American Water Works Association. Chapter 4.

2. Answer: **c.** 1.00 to 3.00 inches.

   Reference: *Water Treatment Operator Handbook,* Revised edition. 2005. Nicholas G. Pizzi. American Water Works Association. Chapter 4.

3. Answer: **c.** Float-type sensor

   Reference: *Water Treatment Operator Handbook,* Revised edition. 2005. Nicholas G. Pizzi. American Water Works Association. Chapter 13.

4. Answer: **a.** electrons.

   Reference: *Basic Science Concepts and Applications*, 4th edition. 2010. Nicholas G. Pizzi, Editor. American Water Works Association. Electricity 1.

5. Answer: **c.** slinger ring.

   Reference: *Pumps & Pumping*, 9th edition. 2010. E.E. "Skeet" Arasmith, Mitch Scheele, & Kimon Zentz. ACR Publications.

6. Answer: **a.** distribute lubrication to the packing.

   Reference: *Pumps & Pumping*, 9th edition. 2010. E.E. "Skeet" Arasmith, Mitch Scheele, & Kimon Zentz. ACR Publications.

7. Answer: **b.** Every year

   Reference: *Water Treatment*, 4th edition. 2010. Nicholas G. Pizzi, Editor. American Water Works Association. Chapter 7.

8. Answer: **c.** Plugging of the well screen

   Reference: *Water Sources*, 4th edition. 2010. Paul Koch, Editor. American Water Works Association. Chapter 2.

9. Answer: **d.** To reduce friction and heat

   Reference: *Pumps & Pumping*, 9th edition. 2010. E.E. "Skeet" Arasmith, Mitch Scheele, & Kimon Zentz. ACR Publications.

10. Answer: **b.** Diaphragm

    Reference: *Pumping: Fundamentals for the Water & Wastewater Maintenance Operator Series.* 2001. Frank R. Spellman and Joanne Drinan. Technomic Publishing Company. Chapter 2.

11. Answer: **b.** 60 to 80°.

    Reference: *Water Treatment,* 4th edition. 2010. Nicholas G. Pizzi, Editor. American Water Works Association. Chapter 3.

12. Answer: **d.** They require periodic flexible tubing replacement

    Reference: *Water Treatment,* 4th edition. 2010. Nicholas G. Pizzi, Editor. American Water Works Association. Chapter 4.

ANSWERS TO ADDITIONAL WATER TREATMENT OPERATOR PRACTICE QUESTIONS 211

13. Answer: **b.** To shut off chlorine gas flow in case of a leak

    Reference: *Water Treatment,* 4th edition. 2010. Nicholas G. Pizzi, Editor. American Water Works Association. Chapter 7.

14. Answer: **a.** Injector

    Reference: *Water Treatment,* 4th edition. 2010. Nicholas G. Pizzi, Editor. American Water Works Association. Chapter 7.

15. Answer: **c.** 400 lb/d

    Reference: *Water Treatment,* 4th edition. 2010. Nicholas G. Pizzi, Editor. American Water Works Association. Chapter 7.

16. Answer: **d.** open only one turn to permit maximum withdrawal rate.

    Reference: *Water Treatment,* 4th edition. 2010. Nicholas G. Pizzi, Editor. American Water Works Association. Chapter 7.

17. Answer: **b.** Diaphragm pump

    Reference: *Water Treatment,* 4th edition. 2010. Nicholas G. Pizzi, Editor. American Water Works Association. Chapter 8.

18. Answer: **b.** a flow meter for measuring total flow such that pacing can be provided.

    Reference: *Water Treatment,* 4th edition. 2010. Nicholas G. Pizzi, Editor. American Water Works Association. Chapter 8.

19. Answer: **b.** 2 ft/sec.

    Reference: *Water Treatment Operator Handbook,* Revised edition. 2005. Nicholas G. Pizzi. American Water Works Association. Chapter 4.

20. Answer: **a.** Rotameter

    Reference: *Water Treatment Operator Handbook,* Revised edition. 2005. Nicholas G. Pizzi. American Water Works Association. Chapter 13.

21. Answer: **a.** an atmospheric vacuum breaker.

    Reference: *Water Quality,* 4th edition. 2010. Joseph A. Ritter, Editor. American Water Works Association. Chapter 3.

22. Answer: **d.** Free chlorine may corrode the mesh

    Reference: *Water Treatment,* 4th edition. 2010. Nicholas G. Pizzi, Editor. American Water Works Association. Chapter 3.

23. Answer: **d.** Induction mixer

    Reference: M20, *Water Chlorination/Chlorination Practices and Principles*, 2nd edition. 2006. American Water Works Association. Chapter 4.

24. Answer: **c.** equal to or greater than

    Reference: *Basic Science Concepts and Applications,* 4th edition. 2010. Nicholas G. Pizzi, Editor. American Water Works Association. Electricity 3.

25. Answer: **a.** Wear rings

    Reference: *Pumping: Fundamentals for the Water & Wastewater Maintenance Operator Series.* 2001. Frank R. Spellman and Joanne Drinan. Technomic Publishing Company. Chapter 4.

26. Answer: **c.** Soft-faced hammer

    Reference: *Pumps & Pumping*, 9th edition. 2010. E.E. "Skeet" Arasmith, Mitch Scheele, & Kimon Zentz. ACR Publications.

27. Answer: **a.** Stored vertically, on end

    Reference: *Pumps & Pumping*, 9th edition. 2010. E.E. "Skeet" Arasmith, Mitch Scheele, & Kimon Zentz. ACR Publications.

28. Answer: **a.** Solvent

    Reference: *Pumps & Pumping*, 9th edition. 2010. E.E. "Skeet" Arasmith, Mitch Scheele, & Kimon Zentz. ACR Publications.

29. Answer: **b.** At least every three months

    Reference: *Pumps & Pumping*, 9th edition. 2010. E.E. "Skeet" Arasmith, Mitch Scheele, & Kimon Zentz. ACR Publications.

30. Answer: **d.** When leakage becomes excessive

    Reference: *Pumping: Fundamentals for the Water & Wastewater Maintenance Operator Series*. 2001. Frank R. Spellman and Joanne Drinan. Technomic Publishing Company. Chapter 4.

31. Answer: **b.** Oil

    Reference: *Pumping: Fundamentals for the Water & Wastewater Maintenance Operator Series*. 2001. Frank R. Spellman and Joanne Drinan. Technomic Publishing Company. Chapter 8.

32. Answer: **a.** Corrosion

    Reference: *Water Treatment,* 4th edition. 2010. Nicholas G. Pizzi, Editor. American Water Works Association. Chapter 3.

33. Answer: **c.** $V^2/2g$

    Reference: *The Water Dictionary: A Comprehensive Reference of Water Terminology*, 2nd edition. 2010. Nancy McTigue, Editor, and James M. Symons, Editor Emeritus. American Water Works Association.

34. Answer: **b.** Split-case pump

    Reference: *Pumps & Pumping*, 9th edition. 2010. E.E. "Skeet" Arasmith, Mitch Scheele, & Kimon Zentz. ACR Publications.

35. Answer: **a.** End suction pump

    Reference: *Pumps & Pumping*, 9th edition. 2010. E.E. "Skeet" Arasmith, Mitch Scheele, & Kimon Zentz. ACR Publications.

36. Answer: **c.** Three stage

    Reference: *Pumps & Pumping*, 9th edition. 2010. E.E. "Skeet" Arasmith, Mitch Scheele, & Kimon Zentz. ACR Publications.

37. Answer: **c.** radial load.

    Reference: *Pumps & Pumping*, 9th edition. 2010. E.E. "Skeet" Arasmith, Mitch Scheele, & Kimon Zentz. ACR Publications.

ANSWERS TO ADDITIONAL WATER TREATMENT OPERATOR PRACTICE QUESTIONS

38. Answer: **c.** Flange coupling

    Reference: *Pumps & Pumping*, 9th edition. 2010. E.E. "Skeet" Arasmith, Mitch Scheele, & Kimon Zentz. ACR Publications.

39. Answer: **b.** pumping too much water.

    Reference: *Pumps & Pumping*, 9th edition. 2010. E.E. "Skeet" Arasmith, Mitch Scheele, & Kimon Zentz. ACR Publications.

40. Answer: **b.** impeller rings.

    Reference: *Pumping: Fundamentals for the Water & Wastewater Maintenance Operator Series*. 2001. Frank R. Spellman and Joanne Drinan. Technomic Publishing Company. Chapter 4.

41. Answer: **b.** Split coupling

    Reference: *Pumping: Fundamentals for the Water & Wastewater Maintenance Operator Series*. 2001. Frank R. Spellman and Joanne Drinan. Technomic Publishing Company. Chapter 4.

42. Answer: **c.** thrust bearings.

    Reference: *Pumping: Fundamentals for the Water & Wastewater Maintenance Operator Series*. 2001. Frank R. Spellman and Joanne Drinan. Technomic Publishing Company. Chapter 4.

43. Answer: **c.** Self-aligning double-row ball bearings

    Reference: *Pumping: Fundamentals for the Water & Wastewater Maintenance Operator Series*. 2001. Frank R. Spellman and Joanne Drinan. Technomic Publishing Company. Chapter 4.

44. Answer: **d.** Re-evaluate the pump's requirements and correct the condition

    Reference: *Pumping: Fundamentals for the Water & Wastewater Maintenance Operator Series*. 2001. Frank R. Spellman and Joanne Drinan. Technomic Publishing Company. Chapter 9.

45. Answer: **c.** Submersible pump

    Reference: *Pumping: Fundamentals for the Water & Wastewater Maintenance Operator Series*. 2001. Frank R. Spellman and Joanne Drinan. Technomic Publishing Company. Chapter 10.

46. Answer: **d.** Positive displacement pump

    Reference: *Pumping: Fundamentals for the Water & Wastewater Maintenance Operator Series*. 2001. Frank R. Spellman and Joanne Drinan. Technomic Publishing Company. Chapter 11.

47. Answer: **a.** 4 feet

    Reference: *Pumping: Fundamentals for the Water & Wastewater Maintenance Operator Series*. 2001. Frank R. Spellman and Joanne Drinan. Technomic Publishing Company. Chapter 11.

48. Answer: **c.** Electromotive principle

    Reference: *Water Quality*, 4th edition. 2010. Joseph A. Ritter, Editor. American Water Works Association. Chapter 5.

49. Answer: **b.** Electrical resistance

    Reference: *Water Quality,* 4th edition. 2010. Joseph A. Ritter, Editor. American Water Works Association. Chapter 5.

50. Answer: **c.** 5 rpm

    Reference: *Water Treatment,* 4th edition. 2010. Nicholas G. Pizzi, Editor. American Water Works Association. Chapter 3.

51. Answer: **b.** a failed injector.

    Reference: *Water Treatment,* 4th edition. 2010. Nicholas G. Pizzi, Editor. American Water Works Association. Chapter 7.

52. Answer: **c.** an empty chlorine container.

    Reference: *Water Treatment,* 4th edition. 2010. Nicholas G. Pizzi, Editor. American Water Works Association. Chapter 7.

53. Answer: **a.** Aluminum

    Reference: *Water Treatment Operator Handbook,* Revised edition. 2005. Nicholas G. Pizzi. American Water Works Association. Chapter 10.

54. Answer: **d.** Loss of head meter

    Reference: *Water Treatment Operator Handbook,* Revised edition. 2005. Nicholas G. Pizzi. American Water Works Association. Chapter 13.

55. Answer: **d.** Pump test to confirm capacity

    Reference: *Water Sources,* 4th edition. 2010. Paul Koch, Editor. American Water Works Association. Chapter 2.

56. Answer: **b.** fluid from a container by suction.

    Reference: *The Water Dictionary: A Comprehensive Reference of Water Terminology,* 2nd edition. 2010. Nancy McTigue, Editor, and James M. Symons, Editor Emeritus. American Water Works Association.

57. Answer: **a.** Automatic residual control gas feeder

    Reference: M20, *Water Chlorination/Chlorination Practices and Principles*, 2nd edition. 2006. American Water Works Association. Chapter 4.

58. Answer: **b.** Stainless steel

    Reference: M20, *Water Chlorination/Chlorination Practices and Principles*, 2nd edition. 2006. American Water Works Association. Chapter 5.

59. Answer: **c.** Number of amperes to be carried

    Reference: *Basic Science Concepts and Applications*, 4th edition. 2010. Nicholas G. Pizzi, Editor. American Water Works Association. Electricity 3.

60. Answer: **b.** iron salts.

    Reference: M37, *Operational Control of Coagulation and Filtration Processes*, 3rd edition. 2011. American Water Works Association. Chapter 2.

61. Answer: **d.** oil drip system.

    Reference: *Pumps & Pumping*, 9th edition. 2010. E.E. "Skeet" Arasmith, Mitch Scheele, & Kimon Zentz. ACR Publications.

ANSWERS TO ADDITIONAL WATER TREATMENT OPERATOR PRACTICE QUESTIONS 215

62. Answer: **c.** calcium.

Reference: *Pumps & Pumping*, 9th edition. 2010. E.E. "Skeet" Arasmith, Mitch Scheele, & Kimon Zentz. ACR Publications.

63. Answer: **b.** 180°F

Reference: *Pumping: Fundamentals for the Water & Wastewater Maintenance Operator Series*. 2001. Frank R. Spellman and Joanne Drinan. Technomic Publishing Company. Chapter 6.

64. Answer: **c.** In the suction and discharge hoses

Reference: *Water Treatment,* 4th edition. 2010. Nicholas G. Pizzi, Editor. American Water Works Association. Chapter 7.

65. Answer: **c.** 1/8 of an inch

Reference: *Pumping: Fundamentals for the Water & Wastewater Maintenance Operator Series*. 2001. Frank R. Spellman and Joanne Drinan. Technomic Publishing Company. Chapter 4.

66. Answer: **d.** low-head, high capacity centrifugal type pump.

Reference: *Water Treatment*, 4th edition. 2010. Nicholas G. Pizzi, Editor. American Water Works Association. Chapter 7.

67. Answer: **a.** Floating proportional control

Reference: *Water Treatment Operator Handbook,* Revised edition. 2005. Nicholas G. Pizzi. American Water Works Association. Chapter 13.

68. Answer: **c.** Float-type sensor

Reference: *Water Treatment Operator Handbook,* Revised edition. 2005. Nicholas G. Pizzi. American Water Works Association. Chapter 13.

69. Answer: **c.** At least 72 hours

Reference: *Water Sources,* 4th edition. 2010. Paul Koch, Editor. American Water Works Association. Chapter 2.

70. Answer: **c.** American Society of Mechanical Engineers

Reference: M20, *Water Chlorination/Chlorination Practices and Principles*, 2nd edition. 2006. American Water Works Association. Chapter 4.

71. Answer: **a.** 5 to 10°C

Reference: M20, *Water Chlorination/Chlorination Practices and Principles*, 2nd edition. 2006. American Water Works Association. Chapter 4.

72. Answer: **b.** 20%

Reference: M20, *Water Chlorination/Chlorination Practices and Principles*, 2nd edition. 2006. American Water Works Association. Chapter 4.

73. Answer: **c.** less expensive to maintain than gravimetric feeders.

Reference: *Water Treatment*, 4th edition. 2010. Nicholas G. Pizzi, Editor. American Water Works Association. Chapter 8.

74. Answer: **a.** Feedback control

Reference: *Water Treatment Operator Handbook,* Revised edition. 2005. Nicholas G. Pizzi. American Water Works Association. Chapter 13.

75. Answer: **d.** Proportional plus reset control

    Reference: *Water Treatment Operator Handbook,* Revised edition. 2005. Nicholas G. Pizzi. American Water Works Association. Chapter 13.

Perform Security, Safety, and Administrative Procedures

Answers to Additional Practice Questions

1. Answer: **b.** Methane

    Reference: *Water Treatment,* 4th edition. 2010. Nicholas G. Pizzi, Editor. American Water Works Association. Chapter 14.

2. Answer: **d.** Chlorine inhalation may lead to delayed reactions such as pulmonary edema

    Reference: M20, *Water Chlorination/Chlorination Practices and Principles*, 2nd edition. 2006. American Water Works Association. Chapter 5.

3. Answer: **a.** for at least 15 minutes.

    Reference: M20, *Water Chlorination/Chlorination Practices and Principles*, 2nd edition. 2006. American Water Works Association. Chapter 5.

4. Answer: **c.** black stain.

    Reference: *Water Treatment Operator Handbook,* Revised edition. 2005. Nicholas G. Pizzi. American Water Works Association. Chapter 10.

5. Answer: **b.** 3 a.m.

    References: *Water Sources,* 4th edition. 2010. Paul Koch, Editor. American Water Works Association. Chapter 5.

6. Answer: **d.** 7 p.m.

    Reference: *Water Sources,* 4th edition. 2010. Paul Koch, Editor. American Water Works Association. Chapter 5.

7. Answer: **b.** Taste and odors

    Reference: *Water Quality,* 4th edition. 2010. Joseph A. Ritter, Editor. American Water Works Association. Chapter 7.

8. Answer: **b.** Chlorine cylinders stored outside should be stored on elevated racks to help prevent corrosion

    Reference: *Water Treatment,* 4th edition. 2010. Nicholas G. Pizzi, Editor. American Water Works Association. Chapter 7.

9. Answer: **c.** inspecting the dome and no product is flowing.

    Reference: M20, *Water Chlorination/Chlorination Practices and Principles*, 2nd edition. 2006. American Water Works Association. Chapter 5.

10. Answer: **b.** Entry supervisor

    Reference: *Water Treatment Operator Handbook,* Revised edition. 2005. Nicholas G. Pizzi. American Water Works Association. Chapter 13.

11. Answer: **b.** Giardiasis

    Reference: *Water Quality,* 4th edition. 2010. Joseph A. Ritter, Editor. American Water Works Association. Chapter 4.

ANSWERS TO ADDITIONAL WATER TREATMENT OPERATOR PRACTICE QUESTIONS

12. Answer: **c.** virus.

    Reference: *Water Treatment,* 4th edition. 2010. Nicholas G. Pizzi, Editor. American Water Works Association. Chapter 7.

13. Answer: **d.** person-to-person contact.

    Reference: *Water Quality,* 4th edition. 2010. Joseph A. Ritter, Editor. American Water Works Association. Chapter 4.

14. Answer: **b.** are fitted with a low-air-pressure alarm that sounds, alerting the wearer to leave the contaminated site.

    Reference: *Water Treatment,* 4th edition. 2010. Nicholas G. Pizzi, Editor. American Water Works Association. Chapter 7.

15. Answer: **d.** 20 to 50 ppm

    Reference: *Water Treatment Operator Handbook,* Revised edition. 2005. Nicholas G. Pizzi. American Water Works Association. Chapter 8.

16. Answer: **c.** Authorized attendant

    Reference: *Water Treatment Operator Handbook,* Revised edition. 2005. Nicholas G. Pizzi. American Water Works Association. Chapter 13.

17. Answer: **c.** gastroenteritis.

    Reference: *Water Quality,* 4th edition. 2010. Joseph A. Ritter, Editor. American Water Works Association. Chapter 4.

18. Answer: **a.** Cholera

    Reference: *Water Quality,* 4th edition. 2010. Joseph A. Ritter, Editor. American Water Works Association. Chapter 4.

19. Answer: **a.** highly volatile.

    Reference: M20, *Water Chlorination/Chlorination Practices and Principles*, 2nd edition. 2006. American Water Works Association. Chapter 4.

20. Answer: **c.** the tank to collapse.

    Reference: M20, *Water Chlorination/Chlorination Practices and Principles*, 2nd edition. 2006. American Water Works Association. Chapter 4.

21. Answer: **c.** Copper

    Reference: M20, *Water Chlorination/Chlorination Practices and Principles*, 2nd edition. 2006. American Water Works Association. Chapter 5.

22. Answer: **c.** 30 minutes

    Reference: M20, *Water Chlorination/Chlorination Practices and Principles*, 2nd edition. 2006. American Water Works Association. Chapter 5.

23. Answer: **a.** reduced demand on supply source.

    Reference: *Water Sources,* 4th edition. 2010. Paul Koch, Editor. American Water Works Association. Chapter 5.

24. Answer: **c.** temperature increases.

    Reference: *Water Sources,* 4th edition. 2010. Paul Koch, Editor. American Water Works Association. Chapter 6.

25. Answer: **b.** 0.5 mg/L

    Reference: *Water Treatment Operator Handbook,* Revised edition. 2005. Nicholas G. Pizzi. American Water Works Association. Chapter 10.

26. Answer: **d.** iron reacting with tannic acid in the coffee.

    Reference: *Water Quality,* 4th edition. 2010. Joseph A. Ritter, Editor. American Water Works Association. Chapter 7.

27. Answer: **c.** household plumbing systems.

    Reference: *Water Treatment,* 4th edition. 2010. Nicholas G. Pizzi, Editor. American Water Works Association. Chapter 9.

28. Answer: **b.** Kit B

    Reference: *Water Treatment,* 4th edition. 2010. Nicholas G. Pizzi, Editor. American Water Works Association. Chapter 7.

29. Answer: **d.** 2,000 to 3,000 ppm.

    Reference: M20, *Water Chlorination/Chlorination Practices and Principles*, 2nd edition. 2006. American Water Works Association. Chapter 5.

30. Answer: **a.** 4 to 6 μm

    Reference: *Water Quality,* 4th edition. 2010. Joseph A. Ritter, Editor. American Water Works Association. Chapter 4.

31. Answer: **a.** *Salmonella*

    Reference: *Water Quality,* 4th edition. 2010. Joseph A. Ritter, Editor. American Water Works Association. Chapter 4.

32. Answer: **d.** Soda ash

    Reference: *Water Treatment,* 4th edition. 2010. Nicholas G. Pizzi, Editor. American Water Works Association. Chapter 11.

33. Answer: **b.** 6 ft

    Reference: *Water Treatment,* 4th edition. 2010. Nicholas G. Pizzi, Editor. American Water Works Association. Chapter 13.

34. Answer: **b.** 2

    Reference: *Water Treatment Operator Handbook,* Revised edition. 2005. Nicholas G. Pizzi. American Water Works Association. Chapter 8.

35. Answer: **c.** 150 to 200 ppm.

    Reference: *Water Treatment Operator Handbook,* Revised edition. 2005. Nicholas G. Pizzi. American Water Works Association. Chapter 8.

36. Answer: **a.** liver.

    Reference: *Basic Microbiology for Drinking Water Personnel*, 2nd edition. 2006. Dennis R. Hill. American Water Works Association. Chapter 3.

37. Answer: **a.** 10 to 25 nanometers

    Reference: *Water Quality,* 4th edition. 2010. Joseph A. Ritter, Editor. American Water Works Association. Chapter 4.

38. Answer: **b.** 35

    Reference: M20, *Water Chlorination/Chlorination Practices and Principles*, 2nd edition. 2006. American Water Works Association. Chapter 2.

39. Answer: **b.** 20%

    Reference: M20, *Water Chlorination/Chlorination Practices and Principles*, 2nd edition. 2006. American Water Works Association. Chapter 4.

40. Answer: **c.** An explosive mixture of nitrogen trichloride

    Reference: M20, *Water Chlorination/Chlorination Practices and Principles*, 2nd edition. 2006. American Water Works Association. Chapter 5.

41. Answer: **a.** 25 to 30%

    Reference: *Water Sources*, 4th edition. 2010. Paul Koch, Editor. American Water Works Association. Chapter 4.

42. Answer: **b.** More than 50%

    Reference: *Water Sources*, 4th edition. 2010. Paul Koch, Editor. American Water Works Association. Chapter 4.

43. Answer: **c.** The Resource Conservation and Recovery Act (RCRA)

    Reference: *Water Sources*, 4th edition. 2010. Paul Koch, Editor. American Water Works Association. Chapter 7.

44. Answer: **b.** *Code of Federal Regulations.*

    Reference: *Water Treatment Operator Handbook,* Revised edition. 2005. Nicholas G. Pizzi. American Water Works Association. Chapter 13.

45. Answer: **d.** Exposure differences to contaminants

    Reference: *Water Treatment Operator Handbook,* Revised edition. 2005. Nicholas G. Pizzi. American Water Works Association. Chapter 1.

46. Answer: **d.** 30 days.

    Reference: *Water Treatment*, 4th edition. 2010. Nicholas G. Pizzi, Editor. American Water Works Association. Chapter 9.

47. Answer: **c.** stomach flu symptoms.

    Reference: *Basic Microbiology for Drinking Water Personnel*, 2nd edition. 2006. Dennis R. Hill. American Water Works Association. Chapter 3.

48. Answer: **a.** Interfere with nutrient absorption

    Reference: *Basic Microbiology for Drinking Water Personnel*, 2nd edition. 2006. Dennis R. Hill. American Water Works Association. Chapter 3.

49. Answer: **b.** 2.0 to 3.0 μm

    Reference: *Basic Microbiology for Drinking Water Personnel*, 2nd edition. 2006. Dennis R. Hill. American Water Works Association. Chapter 3.

50. Answer: **c.** Kit C

    Reference: *Water Treatment,* 4th edition. 2010. Nicholas G. Pizzi, Editor. American Water Works Association. Chapter 7.

51. Answer: **a.** North American Chlorine Emergency Plan

    Reference: *Water Treatment,* 4th edition. 2010. Nicholas G. Pizzi, Editor. American Water Works Association. Chapter 7.

52. Answer: **c.** 5 ppm.

    Reference: M20, *Water Chlorination/Chlorination Practices and Principles*, 2nd edition. 2006. American Water Works Association. Chapter 4.

53. Answer: **b.** 2,500 lb.

    Reference: M20, *Water Chlorination/Chlorination Practices and Principles*, 2nd edition. 2006. American Water Works Association. Chapter 5.

54. Answer: **a.** Clean up the source of contamination caused by the disaster

    Reference: *Water Sources,* 4th edition. 2010. Paul Koch, Editor. American Water Works Association. Chapter 4.

55. Answer: **c.** 50 to 100%.

    Reference: *Water Sources,* 4th edition. 2010. Paul Koch, Editor. American Water Works Association. Chapter 5.

56. Answer: **b.** Possible stimulation of water service growth

    Reference: *Water Sources,* 4th edition. 2010. Paul Koch, Editor. American Water Works Association. Chapter 5.

57. Answer: **d.** Federal Underground Injection Control Program

    Reference: *Water Sources,* 4th edition. 2010. Paul Koch, Editor. American Water Works Association. Chapter 3.

58. Answer: **a.** Riparian doctrine.

    Reference: *Water Sources,* 4th edition. 2010. Paul Koch, Editor. American Water Works Association. Chapter 5.

59. Answer: **b.** absolute ownership.

    Reference: *Water Sources,* 4th edition. 2010. Paul Koch, Editor. American Water Works Association. Chapter 5.

60. Answer: **c.** reasonable use.

    Reference: *Water Sources,* 4th edition. 2010. Paul Koch, Editor. American Water Works Association. Chapter 5.

61. Answer: **a.** Correlative rights

    Reference: *Water Sources,* 4th edition. 2010. Paul Koch, Editor. American Water Works Association. Chapter 5.

62. Answer: **d.** Appropriation-permit systems

    Reference: *Water Sources,* 4th edition. 2010. Paul Koch, Editor. American Water Works Association. Chapter 5.

63. Answer: **a.** 0.1 to 3.0%

    Reference: *Water Sources,* 4th edition. 2010. Paul Koch, Editor. American Water Works Association. Chapter 6.

ANSWERS TO ADDITIONAL WATER TREATMENT OPERATOR PRACTICE QUESTIONS 221

64. Answer: **d.** Coliphages

    Reference: *Basic Microbiology for Drinking Water Personnel*, 2nd edition. 2006. Dennis R. Hill. American Water Works Association. Chapter 1.

65. Answer: **a.** October 1

    Reference: *The Water Dictionary: A Comprehensive Reference of Water Terminology*, 2nd edition. 2010. Nancy McTigue, Editor, and James M. Symons, Editor Emeritus. American Water Works Association.

66. Answer: **c.** fungi.

    Reference: *The Water Dictionary: A Comprehensive Reference of Water Terminology*, 2nd edition. 2010. Nancy McTigue, Editor, and James M. Symons, Editor Emeritus. American Water Works Association.

67. Answer: **a.** USEPA and state primacy programs

    Reference: *Water Quality,* 4th edition. 2010. Joseph A. Ritter, Editor. American Water Works Association. Chapter 2.

68. Answer: **d.** 100%

    Reference: M20, *Water Chlorination/Chlorination Practices and Principles*, 2nd edition. 2006. American Water Works Association. Chapter 4.

69. Answer: **a.** relatively safe compared to using gaseous chlorine.

    Reference: *Water Treatment*, 4th edition. 2010. Nicholas G. Pizzi, Editor. American Water Works Association. Chapter 7.

70. Answer: **c.** *Bacillus*

    Reference: *Basic Microbiology for Drinking Water Personnel*, 2nd edition. 2006. Dennis R. Hill. American Water Works Association. Chapter 2.

Evaluate Characteristics of Source Water

Answers to Additional Practice Questions

1. Answer: **b.** Discolored laundry brown and changed taste of water, coffee, tea, and other beverages

    Reference: *Water Quality,* 4th edition. 2010. Joseph A. Ritter, Editor. American Water Works Association. Chapter 1.

2. Answer: **a.** Hydrogen sulfide

    Reference: *Water Treatment,* 4th edition. 2010. Nicholas G. Pizzi, Editor. American Water Works Association. Chapter 14.

3. Answer: **d.** Radon

    Reference: *Water Treatment,* 4th edition. 2010. Nicholas G. Pizzi, Editor. American Water Works Association. Chapter 14.

4. Answer: **c.** 6.0 to 8.5 units.

    Reference: *Water Treatment Operator Handbook,* Revised edition. 2005. Nicholas G. Pizzi. American Water Works Association. Chapter 4.

5. Answer: **c.** 6.5 to 8.5 units.

   Reference: *Water Treatment Operator Handbook,* Revised edition. 2005. Nicholas G. Pizzi. American Water Works Association. Chapter 4.

6. Answer: **c.** transpiration.

   Reference: *Water Sources,* 4th edition. 2010. Paul Koch, Editor. American Water Works Association. Chapter 1.

7. Answer: **c.** condensation.

   Reference: *Water Sources,* 4th edition. 2010. Paul Koch, Editor. American Water Works Association. Chapter 1.

8. Answer: **b.** algae.

   Reference: *The Water Dictionary: A Comprehensive Reference of Water Terminology,* 2nd edition. 2010. Nancy McTigue, Editor, and James M. Symons, Editor Emeritus. American Water Works Association.

9. Answer: **c.** Radon

   Reference: *Water Quality,* 4th edition. 2010. Joseph A. Ritter, Editor. American Water Works Association. Chapter 7.

10. Answer: **d.** Turbidity

    Reference: *Water Treatment,* 4th edition. 2010. Nicholas G. Pizzi, Editor. American Water Works Association. Chapter 4.

11. Answer: **b.** 0.5 mg/L

    Reference: *Water Treatment,* 4th edition. 2010. Nicholas G. Pizzi, Editor. American Water Works Association. Chapter 2.

12. Answer **b.** blue-green algae.

    Reference: M7, *Problem Organisms in Water: Identification and Treatment,* 3rd edition. 2003. American Water Works Association. Chapter 10.

13. Answer **c.** Coliform bacteria

    Reference: M37, *Operational Control of Coagulation and Filtration Processes,* 3rd edition. 2011. American Water Works Association. Chapter 2.

14. Answer **c.** have cysts that can penetrate filters.

    Reference: M7, *Problem Organisms in Water: Identification and Treatment,* 3rd edition. 2003. American Water Works Association. Chapter 11.

15. Answer **a.** Unappealing appearance and indication that dissolved organics may be present

    Reference: *Water Quality,* 4th edition. 2010. Joseph A. Ritter, Editor. American Water Works Association. Chapter 1.

16. Answer **c.** oxides, carbonates, and hydroxides.

    Reference: *Water Quality,* 4th edition. 2010. Joseph A. Ritter, Editor. American Water Works Association. Chapter 6.

17. Answer: **c.** Humic materials

    Reference: *Water Quality,* 4th edition. 2010. Joseph A. Ritter, Editor. American Water Works Association. Chapter 7.

ANSWERS TO ADDITIONAL WATER TREATMENT OPERATOR PRACTICE QUESTIONS   223

18. Answer: **c.**   saturated zone.

    Reference: *Water Sources,* 4th edition. 2010. Paul Koch, Editor. American Water Works Association. Chapter 1.

19. Answer: **d.**   Radial wells

    Reference: *Water Sources,* 4th edition. 2010. Paul Koch, Editor. American Water Works Association. Chapter 2.

20. Answer: **d.**   groundwater flow.

    Reference: *Water Sources,* 4th edition. 2010. Paul Koch, Editor. American Water Works Association. Chapter 3.

21. Answer: **a.**   0 to 5 mg/L

    Reference: *Water Treatment,* 4th edition. 2010. Nicholas G. Pizzi, Editor. American Water Works Association. Chapter 14.

22. Answer: **c.**   Viruses, bacteria, and protozoans

    Reference: *Water Treatment Operator Handbook,* Revised edition. 2005. Nicholas G. Pizzi. American Water Works Association. Chapter 1.

23. Answer: **b.**   Limestone

    Reference: *Water Treatment Operator Handbook,* Revised edition. 2005. Nicholas G. Pizzi. American Water Works Association. Chapter 2.

24. Answer: **a.**   confined aquifer.

    Reference: *Water Sources,* 4th edition. 2010. Paul Koch, Editor. American Water Works Association. Chapter 1.

25. Answer: **b.**   unsaturated zone.

    Reference: *Water Sources,* 4th edition. 2010. Paul Koch, Editor. American Water Works Association. Chapter 1.

26. Answer: **b.**   Agricultural activities

    Reference: *Water Sources,* 4th edition. 2010. Paul Koch, Editor. American Water Works Association. Chapter 2.

27. Answer: **b.**   Cold water contains more DO than warm water

    Reference: *Water Quality,* 4th edition. 2010. Joseph A. Ritter, Editor. American Water Works Association. Chapter 6.

28. Answer: **d.**   alkalinity.

    Reference: M7, *Problem Organisms in Water: Identification and Treatment,* 3rd edition. 2003. American Water Works Association. Chapter 10.

29. Answer: **a.**   < 50 mg/L

    Reference: *Water Treatment,* 4th edition. 2010. Nicholas G. Pizzi, Editor. American Water Works Association. Chapter 2.

30. Answer: **a.**   8 hours

    Reference: *Water Treatment,* 4th edition. 2010. Nicholas G. Pizzi, Editor. American Water Works Association. Chapter 9.

31. Answer: **c.** Every day

    Reference: *Water Treatment,* 4th edition. 2010. Nicholas G. Pizzi, Editor. American Water Works Association. Chapter 9.

32. Answer: **d.** advection.

    Reference: *Water Sources,* 4th edition. 2010. Paul Koch, Editor. American Water Works Association. Chapter 1.

33. Answer: **d.** hydrogen bonding.

    Reference: *Basic Chemistry for Water & Wastewater Operators*, Revised edition. 2005. Darshan Singh Sarai. American Water Works Association. Chapter 9.

34. Answer: **c.** low nutrient body of water with abundant oxygen.

    Reference: *Water Treatment Operator Handbook*, Revised edition. 2005. Nicholas G. Pizzi. American Water Works Association. Chapter 2.

35. Answer: **a.** In the hypolimnion

    Reference: *Water Treatment,* 4th edition. 2010. Nicholas G. Pizzi, Editor. American Water Works Association. Chapter 2.

36. Answer: **b.** Noncarbonated hardness

    Reference: *Water Treatment,* 4th edition. 2010. Nicholas G. Pizzi, Editor. American Water Works Association. Chapter 11.

37. Answer: **c.** ephemeral streams.

    Reference: *Water Sources,* 4th edition. 2010. Paul Koch, Editor. American Water Works Association. Chapter 3.

38. Answer: **c.** Taste and odors

    Reference: M7, *Problem Organisms in Water: Identification and Treatment*, 3rd edition. 2003. American Water Works Association. Chapter 10.

39. Answer: **c.** Flagellates

    Reference: *Water Treatment,* 4th edition. 2010. Nicholas G. Pizzi, Editor. American Water Works Association. Chapter 2.

40. Answer: **a.** are able to photosynthesize.

    Reference: M7, *Problem Organisms in Water: Identification and Treatment*, 3rd edition. 2003. American Water Works Association. Chapter 3.

41. Answer: **d.** Cyanobacteria

    Reference: *The Water Dictionary: A Comprehensive Reference of Water Terminology*, 2nd edition. 2010. Nancy McTigue, Editor, and James M. Symons, Editor Emeritus. American Water Works Association.

42. Answer: **d.** Undesirable taste and a milky appearance

    Reference: *Water Quality,* 4th edition. 2010. Joseph A. Ritter, Editor. American Water Works Association. Chapter 1.

43. Answer: **d.** Dissolved oxygen may oxidize iron and manganese

    Reference: *Water Quality,* 4th edition. 2010. Joseph A. Ritter, Editor. American Water Works Association. Chapter 6.

ANSWERS TO ADDITIONAL WATER TREATMENT OPERATOR PRACTICE QUESTIONS 225

44. Answer: **a.** Radium

   Reference: *Water Quality,* 4th edition. 2010. Joseph A. Ritter, Editor. American Water Works Association. Chapter 7.

45. Answer: **d.** porosity.

   Reference: *Water Sources,* 4th edition. 2010. Paul Koch, Editor. American Water Works Association. Chapter 1.

46. Answer: **a.** Septic tanks and leech fields

   Reference: *Water Sources,* 4th edition. 2010. Paul Koch, Editor. American Water Works Association. Chapter 2.

47. Answer: **d.** 60 ft.

   Reference: *Water Sources,* 4th edition. 2010. Paul Koch, Editor. American Water Works Association. Chapter 2.

48. Answer: **d.** The pumping rate

   Reference: *Water Sources,* 4th edition. 2010. Paul Koch, Editor. American Water Works Association. Chapter 2.

49. Answer: **d.** algae.

   Reference: *Water Treatment,* 4th edition. 2010. Nicholas G. Pizzi, Editor. American Water Works Association. Chapter 2.

50. Answer: **c.** they commonly adhere to boats so they move around and disperse eggs.

   Reference: *Water Treatment,* 4th edition. 2010. Nicholas G. Pizzi, Editor. American Water Works Association. Chapter 2.

51. Answer: **d.** Low turbidity

   Reference: *Water Treatment,* 4th edition. 2010. Nicholas G. Pizzi, Editor. American Water Works Association. Chapter 4.

52. Answer: **b.** Sand

   Reference: *Water Sources,* 4th edition. 2010. Paul Koch, Editor. American Water Works Association. Chapter 1.

53. Answer: **a.** electron.

   Reference: *Water Quality,* 4th edition. 2010. Joseph A. Ritter, Editor. American Water Works Association. Chapter 7.

54. Answer: **d.** Photon

   Reference: *Water Quality,* 4th edition. 2010. Joseph A. Ritter, Editor. American Water Works Association. Chapter 7.

55. Answer: **c.** Radon

   Reference: *Water Quality,* 4th edition. 2010. Joseph A. Ritter, Editor. American Water Works Association. Chapter 7.

56. Answer: **c.** *Anabaena*

   Reference: M7, *Problem Organisms in Water: Identification and Treatment*, 3rd edition. 2003. American Water Works Association. Chapter 10.

57. Answer: **c.** Powdered activated carbon

    Reference: *Water Treatment,* 4th edition. 2010. Nicholas G. Pizzi, Editor. American Water Works Association. Chapter 2.

58. Answer: **d.** 0.14 mg/L

    Reference: *Water Treatment,* 4th edition. 2010. Nicholas G. Pizzi, Editor. American Water Works Association. Chapter 2.

59. Answer: **d.** Shock treatment using chlorine at a dosage of 10.0 mg/L for 30 minutes

    Reference: M20, *Water Chlorination/Chlorination Practices and Principles,* 2nd edition. 2006. American Water Works Association. Chapter 6.

60. Answer: **c.** pesticides.

    Reference: *Water Quality,* 4th edition. 2010. Joseph A. Ritter, Editor. American Water Works Association. Chapter 7.

61. Answer: **b.** Alpha

    Reference: *Water Quality,* 4th edition. 2010. Joseph A. Ritter, Editor. American Water Works Association. Chapter 7.

62. Answer: **d.** Tritium

    Reference: *Water Quality,* 4th edition. 2010. Joseph A. Ritter, Editor. American Water Works Association. Chapter 7.

63. Answer: **b.** on its porosity and permeability.

    Reference: *Water Sources,* 4th edition. 2010. Paul Koch, Editor. American Water Works Association. Chapter 1.

64. Answer: **c.** Gamma-ray logging

    Reference: *Water Sources,* 4th edition. 2010. Paul Koch, Editor. American Water Works Association. Chapter 2.

65. Answer: **d.** Infiltration of urban runoff

    Reference: *Water Sources,* 4th edition. 2010. Paul Koch, Editor. American Water Works Association. Chapter 2.

66. Answer: **a.** At least 24 hours

    Reference: *Water Sources,* 4th edition. 2010. Paul Koch, Editor. American Water Works Association. Chapter 2.

67. Answer: **b.** 12°C.

    Reference: *Water Treatment,* 4th edition. 2010. Nicholas G. Pizzi, Editor. American Water Works Association. Chapter 2.

68. Answer: **a.** Manage watershed to prevent nitrates from entering the water system

    Reference: *Water Treatment Operator Handbook,* Revised edition. 2005. Nicholas G. Pizzi. American Water Works Association. Chapter 2.

69. Answer: **c.** oxygen is removed by the decomposition of organic materials making the water capable of dissolving iron and manganese.

    Reference: *Water Treatment Operator Handbook,* Revised edition. 2005. Nicholas G. Pizzi. American Water Works Association. Chapter 10.

ANSWERS TO ADDITIONAL WATER TREATMENT OPERATOR PRACTICE QUESTIONS

70. Answer: **b.** ridge.

    Reference: *Water Sources,* 4th edition. 2010. Paul Koch, Editor. American Water Works Association. Chapter 7.

71. Answer: **b.** The Comprehensive Environmental Response, Compensation, and Liability Act (Superfund)

    Reference: *Water Sources,* 4th edition. 2010. Paul Koch, Editor. American Water Works Association. Chapter 7.

72. Answer: **b.** colloidal solids.

    Reference: *Water Treatment,* 4th edition. 2010. Nicholas G. Pizzi, Editor. American Water Works Association. Chapter 4.

73. Answer: **a.** increases; decreases

    Reference: *Water Treatment,* 4th edition. 2010. Nicholas G. Pizzi, Editor. American Water Works Association. Chapter 4.

74. Answer: **d.** iron.

    Reference: *Basic Microbiology for Drinking Water Personnel*, 2nd edition. 2006. Dennis R. Hill. American Water Works Association. Chapter 3.

75. Answer: **a.** It is naturally occurring.

    Reference: *Water Treatment Plant Operation*, Volume I, 6th edition. 2008. Ken Kerri. California State University–Sacramento. Page 29.

Answers to Math Questions for Water Treatment Operators Levels I & II

1. Answer: **b.** 315.7 mg/L

   Equation: Hardness, mg/L = $\dfrac{(\text{Hardness, gpg})(17.12 \text{ mg/L})}{1 \text{ gpg}}$

   Hardness, mg/L = $\dfrac{(18.44, \text{gpg})(17.12 \text{ mg/L})}{1 \text{ gpg}}$ = 315.69 mg/L,

   round to **315.7 mg/L**

2. Answer: **b.** 6.44 gpm/ft

   Equation: Specific yield, gpm/ft = $\dfrac{\text{Well yield, gpm}}{\text{Drawdown, ft}}$

   Specific yield, gpm/ft = $\dfrac{105 \text{ gpm}}{16.3 \text{ ft}}$ = **6.44 gpm/ft**

3. Answer: **d.** 159.7 feet

   Equation: Drawdown, ft = Pumping water level, ft − Static water level, ft

   Rearrange the equation to solve for pumping water level.

   Pumping Water Level, ft = Drawdown, ft + Static water level

   Substitute known values and solve.

   Pumping water level, ft = 21.1 ft + 138.6 ft = **159.7 ft**

4. Answer: **c.** 11,790 lb/ft^3

   Equation: Pressure = (Depth, ft)(Density, 62.4 lb/ft^3)

   Pressure = (189 ft)(62.4 lb/ft^3) = 11,793.6 lb/ft^2, round to **11,790 lb/ft^3**

5. Answer: **d.** 14,380 gpd/ft

   Equation: Weir overflow rate = $\dfrac{\text{Flow, gpd}}{\text{Weir length, ft}}$

   Weir overflow rate = $\dfrac{2{,}330{,}000 \text{ gpd}}{162 \text{ ft}}$ = 14,382.7 gpd/ft,

   round to **14,380 gpd/ft**

6. Answer: **b.** 15,800 gpm

   (35.1 cfs)(60 sec/min)(7.48 gal/ft^3) = 15,752.88 gpm, round to **15,800 gpm**

7. Answer: **b.** 12 cfs

   $\dfrac{(7.7 \text{ mgd})(1{,}000{,}000 \text{ gal})(1 \text{ ft}^3)(1 \text{ day})(1 \text{ min})}{(1 \text{ mil gal})(7.48 \text{ gal})(1440 \text{ min})(60 \text{ sec})}$ = 11.91 cfs, round to **12 cfs**

8. Answer: **a.** 104 mil gal

   [(318 acre-ft)(43,560 ft^3/acre-ft)(7.48 gal/ft^3)] ÷ 1,000,000 = 103.61, round to **104 mil gal**

9. Answer: **a.** 20°C

   Equation: °C = (°F − 32) × 5/9, or easier °C = (°F − 32) ÷ 1.8

   °C = (68°F − 32) × 5/9 = (36 × 5) ÷ 9 = 180 ÷ 9 = **20°C** or

   °C = (68°F − 32) ÷ 1.8 = (36°F) ÷ 1.8 = **20°C**

10. Answer: **c.** 258

    Number, x = (316)(81.5% ÷ 100%) = 257.54, round to **258**

11. Answer: **c.** 2.2% soda ash slurry

    Know: 1 gal of water = 8.34 lb

    $$\text{Percent soda ash} = \frac{(8.25 \text{ lb})(100\%)}{8.25 \text{ lb} + 45 \text{ gal}(8.34 \text{ lb/gal})} = \frac{(8.25 \text{ lb})(100\%)}{8.25 \text{ lb} + 375.3 \text{ lb}}$$

    $$\frac{(8.25 \text{ lb})(100\%)}{383.55 \text{ lb}} = 2.15\%, \text{ round to } \mathbf{2.2\% \text{ soda ash slurry}}$$

12. Answer: **d.** 133,000 ft²

    Equation: Area = πr^2, where π = 3.14

    First find the radius: Radius = Diameter ÷ 2 = 411 ÷ 2 = 205.5 ft

    Area of tank, ft² = (3.14)(205.5 ft)(205.5 ft) = 132,602.985 ft², round to **133,000 ft²**

13. Answer: **d.** 600 ft

    Equation: Circumference = 2π(radius) or $2\pi r$

    Circumference = 2(3.14)(95 ft) = 596.6 ft, round to **600 ft**

14. Answer: **d.** 76,000 gal

    First, determine the number of feet in 1.09 miles.

    (5,280 ft/mile)(1.09 miles) = 5,755.2 ft

    Volume, gal = (0.785)(1.5 ft)(1.5 ft)(5,755.2 ft)(7.48 gal/ft³) = 76,035 gal, round to **76,000 gal**

15. Answer: **b.** 1.5 mL alum

    Alum Dosage, mg/L

    $$= \frac{(\text{stock, mL})(1{,}000 \text{ mg/gram})(\text{Concentration in grams/liter})}{\text{Sample size, mL}}$$

    $$= \frac{(5.0 \text{ mL})(1{,}000 \text{ mg/gram})(0.30 \text{ grams/liter})}{1{,}000 \text{ mL}}$$

    = **1.5 mg/L alum**

16. Answer: **c.** 18 mg/L as $CaCO_3$

    Phenolphthalein alkalinity, mg/L as $CaCO_3$

    $$= \frac{(A, \text{ titrant, mL})(0.02 \text{ N})(50{,}000)}{\text{Sample size, mL}}$$

    $$= \frac{(1.8 \text{ mL})(0.02 \text{ N})(50{,}000)}{100 \text{ mL}} = \mathbf{18 \text{ mg/L as } CaCO_3}$$

ANSWERS TO MATH QUESTIONS FOR WATER TREATMENT OPERATORS LEVELS I & II 231

17. Answer: **c.** 9,335,000 gal

    First, convert acre-ft into ft^3

    (28.65 ac-ft)(43,560 ft^3/acre-ft) = 1,247,994 ft^3

    Then multiple this result by the number of gallons in a cubic foot.

    (1,247,994 ft^3)(7.48 gal/ft^3) = 9,334,995.12 gal, round to **9,335,000 gal**

18. Answer: **b.** 2.4% slurry
    Know: 1 gal of water = 8.34 lb

    First determine the total weight of the water.

    Water, lb = (8.34 lb/gal)(35 gal) = 291.9

    $$\% \text{ Polymer} = \frac{(7.3 \text{ lb})(100\%)}{7.3 \text{ lb} + 291.9 \text{ lb}} = \frac{(7.3 \text{ lb})(100\%)}{299.2 \text{ lb}} = 2.4398\%$$

    round to **2.4% slurry**

19. Answer: **d.** 93.4% ntu removed
    Equation: $\text{Percent ntu removed} = \frac{(\text{influent ntu} - \text{effluent ntu})(100\%)}{\text{influent ntu}}$ or

    $$\frac{(\text{In} - \text{Out})(100\%)}{\text{In}}$$

    $$\text{Percent ntu removed} = \frac{(17.1 \text{ ntu} - 1.13 \text{ ntu})(100\%)}{17.1 \text{ ntu}} = \frac{(15.97 \text{ ntu})(100\%)}{17.1 \text{ ntu}}$$

    = 93.39%, round to **93.4% ntu removed**

20. Answer: **c.** 2.52 hr

    First determine the volume in gallons for the clarifier.

    Volume, gal = (0.785)(Diameter)^2(Depth)(7.48 gal/ft^3)

    Volume, gal = (0.785)(112.2 ft)(112.2 ft)(10.33 ft)(7.48 gal/ft^3) = 763,584.83 gal

    Then, convert mgd to gallons per hour as detention time is asked for in hours.

    (7.26 mgd)(1,000,000/1 M)(1 day/24 hrs) = 302,500 gph

    Equation: $\text{Detention time, hrs} = \frac{\text{Volume, gal}}{\text{Flow rate, gal/hour (gph)}}$

    $$\text{Detention time, hrs} = \frac{763,584.83 \text{ gal}}{302,500 \text{ gph}} = \mathbf{2.52 \text{ hr}}$$

21. Answer: **a.** 6.9 mg/L lime

    First calculate the alkalinity that will react with the alum by using a ratio.

$$\frac{x \text{ mg/L alkalinity}}{8.5 \text{ mg/L, Alum (dosage)}} = \frac{0.45 \text{ mg/L, Alkalinity}}{1 \text{ mg/L, Alum}}$$

$$x \text{ mg/L alkalinity} = \frac{(0.45 \text{ mg/L, Alkalinity})(8.5 \text{ mg/L Alum})}{1 \text{ mg/L, Alum}} = 3.825 \text{ mg/L}$$

Next calculate the total alkalinity required.

Total alkalinity, mg/L = Residual alkalinity, mg/L + Alkalinity consumed by alum, mg/L

Total alkalinity, mg/L = 14 mg/L + 3.825 mg/L = 17.825 mg/L total alkalinity

Next, determine the amount of alkalinity that must be added to the water.

Added alkalinity, mg/L = Total alkalinity required, mg/L − Raw water alkalinity, mg/L

Added alkalinity, mg/L = 17.825 mg/L − 9.0 mg/L = 8.825 mg/L added alkalinity

Finally, calculate the lime dosage needed to make this alkalinity.

$$\frac{x \text{ mg/L, Lime}}{8.825 \text{ mg/L, Alkalinity}} = \frac{0.35 \text{ mg/L, Lime}}{0.45 \text{ mg/L, Alkalinity}}$$

$$x \text{ mg/L, Lime} = \frac{(0.35 \text{ mg/L, Lime})(8.825 \text{ mg/L, Alkalinity})}{0.45 \text{ mg/L, Alkalinity}}$$

= 6.864 mg/L, round to **6.9 mg/L lime**

22. Answer: **d.** 33,672 lb copper sulfate
    (43,560 ft^3/acre-ft)(8,850 acre-ft) = 385,506,000 ft^3 × 7.48 gal ÷ 1 ft^3 = 2,883,584,880 gal ÷ 1 million = 2883.6

    Write the equation: Copper sulfate, lb

    $$= \frac{(\text{mil gal})(\text{Dosage, mg/L})(8.34 \text{ lb/gal})}{\text{Percent available copper} \div 100\%}$$

    Copper sulfate, lb = $\frac{(2884 \text{ mil gal})(0.35 \text{ mg/L})(8.34 \text{ lb/gal})}{25\% \div 100\%} = \frac{8{,}418}{0.25}$

    = **33,672 lb copper sulfate**

    In reality, only the upper 20 ft of the reservoir would be treated. Large reservoirs are also treated one section at a time.

23. Answer: **c.** 1.42 SG

    Know: The density of water can also be expressed as lb/gal, or 8.34 lb/gal.

    $$SG = \frac{11.87 \text{ lb/gal}}{8.34 \text{ lb/gal}} = \mathbf{1.42 \text{ SG}}$$

24. Answer: **b.** 4.8 gpm/ft^2
    First, find the surface area of the filter:
    Filter surface area = (24 ft)(28 ft) = 672 ft^2

    Equation: Filtration rate = $\frac{\text{Flow rate, gpm}}{\text{Filter surface area, ft}^2}$

Filtration rate = $\dfrac{3{,}250 \text{ gpm}}{672 \text{ ft}^2}$ = 4.836 gpm/ft², round to **4.8 gpm/ft²**

25. Answer: **c.** 13 gpm/ft²

    First, calculate the area of the filter in square feet.

    Equation: Number of ft² = (Length, ft)(Width, ft)

    Number of ft² = (25 ft)(18.2 ft) = 455 ft²

    Now, convert cfs to gpm.

    Equation: Number of gpm = (ft³/sec)(60 sec/min)(7.48 gal/ft³)

    Number of gpm = (13 ft³/sec)(60 sec/min)(7.48 gal/ft³) = 5,834.4 gpm

    Then determine backwash rate.

    Equation: Backwash rate, gpm/ft² = $\dfrac{\text{Flow rate, gpm}}{\text{Filter surface area, ft}^2}$

    Backwash rate, gpm/ft² = $\dfrac{5{,}834.4, \text{ gpm}}{455, \text{ ft}^2}$ = 12.8 gpm/ft², round to **13 gpm/ft²**

26. Answer: **b.** 8,640 gpm

    Equation: Backwash pumping rate = (Desired backwash rate, gpm/ft²)(Filter area, ft²)

    Backwash pumping rate = (18 gpm/ft²)(20.0 ft)(24.0 ft) = **8,640 gpm**

27. Answer: **c.** 114 mg/L as $CaCO_3$
    Solution: Total alkalinity, mg/L as $CaCO_3$ = $\dfrac{(\text{Titrant, mL})(0.02 \text{ N})(50{,}000)}{\text{Sample size, mL}}$

    Total alkalinity, mg/L as $CaCO_3$ = $\dfrac{(11.4 \text{ mL})(0.02 \text{ N})(50{,}000)}{100 \text{ mL}}$ = **114 mg/L as $CaCO_3$**

28. Answer: **c.** 69.1 mL/min

    Know: 1 day = 1,440 min

    (26.3 gal/day)(3,785 mL/gal)(1 day/1,440 min) = **69.1 mL/min**

29. Answer **a.** 1.81

    Divide the weight of the substance by 62.4 lb/ft³ to compare the substance to water.

    Density = 112.7 lb/ft³ / 62.4 lb/ft³ = **1.81**

30. Answer: **b.** 11.4 ft

    Write the equation, arranging it to solve for the unknown, drawdown.

    Drawdown, ft = Well yield, gpm ÷ Specific yield, gpm/ft

    Drawdown, ft = 325 gpm ÷ 28.4 gpm/ft = **11.4 ft**

31. Answer: **a.** 33.7 hr

First, calculate the number of gallons per foot in the tank.

Equation: gal/ft = (0.785)(Diameter, ft)^2(1 ft)(7.48 gal/ft)

gal/ft = (0.785)(119.8 ft)(119.8 ft)(1 ft)(7.48 gal/ft) = 84,272 gal/ft

Next, determine how many feet of water can be used from the tank.

Number of usable feet of water = 27.6 ft − 16.0 ft = 11.6 ft

Next, multiply the number of usable feet of water by the gal/ft to get the total number of usable gallons.

Total usable gallons = (11.6 ft)(84,272 gal/ft) = 977,555.2 gal

Lastly, determine the hours of supply left.

Supply, hr = $\dfrac{(977,555.2 \text{ gal})}{(483 \text{ gpm})(60 \text{ min/hr})}$ = **33.7 hr**

32. Answer: **d.** 4,300 ft^3

First, find the volume of the cone in cubic feet.

Volume, ft^3 = 1/3(0.785)(Diameter, ft)^2(Depth)

Volume, ft^3 = 1/3(0.785)(14 ft)(14 ft)(12.5 ft) = 641.08 ft^3

Next, find the volume of the cylindrical part of the tank.

Volume = (0.785)(Diameter, ft)^2(Depth) = (0.785)(14 ft)(14 ft)(24 ft) = 3,692.64 ft^3

Lastly, add the two volumes for the answer.

Total volume, ft^3 = 641.08 ft^3 + 3,692.64 ft^3 = 4,333.72 ft^3, round to **4,300 ft^3**

33. Answer: **b.** 2.5 ft

Equation: Q (Flow) = (Area)(Velocity)

20.3 cfs = (5.8 ft)(x ft, Depth)(1.4 fps)

Now solve for depth:

x ft, Depth = $\dfrac{20.3 \text{ cfs}}{(5.8 \text{ ft})(1.4 \text{ fps})}$ = **2.5 ft**

34. Answer: **d.** 46.3 psi

Know: There are 0.433 psi/ft or 2.31 ft/psi.

Use 0.433 psi/ft: psi = (Depth, ft)(0.433 psi/ft)

psi = (107 ft)(0.433 psi/ft) = 46.331 psi, round to **46.3 psi**

ANSWERS TO MATH QUESTIONS FOR WATER TREATMENT OPERATORS LEVELS I & II

35. Answer: **b.** 18,800,000 gal/day

    Equation: Flow, gal/day = (Basin volume, gal)(24 hr/day) ÷ (Detention time, hr)

    First, convert mil gal of the sedimentation basin to gallons.

    Number of gal = 2.18 mil gal × 1,000,000 = 2,180,000 gal

    Substitute known values and solve.

    Flow, gal/day = $\frac{(2,180,000 \text{ gal})(24 \text{ hr/day})}{2.79 \text{ hr}}$ = 18,752,688 gal/day,

    round to **18,800,000 gal/day**

36. Answer: **c.** 29.5 hr

    Equation: Operating time, hr = Treated water, gal ÷ Flow rate, gph

    First, convert flow rate from gpm to gph.

    (245 gpm)(60 min/hr) = 14,700 gph

    Operating time, hr = 434,000 gal ÷ 14,700 gph
    = 29.52 hr, round to **29.5 hr**

37. Answer: **a.** 2.9 mL/min of ZOP

    First, calculate the lb/gal of ZOP.

    (1.63 S.G.)(8.34 lb/gal) = 13.5942 lb/gal

    Next, find the ZOP usage in lb/day using the "pounds" formula.

    ZOP, lb/day = (12.1 mgd)(0.15 mg/L)(8.34 lb/gal) = 15.1371 lb/day

    Then convert lb/day ZOP to gal/day.

    15.1371 lb/day ÷ 13.5942 lb/gal = 1.11 gal/day ZOP

    Lastly, convert gal/day ZOP to mL/min.

    (1.11 gal/day × 3,785 mL/gal) ÷ 1,440 = **2.9 mL/min of ZOP**

38. Answer: **c.** 1,330 mL/min

    First, find the lb/gal for alum.

    Equation: lb/gal Alum = (Specific gravity)(8.34 lb/gal)

    Alum, lb/gal = (1.31)(8.34 lb/gal) = 10.9254 lb/gal

    Equation: Dosage, mg/L = (mL/min)(1,440 min/day)(Alum, lb/gal)(% Purity)

    (mgd)(8.34 lb/gal)(3,785 mL/gal)

Rearrange to solve for unknown (mL/min).

$$\text{Alum, mL/min} = \frac{(\text{Dosage, mg/L})(\text{mgd})(8.34 \text{ lb/gal})(3{,}785 \text{ mL/gal})}{(\text{Alum, lb/gal})(\% \text{ Purity})(1{,}440 \text{ mL/min})}$$

$$= \frac{(10.4 \text{ mg/L})(30.9 \text{ mgd})(8.34 \text{ lb/gal})(3{,}785 \text{ mL/gal})}{(10.9254 \text{ lb/gal})(48.4\%/100\% \text{ Purity})(1{,}440 \text{ mL/min})}$$

$$= \mathbf{1{,}330 \text{ mL/min alum}}$$

39. Answer: **c.** 50.0 mg/L

First, find the number of feet in 1.62 miles.

Number of ft = (5,280 ft/mile)(1.62 miles) = 8,553.6 ft

Next, find the volume in cubic feet for the pipe.

Equation: Volume, ft^3 = (0.785)(Diameter, ft)^2(Length, ft)

Volume, ft^3 = (0.785)(1.50 ft)(1.50 ft)(8,553.6 ft) = 15,108 ft^3

Then, determine the number of gallons.

Number of gal = (15,108 ft^3)(7.48 gal/ft^3) = 113,008 gal

Convert the number of gallons to mil gal.

mil gal = 113,008 gal ÷ 1,000,000 = 0.113 mil gal

Lastly, find the dosage in mg/L.

Dosage, mg/L = lb of chlorine ÷ (mil gal × 8.34 lb/gal)
 = 47.2 lb ÷ (0.113 mil gal × 8.34 lb/gal)
 = 47.2 ÷ 0.94
 = 50.21 mg/L, round to **50.0 mg/L**

40. Answer: **a.** 1.40 SG

Know: Water has a density of 62.4 lb/ft^3.

Divide the density of the unknown by the density of water.

Equation: SG = Density of substance ÷ Density of water

SG of unknown substance = 87.6 lb/ft^3 ÷ 62.4 lb/ft^3 = **1.40 SG**

Answers to Math Questions for Water Treatment Operators Levels III & IV

1. Answer: **d.** 457 mL/min

   Equation: $\dfrac{655 \text{ mL/min}}{26.5 \text{ cfs}} = \dfrac{x, \text{ mL/min}}{18.5 \text{ cfs}}$

   Solve for x:
   $x, \text{mL/min} = \dfrac{(655 \text{ mL/min})(18.5 \text{ cfs})}{26.5 \text{ cfs}} = 457.26 \text{ mL/min},$
   round to **457 mL/min**

2. Answer: **a.** 79.7%

   First, convert mgd to gpm or vice versa.

   Know: 1 day = 1,440 min

   Number of gpm = (mgd × 1,000,000) ÷ 1,440 min

   Number of gpm = (1.03 mgd × 1,000,000) ÷ 1,440 min = 715 gpm

   Now calculate the percent recovery.

   Formula is: Percent recovery = (Product Flow × 100 %) ÷ Feed Flow

   Percent Recovery = (570 gpm × 100 %) ÷ 715 gpm = **79.7%**

3. Answer: **b.** 37 lb/day, F

   First convert gpm to mgd.

   mgd = [(1,750 gpm)(1,440 min/day)] ÷ (1,000,000) = 2.52 mgd

   Next, determine the fluoride (F) required.

   F, required = F desired − F raw water: F, required = 1.20 mg/L − 0.15 mg/L = 1.05 mg/L F

   Equation: Fluoride feed rate $= \dfrac{(\text{Dosage, mg/L})(\text{mgd})(8.34 \text{ lb/gal})}{(\text{Percent F ion})(\text{Chemical purity})}$

   $\dfrac{(1.05 \text{ mg/L F})(2.52 \text{ mgd})(8.34 \text{ lb/gal})}{(60.6\%/100\%)(98.1\%/100\%)} = 37.12$ lb/day, round to **37 lb/day, F**

4. Answer: **b.** 39.6 hrs

   Equation: Operating time, hrs = Treated water, gal ÷ Flow rate, gph

   First, convert flow rate from gpm to gph.

   (125 gpm)(60 min/hr) = 7,500 gph

   Operating time, hrs = 297,000 gal ÷ 7,500 gph = **39.6 hrs**

5. Answer: **a.** 375 ft²/filter

   First, calculate the number of gpm.

   (32.7 cfs)(7.48 gal/ft³)(60 s/min) = 14,675.76 gpm

   Filtration rate, gpm/ft² = Flow rate, gpm ÷ Filter surface area, ft²

238 WATER OPERATOR CERTIFICATION STUDY GUIDE

Rearrange the formula:

Filter surface area, ft² = Flow rate, gpm ÷ Filtration rate, gpm/ft²

Filter surface area, ft² = 14,675.76 gpm ÷ 4.89 gpm/ft²
= 3,001.178 ft² (area for all 8 filters)

Filter area for each filter = 3,001.178 ft² ÷ 8 filters = 375.15 ft², round to **375 ft²/filter**

6. Answer: **b.** 27,000 lb/yr of Fe removed

First, calculate the iron removal in ppm.

(0.52 mg/L)(84%/100%) = 0.52 mg/L(0.84) = 0.4368 mg/L

Determine the amount of water in mil gal produced for the year (yr).

(20.2 mgd)(365 days/year) = 7,373 mil gal/year

Next, using the "pounds" equation, solve for the number of lb/year (yr).

Fe, lbs/yr = (mil gal/yr)(Dosage, mg/L)(8.34 lb/gal)

(7,373 mil gal/yr)(0.4368 mg/L)(8.34 lb/gal) = 26,859 lb/yr, round to **27,000 lb/yr**

7. Answer: **b.** 89% Mineral rejection

Equation:

Percent mineral rejection = [1 − (Product TDS, mg/L ÷ Feed TDS, mg/L)] × 100%

Percent mineral rejection = [1 − (135 mg/L ÷ 1,230 mg/L)](100%)

Percent mineral rejection = [1 − (0.1098 mg/L)](100%)

Percent mineral rejection = [0.8902](100%) = 89.02%, round to
89% mineral rejection

8. Answer: **d.** 2.2 Log removal

First determine % removal.

% Removal = [(In − Out)(100%)] ÷ In

% Removal = [(295 − 2.0)(100%)] ÷ 295 = 99.322%

Next, change percent to decimal form for 99.322%.

100% = 1.0 and 99.322% = 0.99322

Then, calculate the log removal.

Equation: Log Removal = $(\log_{10})(-)(1.0 - \%$ Removed in decimal form)

Log Removal = $(\log_{10})(-)(1.0 - 0.99322)$ =

$(\log_{10})(-)(0.00678) = (-)(-2.17) = 2.17$ log, round to **2.2 log removal**

9. Answer: **c.** 36.75 grams

Equation: Normality (N) = (number of gram-equivalents of solute) ÷ (number of liters of solution)

0.25 N = number of gr-eq of solute ÷ 4.5 Liters of solution
= (0.25 N)(4.5 Liters of solution)
= 1.125 gr-eq required

The formula weight for H_3PO_4 = 1/3 the gram equivalent weight because 3 hydrogen atoms (H+) combine with 1 PO_4^{-3}. Therefore, divide the gram formula weight 98.00 by 3.

98.00 gram formula weight/3 = 32.67 grams per equivalent weight.

Number of grams = (# of gr-eq)(# g/gr-eq)= (1.125 gr-eq)(32.67 g/gr-eq) = **36.75 grams**

10. Answer: **c.** 18.6 mL polymer

First, find the number of lb/gal of polymer.

Polymer, lb/gal = (SG)(8.34 lb/gal) = (1.27)(8.34 lb/gal) = 10.5918 lb/gal

Next, determine the number of grams/mL.

$$\text{Polymer, grams/mL} = \frac{(10.5918 \text{ lb/gal})(84.5\% \text{ Purity})(454 \text{ grams/lb})}{(3,785 \text{ mL/gal})(100\%)}$$

Polymer, grams/mL = 1.0735 grams/mL

Convert grams/mL to milligrams/mL.

Number of mg/mL = (1.0735 grams/mL)(1,000 mg/g) = 1,073.5 mg/mL

Next, convert mL to liters by multiplying by 1,000 mL/L.

Number of mg/L = (1,073.5 mg/mL)(1,000 mL/L) = 1,073,500 mg/L

Lastly, determine the number of mL required. Equation: $C_1V_1 = C_2V_2$

(1,073,500 mg/L)(x, mL) = (20,000 mg/Liter)(1,000 mL)

$$x, \text{mL} = \frac{(20,000 \text{ mg/Liter})(1,000 \text{ mL})}{1,073,500 \text{ mg/mL}} = 18.63 \text{ mL},$$

round to **18.6 mL Polymer**

11. Answer: **d.** 8.85 hr

Write the equation with units asked for in question:

Detention time, hrs = (Volume, gal) ÷ (Flow rate, gph)

Next, determine the capacity in gallons for each basin by converting from mil gal to gallons for the clearwell. Then add all basins for total volume in gallons.

Volume, gal floc basins = (45.0 ft)(10.0 ft)(11.0 ft)(7.48 gal/ft³)(4 basins)
 = 148,104 gal

Volume of sedimentation basin = (285 ft)(65.0 ft)(11.4 ft)(7.48 gal/ft³)
 = 1,579,664 gal

Volume of filters = (35.0 ft)(28.0 ft)(12.3 ft)(7.48 gal/ft3)(8 filters)
 = 721,311 gal

Volume of clearwell = (2.05 mil gal)(1,000,000/1 M) = 2,050,000 gal

Add the volumes together.

148,104 + 1,579,664 + 721,311 + 2,050,000 = 4,499,079 gal

Then, convert the flow rate of 12.2 mgd to gph.

$$\text{gph} = \frac{(12.2 \text{ mgd})(1,000,000 \text{ gal})(1 \text{ day})}{(1 \text{ M})(24 \text{ hr})} = 508,333 \text{ gph}$$

Substitute known values and solve.

Detention time, hr = 4,499,079 gal ÷ 508,333 gph = **8.85 hr**

12. Answer: **a.** 0.86 inactivation ratio

First determine the CT table value (Appendix B).

The CT table value is found by going to a 1.5-log removal table, finding the chart with a pH of 7.6, going down the left column and finding a temperature of 12°C, and then going over to 0.60 mg/L free chlorine residual. This number that intersects the temperature and the chlorine residual is the CT table value. In this case the CT table value is 57.80.

Next, calculate the CT.

CT calculated = (Chlorine concentration, mg/L)(Time, min)

CT calculated = (0.60, mg/L)(83 min) = 49.8 is CT calculated

Next, calculate the inactivation ratio.

Inactivation Ratio = Calculated CT Value ÷ CT table value

Inactivation Ratio = 49.8 ÷ 57.8 = **0.86 inactivation ratio**

Because the inactivation ratio value is less than 1.0, this system does not meet the CT criteria and is not in compliance.

13. Answer: **d.** 163 mL/min, H_2SiF_6

    Solution: F required = F desired − F in raw water

    F required = 1.20 mg/L − 0.20 mg/L = 1.00 mg/L

    $$\frac{(\% \text{ Solution})(10{,}000 \text{ mg/L})}{1\%} = \# \text{ of mg/L}$$

    (20.5% solution of H_2SiF_6)(10,000 mg/L per 1%) = 205,000 mg/L

    Equation: (F, mg/L)(mgd)(8.34 lb/gal) = (Percent, mg/L)(mgd)(Solution, lb/gal)(% F ion/100%)

    (1.0 mg/L F)(11.8 mgd)(8.34 lb/gal)
    = (205,000 mg/L)(x mgd Flow)(9.8 lb/gal)(79.0%/100% F)

    Solve for x: x mgd Flow = $\dfrac{(0.20 \text{ mg/L F})(11.8 \text{ mgd})(8.34 \text{ lb/gal})}{(205{,}000 \text{ mg/L})(9.8 \text{ lb/gal})(79.0\%/100\% \text{ F})}$

    = 0.000062 mgd

    Convert mgd to gallons:
    (0.000062 mgd)(1,000,000) = 62 gal/day H_2SiF_6

    Then convert gal/day to mL/min.

    H_2SiF_6, mL/min = (62 gal/day)(3,785 mL/gal)(1 day/1,440 min)
    = 162.98 mL/min, round to **163 mL/min, H_2SiF_6**

14. Answer: **a.** 190 mg/L, $Ca(OH)_2$

    Equation: Hydrated Lime Feed, mg/L = [(A + B + C + D)(1.15 - excess)] ÷ (Lime purity)

    Where A is CO_2 in source water: A = (mg/L as CO_2)(74.1/44): where 74.1 = molecular weight (MW) of $Ca(OH)_2$. Substitute known values and solve:

    A = (13 mg/L)(74.1/44) = 21.89 mg/L

    Where: B is Bicarbonate (total) Alkalinity removed in softening:

    B = (mg/L as $CaCO_3$ removed)(74.1/100.1): where 100.1 = MW as $CaCO_3$.

    B = (165 mg/L − 39 mg/L)(74.1/100.1) = 93.27 mg/L

    Where: C is Hydroxide Alkalinity in softened effluent:
    C = (mg/L as $CaCO_3$)(74.1/100.1).

    In this case there is no Hydroxide Alkalinity, therefore C = 0

    Where: D is Magnesium removed in softening: D = (mg/L as Mg^{2+})(74.1/24.3): where 24.3 = MW of Mg^{2+}

$D = (21 \text{ mg/L} - 7.8 \text{ mg/L})(74.1/24.3) = 40.25 \text{ mg/L}$

Hydrated Lime Feed, mg/L = $[(21.89 \text{ mg/L} + 93.27 \text{ mg/L} + 0 + 40.25 \text{ mg/L})(1.15)]$
$\div (92\%/100\%) = 194.26$ mg/L, round to **190 mg/L, Ca(OH)$_2$**

15. Answer: **b.** 10.1 gpd/ft^2

    Equation: Flow, gpd/ft^2
    $$= \frac{(\text{Water flux, gm/cm}^2/\text{s})(2.54 \text{ cm/in.})^2(12 \text{ in./ft})^2(60 \text{ s/min})(60 \text{ min/hr})(24 \text{ hr/d})}{(1{,}000 \text{ gm/L})(3.785 \text{ L/gal})}$$

    First convert the water flux to decimal form: $4.75 \times 10^{-4} = 0.000475$

    Substitute water flux in decimal form and solve.

    Flow, gpd/ft^2
    $$= \frac{(0.000475 \text{ gm/cm}^2/\text{s})(2.54 \text{ cm/in.})^2(12 \text{ in./ft})^2(60 \text{ s/min})(60 \text{ min/hr})(24 \text{ hr/d})}{(1{,}000 \text{ gm/L})(3.785 \text{ L/gal})}$$

    Flow, gpd/ft^2
    $$= \frac{(0.000475 \text{ gm/cm}^2/\text{s})(6.4516 \text{ cm}^2/\text{in.}^2)(144 \text{ in.}^2/\text{ft}^2)(86{,}400 \text{ s/d})}{3{,}785 \text{ gm/gal}}$$

    Flow, gpd/ft^2 = $\dfrac{38127.407616 \text{ gm/d-ft}^2}{3{,}785 \text{ gm/gal}}$ = 10.07 gpd/ft^2,

    round to **10.1 gpd/ft^2**

16. Answer: **b.** 1.1 inactivation ratio, plant is in compliance

    First calculate the CT for each.

    Filtration CT = (12 min)(0.45 mg/L) = 5.4

    Piping CT = (4.5 min)(0.40 mg/L) = 1.8

    Clearwell CT = (51 min)(1.20 mg/L) = 61.2

    Extrapolate CT values from table in Appendix A.

    Filtration 0.4 mg/L chlorine residual = 47.1 and 0.6 mg/L chlorine residual = 48.7 from tables. Use these because 0.45 mg/L falls between these two values

    The difference between these two values is 0.2 mg/L. The difference between 0.45 mg/L and 0.40 mg/L is 0.05 mg/L.

    Thus (0.05 mg/L / 0.2 mg/L)(100%) = (0.25)(100%) = 25%

    Extrapolate equation for this problem:

    CT value = $\dfrac{(48.7 - 47.1)(25\%)}{100\%} + 47.1 = 0.4 + 47.1 = 47.5$ CT value

    Next, calculate the inactivation ratio (IR).

    Equation: IR = Calculated CT ÷ CT Value, Table

    Filtration IR = (5.4 Calc., CT) ÷ (47.5, CT Value, Table) = 0.114

    Piping IR = (1.8 Calc., CT) ÷ (47.1, CT Value, Table) = 0.038

    Clearwell IR = (61.2 Calc., CT) ÷ (62.3 CT Value, Table) = 0.982

Lastly, add all three inactivation ratios.

IR, total = 0.114 + 0.038 + 0.982 = 1.134, round to **1.1 inactivation ratio**.

Because the inactivation ratio is greater than 1.0, the water plant is in compliance with the required CT criteria.

17. Answer: **b.** 2.19 mg/L polymer aid

    First, determine the number of ml/min for the polymer aid (Polymer):

    (106 mL) ÷ 5 min = 21.2 mL/min

    Next, calculate the lb/gal for the polymer = (SG)(8.34 lb/gal)

    Polymer, lb/gal = (1.26)(8.34 lbs/gal) = 10.5084 lb/gal

    Then, find the number of mgd:

    (3,225 gpm × 1,440 min/day) ÷ 1,000,000 = 4.644 mgd

    Or, use the dosage equation with conversions added for convenience (dosage/conversion equation):

    $$\text{Polymer, mg/L} = \frac{(\text{mL/min})(1,440 \text{ min/day})(\text{Polymer, lb/gal})}{(3,785 \text{ mL/gal})(\text{mgd})(8.34 \text{ lb/gal})}$$

    $$\text{Polymer, mg/L} = \frac{(21.2)(1,440)(10.5084)}{(3,785)(4.644)(8.34)} = 2.188 \text{ mg/L, round to } \mathbf{2.19 \text{ mg/L}}$$

18. Answer: **c.** 5,650 ft³

    First find the volume of the cone in cubic feet.

    Volume, ft³ = 1/3(0.785)(Diameter, ft)²(Depth, ft)

    Volume, ft³ = 1/3(0.785)(15 ft)(15 ft)(12 ft) = 706.5 ft³

    Next find the volume of the cylindrical part of the tank.

    Volume = (0.785)(Diameter)²(Depth) = (0.785)(15 ft)(15 ft)(28 ft)
    = 4,945.5 ft³

    Lastly, add the two volumes for the answer.

    Total volume, ft³ = 706.5 ft³ + 4,945.5 ft³ = 5,652 ft³, round **to 5,650 ft³**

19. Answer: **b.** 4,190 mil gal/year

    First, find the number of inches collected from the rainfall.

    (22.6 inches)(6.75% ÷ 100%) = 1.5255 inches

    Convert the number of inches of rain collected to feet:

    Collected rainwater, ft = (1.5255 inches)(1 ft/12 in.) = 0.127125 ft

    Then, calculate the number of gallons in 1 acre that is covered by 0.127125 ft of water.

    (43,560 ft²/acre)(7.48 gal/ft³)(0.127125 ft) = 41,421 gal/acre

    Convert this to mil gal: (41,421 gal/acre ÷ 1,000,000) = 0.04142 mil gal/acre

    Next, find the number of acres in 158 square miles:

    Number of acres = (640 acres/mile)(158 miles) = 101,120 acres

    Then, calculate the number of mil gal collected from the watershed.

    (0.04142 mil gal/acre)(101,120 acres) = 4,188.39 mil gal,
    round to **4,190 mil gal/year**

20. Answer: **c.** 211,900 gal

    First, convert the capacity in kilograins to grains:

    (25.5 kilograins/ft³)(1,000 grains/kilograins) = 25,500 grains/ft³

    Next, find the capacity of the unit in grains.

    (118 ft³)(25,500 grains/ft³) = 3,009,000 grains

    Equation: Water treatment capacity, gal
    = (Exchange capacity in grains) ÷ Hardness, gpg
    = (3,009,000 grains) ÷ 14.2 gpg = 211,901 gal, round to **211,900 gal**

21. Answer: **c.** 10 g of CaO

    Solution: Wt-volume (vol.) percent
    = [(Wt. of solute, g)(100%)] ÷ (Volume of solution, mL)

    First, determine volume in mL. Know: 1 Liter = 1,000 mL

    1.0% = (Wt of solute, g)(100%) ÷ 1,000 mL

    Solve for weight of solute, g, by multiplying both side of the equation by 1,000 mL and dividing both sides by 100. Thus:

    Wt of solute, g = [(1.0%)(1,000 mL)] ÷ 100 = **10 g of CaO**

22. Answer: **d.** 99.7%

    Equation: Percent ntu removal = [(Influent ntu − Effluent ntu)(100%)]
    ÷ (Influent ntu)

    % ntu removal = (5.45 ntu − 0.018 ntu)(100%) ÷ 5.45 ntu
    = (5.432 ntu)(100%) ÷ 5.45 ntu = **99.7%**

23. Answer: **b.** 5.26 psi

    Equation: psi = (Depth, ft) ÷ 2.31 ft/psi

    psi = 8.95 ft ÷ 2.31 ft/psi = 3.874 psi

    Next, find the specific gravity (SG) of the alum.

    Alum, SG = 11.32 lb/gal ÷ 8.34 lb/gal = 1.357 SG of Alum

    Then, multiply the psi by the alum's SG to determine the psi.

    psi = (3.874 psi)(1.357) = 5.257 psi, round to **5.26 psi**

24. Answer: **b.** 680 min

    First, determine the volume in gallons for the clarifier.

    Equation: Volume, gal = (0.785)(Diameter, ft)²(Depth, ft)(7.48 gal/ft³)

    Volume, gal = (0.785)(152 ft)(152 ft)(14.8 ft)(7.48 gal/ft³) = 2,007,799 gal

    Then convert mgd to gallons per minute as detention time is asked for in minutes.

    Know: 1 day = 1,440 min

    (4.25 mgd × 1,000,000) ÷ 1,440 min = 2,951.4 gpm

    Equation: Detention time, min = (Volume, gal) ÷ Flow rate, gpm

    Detention time, min = 2,007,799 gal ÷ 2,951.4 gpm = 680.29 min, round to **680 min**

25. Answer: **c.** 6.87 mg/L

    First, determine the number of mL/min for the polymer:

    191 mL ÷ 3 min = 63.667 mL/min

    Next, calculate the lb/gal for the polymer:

    Polymer, lb/gal = (SG)(8.34 lb/gal) = (1.34)(8.34 lb/gal) = 11.1756 lb/gal

    Then, find the number of mgd:

    mgd = (3,280 gpm × 1,440) ÷ 1,000,000 = 4.7232 mgd

    Use the dosage equation with conversions added for convenience (dosage/conversion equation):

    $$\text{Polymer, mg/L} = \frac{(\text{mL/min})(1{,}440 \text{ min/day})(\text{lb/gal, Polymer})}{(3{,}785 \text{ mL/gal})(\text{mgd})(8.34 \text{ lb/gal})}$$

    $$\text{Polymer, mg/L} = \frac{(63.667 \text{ mL/min})(1{,}440 \text{ min/day})(11.1756 \text{ lb/gal})}{(3{,}785 \text{ mL/gal})(4.7232 \text{ mgd})(8.34 \text{ lb/gal})}$$

    = **6.87 mg/L**

26. Answer: **a.** 11.1 microliters

    Equation:

    Polymer, microliters = Polymer desired, mg/L ÷ [(SG)(Percent conc./100%)]

    Polymer, microliters = 6 mg/L ÷ [(1.35)(80%/100%)] = 5.556 microliters

    However, because 2-liter jars are being used, the result has to be multiplied by 2.

    Polymer, microliters = (5.556 microliters)(2) = **11.1 microliters**

27. Answer: **b.** 0.525 lb/min

    Equation: Soda ash, lb/day = (mgd)(Dosage, mg/L)(8.34 lb/gal)

    Soda ash, lb/day = (12.5 mgd)(7.25 mg/L)(8.34 lb/gal) = 755.81 lb/day

    Then convert to lb/min:

    755.81 lb/day ÷ 1,440 min/day = **0.525 lb/min**

28. Answer: **c.** 120 lb/day

    First, determine the number of mgd being treated:

    mgd = (2,575 gpm × 1,440) ÷ 1,000,000 = 3.708 mgd

    Calcium hypochlorite, lb/day = [(mgd)(Dosage, mg/L)(8.34 lb/gal)] ÷ (Percent available chlorine/100%)

    Substitute known values and solve:

    [(3.708 mgd)(2.5 mg/L)(8.34 lb/gal)] ÷ 65% Available chlorine/100%
      = 118.9 lb/day, round to **120 lb/day**

29. Answer: **d.** 78.1 mL/min

    Find the number of pounds per day of polymer required by using the "pounds" equation.

    Polymer, lb/day = (mgd)(Dosage, mg/L)(8.34 lb/gal)

    Polymer, lb/day = (10.0 mgd)(3.95 mg/L)(8.34 lb/gal) = 329.43 lb/day

    Next, determine the lb/gal of the polymer solution.

    Polymer, lb/gal = (SG)(8.34 lb/gal) = (1.33)(8.34 lb/gal) = 11.0922 lb/gal

Convert the number of lb/day to number of gal/day.

Polymer, gal/day = 329.43 lb/day ÷ 11.0922 lb/gal = 29.699 gal/day

Then convert gal/day to mL/min.

Polymer, mL/min = [(29.699 gal/day)(3,785 mL/gal)] ÷ (1,440 min/day)
= **78.1 mL/min**

30. Answer: **c.** 5.36 gpm/ft^2

    First, convert the number of cfs (ft^3/s) to gpm.

    (21.5 ft^3/s)(7.48 gal/ft^3)(60 sec/min) = 9,649.2 gpm

    Next, determine the combined filtration area of the four filters.

    Total filter area = (4 filters)(450 ft) = 1,800 ft

    Equation: Filtration rate = Flow rate, gpm ÷ Filter surface area, ft^2

    Filtration rate = 9,649.2 gpm ÷ 1,800 ft^2 = **5.36 gpm/ft^2**

31. Answer: **a.** 332 mg/L

    Equation: Hardness, mg/L = [(Hardness, gpg)(17.12 mg/L)] ÷ (1 gpg)

    Hardness, mg/L = (19.4 gpg)(17.12 mg/L) ÷ (1 gpg) = 332.128 mg/L,
    round to **332 mg/L**

32. Answer: **b.** 437,000 gal

    First, determine the hardness of the water in gpg.

    (347 mg/L) ÷ (17.12 mg/L per grains/gal) = 20.269 gpg

    Equation: Water treatment capacity, gal = (Exchange capacity in grains)
    ÷ (Hardness, gpg)

    Water treatment capacity, gal = (8,850,000 grains) ÷ 20.269 gpg
    = 436,627 gal, round to **437,000 gal**

33. Answer: **d.** 9.8 gpm/ft^2

    Equation: Backwash (BW) rate, gpm/ft^2 = BW pumping rate, gpm ÷ Filter area, ft^2

    First, convert the BW rate in cfs into gpm.

    gpm = (13.5 cfs)(7.48 gal/ft^2)(60 sec/min) = 6,058.8 gpm

    BW rate, gpm/ft^2 = 6,058.8 gpm ÷ 620 ft^2 = 9.77 gpm/ft^2, round to **9.8 gpm/ft^2**

34. Answer: **d.** 2,140 gpm

    First, find the water production during the 12-hr interval.

    Gallons of water treated in 12-hr interval = (2.95 mgd)(1,000,000)
    (12 hr)/24 hr

    Gallons of water treated in 12-hr interval = 1,475,000 gal

    Next, find the gallons contained in the 4.25-ft drop in water level.

    Equation: Volume, tank = (0.785)(Diameter, ft)^2(Height, ft)

    Volume of 4.25 ft in 50.0-ft diameter tank
    = (0.785)(50.0 ft)(50.0 ft)(4.25 ft)(7.48 gal/ft^3) = 62,388 gal

    Production plus the loss in level is the amount the discharge pump had to send into the distribution system, but first find the number of minutes in 12 hr.
    (12 hr)(60 min/hr) = 720 min

246   WATER OPERATOR CERTIFICATION STUDY GUIDE

Then determine total gallons discharge pumps moved.

Total gal discharge pumps moved in 12 hr = 1,475,000 gal + 62,388 gal
= 1,537,388 gal

Lastly, divide the number of gallons the discharge pumps moved by the time in minutes.

Discharge pumps, gpm = 1,537,388 gal ÷ 720 min = 2,135 gpm, round to **2,140 gpm**

35. Answer: **c.** 1.5 gpm/ft^2

First, convert the number of ft^3/sec to gpm.

(16.5 ft^3/sec)(7.48 gal/ft^3)(60 sec/min) = 7,405.2 gpm

Next, calculate the total surface area for all four filters:

(480 ft^2)(10 filters) = 4,800 ft^2

Equation: Filtration rate = Flow rate, gpm ÷ Filter surface area, ft^2

Filtration rate = 7,405.2 gpm ÷ 4,800 ft^2 = 1.54275 gpm/ft^2, round to **1.5 gpm/ft^2**

36. Answer: **c.** 8,600 lb/yr

Equation: Algae removal, lb/yr = (mgd)(365 days/yr)(8.34 lb/gal)(Dosage, mg/L)(Percent removal efficiency)

Algae removal, lb/yr = (14.1 mgd)(365 days/yr)(8.34 lb/gal)(0.21 mg/L)(0.957)

Algae removal, lb/yr = 8,625.996 lb/yr, round to **8,600 lb/yr**

Water Distribution Operator Requirements

	Level I	Level II	Level III	Level IV
System Information/Components	9%	9%	9%	9%
Monitor, Evaluate, and Adjust Disinfection	11%	11%	10%	10%
Laboratory Analysis	21%	21%	21%	20%
Install Equipment	5%	5%	5%	5%
Operate Equipment	18%	18%	18%	18%
Perform Maintenance	20%	20%	20%	21%
Perform Security, Safety, and Administrative Procedures	16%	16%	17%	17%

Distribution System Information/Components

System Information/Components	Level I	Level II	Level III	Level IV
Assess system demand	Application	Application	Analysis	Analysis
Install joint restraints	Application	Application	Application	Analysis
Install shoring	Comprehension	Application	Application	Analysis
Install thrust blocks	Comprehension	Application	Application	Analysis
Layout system	N/A	N/A	Application	Application
Map system	Comprehension	Comprehension	Comprehension	Comprehension
Perform pressure readings	Application	Application	Application	Analysis
Preparedness contingency/Contingency plan	Comprehension	Comprehension	Application	Analysis
Read blueprints, readings, and maps	Application	Application	Application	Analysis
Select materials	Comprehension	Comprehension	Application	Application
Select type of pipes	Comprehension	Comprehension	Application	Analysis
Size mains	Comprehension	Comprehension	Application	Analysis
Write plans	Comprehension	Application	Application	Analysis

Required Capabilities

Knowledge of:

- Approved backflow methods and devices
- Biological science
- Blueprint readings
- Building codes
- Corrosion control process (including cathodic protection)
- Fire flow requirements
- Function of recordkeeping system
- General hydraulic principles
- Hydrology
- Local codes and ordinances
- Measuring instruments
- Mechanical drafting
- Operation and maintenance practices
- Pipe fittings and joining methods
- Piping material, type, and size
- Potential causes of disasters in facility
- Potential impact of disasters in facility
- Regulations
- Standards
- Watershed management

Ability to:

- Adjust equipment
- Assess likelihood of disaster occurring
- Generate a written safety program
- Generate capital plans
- Generate long- and short-term plans
- Interpret data
- Organize information
- Perform distribution math
- Perform impact assessments
- Perform physical measurements
- Record information
- Write policies and procedures
- Review reports

System Information/Components

Sample Questions for Level I

Answers on page 339

1. Head is measured in

    a. absolute pressure.
    b. gauge pressure.
    c. feet.
    d. foot-pounds.

2. A plat is

    a. a map.
    b. a corrosion point on a pipe.
    c. organelle found in some protozoans.
    d. a highly corrosive soil type.

3. Which type of valve will prevent the collapse of a pipe?

    a. Pressure-relief valve
    b. Needle valve
    c. Pinch valve
    d. Air-and-vacuum relief valve

4. The highest degree of protection for the exterior of a coated steel pipe is

    a. cathodic protection.
    b. bituminous materials.
    c. plastic coatings.
    d. polyethylene tapes.

5. At which time of day is the age of the water stored in the distribution system the highest?

    a. Early morning
    b. Late morning
    c. Early afternoon
    d. Late evening

Sample Questions for Level II

Answers on page 339

1. The amount of liquid that can be raised vertically by a given pressure is called

   a. pressure head.
   b. total head.
   c. velocity head.
   d. pump head.

2. The amount of energy in feet that a pump supplies to a fluid is called

   a. velocity head.
   b. pump head.
   c. total head.
   d. pressure head.

3. The C-value is a measure of a pipe's wall

   a. smoothness.
   b. smoothness giving even flow.
   c. smoothness that retards turbulent flow.
   d. roughness that retards flow due to friction.

4. Which one of the following is a type of joint for ductile iron piping?

   a. Expansion joint
   b. Push-on joint
   c. Bell and spigot with rubber o-ring
   d. Rubber gasket joint

5. The correct protective methods for backflow-prevention devices in order of decreasing effectiveness are

   a. air gap, VB, RPZ, and DCVA.
   b. air gap, VB, DCVA, and RPZ.
   c. air gap, RPZ, VB, and DCVA.
   d. air gap, RPZ, DCVA, and VB.

Sample Questions for Level III

Answers on page 340

1. As the total head on a system increases, the volume that a centrifugal pump delivers is reduced

   a. directly.
   b. indirectly.
   c. proportionately.
   d. disproportionately.

2. Which type of distribution system configuration surrounds the distribution area with large-diameter mains?

   a. Grid system
   b. Dendritic system
   c. Arterial-loop system
   d. Tree system

3. Which type of distribution system configuration has interconnected mains?

   a. Grid system
   b. Dendritic system
   c. Arterial-loop system
   d. Tree system

4. In general, areas that require high fire flow capacity require the minimum static pressure to be

   a. 25 psi or greater.
   b. 30 psi or greater.
   c. 35 psi or greater.
   d. 40 psi or greater.

5. Which type of joint should be used for ductile iron piping in unstable soils?

   a. Flanged joint
   b. Restrained joint
   c. Flexible ball joint
   d. Push-on joint

Sample Questions for Level IV

Answers on page 340

1. The arrangement of pumps in parallel is typically used

   a. when head is not sufficient and more flow is required.
   b. to increase discharge head without increasing flow.
   c. to increase flow.
   d. to increase pump efficiency.

2. Which of the following is a disadvantage for using steel pipe?

   a. It has the potential for both internal and external corrosion
   b. It is difficult to fabricate but inexpensive to transport
   c. It is easy to install due to weight
   d. It is very inexpensive

3. Friction head loss can be calculated using tables based on

   a. Charles' equation.
   b. Graham's formula.
   c. the Hazen-Williams formula.
   d. Russell's calculations.

4. Water behind a dam and above a water treatment plant has energy by virtue of its elevation. This difference in elevation is called elevation head or

   a. kinetic energy.
   b. velocity head.
   c. potential energy.
   d. pressure head.

5. Although velocity head can usually be ignored, it should be considered when it is ____ to ____ percent of the pressure head.

   a. 1; 2
   b. 3; 4
   c. 5; 8
   d. 5; 10

Monitor, Evaluate, and Adjust Disinfection

Monitor, Evaluate, and Adjust Disinfection	Level I	Level II	Level III	Level IV
Monitor Disinfection				
Calcium hypochlorite disinfection	Application	Application	Application	Application
Chlorine gas disinfection	Comprehension	Comprehension	Application	Analysis
Sodium hypochlorite disinfection	Application	Application	Application	Application
Calcium hypochlorite disinfection	Application	Application	Application	Application
Chlorine gas disinfection	Comprehension	Comprehension	Application	Analysis
Sodium hypochlorite disinfection	Application	Application	Application	Application
Evaluate Disinfection				
Calcium hypochlorite disinfection	Analysis	Analysis	Analysis	Analysis
Chlorine gas disinfection	Analysis	Analysis	Analysis	Analysis
Sodium hypochlorite disinfection	Analysis	Analysis	Analysis	Analysis
Adjust Disinfection				
Calcium hypochlorite disinfection	Analysis	Analysis	Analysis	Analysis
Chlorine gas disinfection	Analysis	Analysis	Analysis	Analysis
Sodium hypochlorite disinfection	Analysis	Analysis	Analysis	Analysis
Calcium hypochlorite disinfection	Analysis	Analysis	Analysis	Analysis
Inspect Source Water				
Identify and evaluate potential sources of source water contamination	Analysis	Analysis	Analysis	Analysis
Wells	Application	Application	Application	Application

Required Capabilities

Knowledge of:

- Proper chemical handling and storing
- Regulations
- Reporting requirements
- Safe Drinking Water Act (SDWA)
- Safety procedures
- Sampling requirements
- Testing instruments
- Watershed management
- 40 CFR 141 Subpart C: Monitoring and Analytical Requirements (turbidity, coliforms, organic contaminants, organic contaminants)
- 40 CFR 141 Subpart D: Reporting and Recordkeeping Requirements
- 40 CFR 141 Subpart F: Maximum Contaminant Level Goals and Maximum Residual Disinfectant Level Goals
- 40 CFR 141 Subpart G: National Primary Drinking Water Regulations: Maximum Contaminant Levels and Maximum Residual Disinfectant Levels
- 40 CFR 141 Subpart H: Filtration and Disinfection
- 40 CFR 141 Subpart L: Disinfectant Residuals, Disinfection Byproducts, and Disinfection Byproduct Precursors
- 40 CFR 141 Subpart P: Enhanced Filtration and Disinfection Systems Serving 10,000 or More People
- 40 CFR 141 Subpart T: Enhanced Filtration and Disinfection Systems Serving Fewer Than 10,000 People
- 40 CFR 141 Subpart U: Initial Distribution System Evaluations
- 40 CFR 141 Subpart V: Stage 2 Disinfection Byproducts Requirements

Ability to:

- Adjust equipment
- Adjust flow patterns
- Adjust system units
- Calibrate equipment
- Calibrate instruments
- Interpret data
- Perform distribution math
- Recognize normal and abnormal analytical results

Monitor, Evaluate, and Adjust Disinfection

Sample Questions for Level I

Answers on page 341

1. One of chlorine's advantages is that it
    a. is not influenced much by pH changes.
    b. does not produce chlorinated by-products.
    c. has a persistent residual.
    d. does not cause taste and odor problems.

2. Chlorine gas is _____ times heavier than air.
    a. 1.5
    b. 2.5
    c. 3.5
    d. 4.5

MONITOR, EVALUATE, AND ADJUST DISINFECTION   257

3. After a water storage tank has been chlorinated, which bacteriological test must prove negative before the tank is put back into service?

   a. Gram negative test
   b. HPC test
   c. Coliform test
   d. Chloramine test

4. Which is the best and most reliable method for finding a very small chlorine leak?

   a. Use soap on possible areas of leak and watch for bubbles
   b. Use a strong ammonia solution on a cloth swab and place it by the suspected leak
   c. Use methane gas and watch for a dark brown smoke
   d. Use a chlorine gas detector and place it next to any area suspected of leaking

5. The minimum required free chlorine residual anywhere in the distribution system is

   a. a detectable level.
   b. 0.2 mg/L.
   c. 0.4 mg/L.
   d. 0.5 mg/L.

Sample Questions for Level II

Answers on page 341

1. Sodium hypochlorite (NaOCl) solution is available with _____ available chlorine.

   a. 2 to 5%
   b. 5 to 20%
   c. 25 to 50%
   d. 50 to 70%

2. Booster chlorination is chlorine added

   a. in the coagulation mixing chamber.
   b. before the filters.
   c. at the clearwell.
   d. somewhere in the distribution system.

3. According to AWWA Standard C651, disinfection of water mains requires 24-hour exposure to which minimum free chlorine residual?

   a. 10 mg/L
   b. 25 mg/L
   c. 50 mg/L
   d. 100 mg/L

4. Which must be measured at least at the same time and at the same sampling points in the distribution system that total coliforms are sampled for a water system that only uses surface water?

   a. pH
   b. Langelier Index
   c. Residual disinfectant concentration
   d. Heterotrophic bacteria

5. Hypochlorous acid is

   a. a weak acid.
   b. a strong acid.
   c. easily dissociatable.
   d. hydrophobic.

Sample Questions for Level III

Answers on page 341

1. Which chemical oxidant would be most effective for controlling biological growth?

   a. Chloramines
   b. Chlorine
   c. Ozone
   d. Potassium permanganate

2. Which is the Maximum Residual Disinfectant Level for chlorine?

   a. 4.0 mg/L
   b. 4.5 mg/L
   c. 5.0 mg/L
   d. 6.0 mg/L

3. Which is the most common hypochlorinator problem?

   a. Clogged equipment
   b. Cracked pump head
   c. Corrosion
   d. Broken plunger

4. Which reaction will take longer with chlorine?

   a. Hydrogen sulfide
   b. Ammonia
   c. Organic material
   d. Calcium sulfide

5. If the continuous monitoring of the residual disinfectant concentration for water entering the distribution system fails, how often must grab samples be collected in lieu of continuous monitoring until the problem is repaired?

   a. Every hour
   b. Every 4 hours
   c. Every 8 hours
   d. Every 12 hours

Sample Questions for Level IV

Answers on page 342

1. Which organisms have the greatest resistance to chlorine?

   a. Viruses
   b. Bacteria
   c. Nematodes
   d. *Giardia* cysts

2. Which is the maximum residual for chlorine dioxide to avoid taste and odor complaints?

   a. 0.4 to 0.5 mg/L
   b. 0.9 to 1.0 mg/L
   c. 1.3 to 1.5 mg/L
   d. 1.9 to 2.0 mg/L

3. If secondary chlorine is added to a distribution system that uses chloramines, the chlorine may be added to a desired chlorine to ammonia-nitrogen ratio of

   a. 3.0 : 1 to 3.5 : 1
   b. 3.5 : 1 to 4.0 : 1
   c. 4.0 : 1 to 4.5 : 1
   d. 4.5 : 1 to 5.0 : 1

4. Gaseous chlorine dioxide is generated from

   a. oxygen + chlorine gas + heat (energy).
   b. aqueous chlorine + oxygen + heat (energy).
   c. sodium chlorite + chlorine at low pH values.
   d. hypochlorous acid + oxygen + organic catalyst.

5. Which is the most probable solution if sulfur bacteria are causing corrosion in the distribution system?

   a. Minimize pump activity
   b. Acidification and cleaning
   c. Optimize coagulation, flocculation, and filtration
   d. Routine use of disinfectant and penetrant

Laboratory Analysis

Laboratory Analysis	Level I	Level II	Level III	Level IV
Collect and Preserve Samples				
Chlorine demand	Application	Application	Analysis	Analysis
Chlorine residual	Application	Application	Analysis	Analysis
Coliforms	Application	Application	Analysis	Analysis
Lead/copper	Application	Application	Analysis	Analysis
Nitrate	Application	Application	Analysis	Analysis
Nitrite	Application	Application	Analysis	Analysis
pH	Application	Application	Analysis	Analysis
Radionuclides	Application	Application	Analysis	Analysis
Synthetic organic chemicals (SOC)	Application	Application	Analysis	Analysis
Temperature	Application	Application	Analysis	Analysis
Volatile organic chemicals (VOC)	Application	Application	Analysis	Analysis
Perform Laboratory Analysis				
Chlorine demand	Application	Application	Analysis	Analysis
Chlorine residual	Application	Application	Analysis	Analysis
pH	Application	Application	Analysis	Analysis
Temperature	Application	Application	Analysis	Analysis
Interpret Laboratory Analyses				
Chlorine demand	Application	Application	Analysis	Analysis
Chlorine residual	Application	Application	Analysis	Analysis
Coliforms	Application	Application	Analysis	Analysis
Hardness	Application	Application	Analysis	Analysis
Iron	Application	Application	Analysis	Analysis
Lead/copper	Analysis	Analysis	Analysis	Analysis
Nitrates	Application	Application	Analysis	Analysis
Nitrites	Application	Application	Analysis	Analysis
pH	Application	Application	Application	Analysis
Radionuclides	Application	Application	Analysis	Analysis

Synthetic organic chemicals (SOC)	Application	Application	Analysis	Analysis
Temperature	Application	Application	Application	Analysis
Turbidity	Application	Application	Application	Analysis
Volatile organic chemicals (VOC)	Application	Application	Analysis	Analysis

Required Capabilities

Knowledge of:

- Biological science
- Disinfection concepts
- Disinfection design parameters
- Disinfection process
- General chemistry
- Laboratory equipment
- Monitoring requirements
- Normal characteristics of water
- Physical science
- Proper chemical handling and storing
- Proper sampling procedures
- Quality control/quality assurance practices
- Record keeping policies
- Regulations
- Reporting requirements
- Safe Drinking Water Act (SDWA)
- Safety procedures
- Sampling requirements
- Testing instruments
- 40 CFR 141 Subpart B: Maximum Contaminant Levels (arsenic, nitrate, turbidity)
- 40 CFR 141 Subpart C: Monitoring and Analytical Requirements (turbidity, coliforms, organic contaminants, organic contaminants)
- 40 CFR 141 Subpart D: Reporting and Recordkeeping Requirements
- 40 CFR 141 Subpart E: Special Regulations, Including Monitoring Regulations and Prohibition on Lead Use
- 40 CFR 141 Subpart F: Maximum Contaminant Level Goals and Maximum Residual Disinfectant Level Goals
- 40 CFR 141 Subpart G: National Primary Drinking Water Regulations: Maximum Contaminant Levels and Maximum Residual Disinfectant Levels
- 40 CFR 141 Subpart H: Filtration and Disinfection
- 40 CFR 141 Subpart I: Control of Lead and Copper
- 40 CFR 141 Subpart L: Disinfectant Residuals, Disinfection By-products, and Disinfection By-product Precursors
- 40 CFR 141 Subpart S: Ground Water Rule
- 40 CFR 141 Subpart V: Stage 2 Disinfection By-products Requirements

Ability to:

- Calibrate equipment
- Calibrate instruments
- Determine what information needs to be recorded
- Diagnose/troubleshoot equipment
- Follow written procedures
- Interpret data
- Interpret Material Safety Data Sheets
- Organize information
- Recognize normal and abnormal analytical results
- Record information
- Review reports
- Transcribe data
- Translate technical language into common terminology

Laboratory Analysis

Sample Questions for Level I

Answers on page 343

1. First draw samples for the analysis of lead and copper water must be collected from taps where the water has stood motionless in the plumbing for at least

    a. 4 hours.
    b. 6 hours.
    c. 8 hours.
    d. 24 hours.

2. Samples to be tested for coliforms are collected in plastic bottles that must contain

    a. sodium thiocarbonate.
    b. sodium thiooxalate.
    c. sodium thiosulfate.
    d. sodium thiocyanate.

3. The volume of a sample for coliform compliance is

    a. 100 mL.
    b. 200 mL.
    c. 300 mL.
    d. 0; there is no volume compliance for coliforms.

4. If a water sample is not analyzed immediately for chlorine residual, it is acceptable if it is analyzed within

    a. 10 minutes.
    b. 15 minutes.
    c. 20 minutes.
    d. 30 minutes.

5. Which may be substituted for the analysis of residual disinfectant concentration, when total coliforms are also sampled at the same sampling point?

    a. Heterotrophic plate count (HPC)
    b. Fecal coliforms
    c. Giardia lamblia
    d. Combined chlorine

Sample Questions for Level II

Answers on page 343

1. Which water quality parameter requires a grab sample because it cannot be collected as a composite sample?

    a. pH
    b. Iron
    c. Nitrate
    d. Zinc

2. One chemical characteristic of an acid is that it will

   a. accept an electron pair.
   b. donate an electron pair.
   c. accept a proton.
   d. accept a neutron.

3. Which chemical contains calcium?

   a. Soda ash
   b. Lime
   c. Caustic soda
   d. Sodium bicarbonate

4. The best choice to collect a water sample from a customer's faucet in regards to a complaint would be a

   a. faucet without threads.
   b. faucet that can swivel.
   c. single-lever handle faucet.
   d. faucet with an aerator.

5. When measuring for free chlorine residual, which method is the quickest and simplest?

   a. DPD color comparater
   b. Orthotolidine method
   c. Amperometric titration
   d. 1, 2 nitrotoluene di-amine method

Sample Questions for Level III

Answers on page 343

1. Water that is to be analyzed for inorganic metals should be acidified with

   a. dilute hydrochloric acid.
   b. concentrated hydrochloric acid.
   c. dilute nitric acid.
   d. concentrated nitric acid.

2. If a public water system does not continuously monitor turbidity, how often must it collect a grab sample to measure turbidity from the system's filtered water?

   a. Every 2 hours
   b. Every 4 hours
   c. Every 8 hours
   d. Every 12 hours

3. If a water system uses chlorine for disinfection, but does not provide filtration, which daily parameters following each point of disinfection are required for determining CT calculations?

   a. Lowest value for chlorine residual, pH and temperature, and highest alkalinity
   b. Lowest value for chlorine residual and contact time, lowest temperature, and highest pH
   c. Lowest value for chlorine residual, lowest pH, and lowest temperature
   d. Lowest value for chlorine residual and contact time, pH and alkalinity, and highest temperature

4. A water system exceeds the lead action level if the concentration in more than 10% of tap water samples collected during any monitoring period conducted is greater than

   a. 0.010 mg/L.
   b. 0.012 mg/L.
   c. 0.015 mg/L.
   d. 0.020 mg/L.

5. A water system can reduce its monitoring of total organic carbon (TOC), if it achieves which of the following?

   a. TOC < 1.0 mg/L for 1 year or < 0.5 mg/L for 6 months
   b. TOC < 2.0 mg/L for 1 year or < 1.0 mg/L for 6 months
   c. TOC < 2.0 mg/L for 2 years or < 1.0 mg/L for 1 year
   d. TOC < 3.0 mg/L for 2 years or < 2.0 mg/L for 1 year

Sample Questions for Level IV

Answers on page 344

1. Under the Lead and Copper Rule of the Safe Drinking Water Act, how is the 90th percentile of lead determined for a water system serving fewer than 100 people if the system collects only 5 samples per monitoring period?

   a. By taking the average of the highest and second highest concentrations
   b. By allowing the sample result with the highest concentration to be the 90th percentile value
   c. By allowing the system 2 monitoring periods before reporting
   d. By multiplying the two highest values by 0.9, then taking the average of this result

2. A surface water system can reduce the number of samples for total trihalomethanes (TTHM) and the sum of five haloacetic acids (HAA5), if it has monitored for at least one year and

   a. TTHM ≤ 0.020 mg/L and HAA5 ≤ 0.010 mg/L.
   b. TTHM ≤ 0.030 mg/L and HAA5 ≤ 0.020 mg/L.
   c. TTHM ≤ 0.040 mg/L and HAA5 ≤ 0.030 mg/L.
   d. TTHM ≤ 0.060 mg/L and HAA5 ≤ 0.040 mg/L.

3. A water system that uses chlorine dioxide may reduce the sampling frequency in the distribution system from one three-sample set per month to _____ after one year of monitoring where no individual chlorite sample collected in the distribution system exceeded the chlorite MCL.

   a. one three-sample set per quarter
   b. one six-sample set semi-annually
   c. one three-sample set annually
   d. one six-sample set annually

4. If a water system collects at least 40 samples per month for the analyses of total coliforms, which percent of total coliform positive samples are acceptable for the system to remain in compliance with the maximum contaminant level for total coliforms?

   a. No more than 2%
   b. No more than 3%
   c. No more than 4%
   d. No more than 5%

5. If a water system changes its disinfection sampling point, disinfection type, process, or any other modification identified by the state, it must

   a. perform a sanitary survey.
   b. develop a disinfection benchmark.
   c. develop a disinfection profile.
   d. start over or go back to routine monitoring of disinfection by-products.

Install Equipment

Install Equipment	Level I	Level II	Level III	Level IV
Backflow prevention devices	Comprehension	Comprehension	Comprehension	Comprehension
Hydrants	Comprehension	Application	Application	Application
Meters	Application	Application	Application	Application
Piping	Comprehension	Application	Application	Application
Service connections	Comprehension	Application	Application	Application
Taps	Comprehension	Application	Application	Analysis
Valves	Comprehension	Application	Application	Analysis
Water mains	Comprehension	Application	Application	Analysis

Required Capabilities

Knowledge of:

- Approved backflow methods and devices
- Blueprint readings
- Building codes
- Corrosion control process (including cathodic protection)
- Dechlorination process
- Different types of cross-connections
- Different types of joints, restraints, and thrust blocks
- Function of tools
- Personal protective equipment
- Pipe fittings and joining methods
- Piping material, type, and size
- Pneumatics
- Proper lifting techniques
- Protective coatings and paints
- Safety procedures

Ability to:

- Demonstrate safe work habits
- Diagnose/troubleshoot equipment
- Identify potential safety hazards
- Inspect pumps
- Operate safety equipment
- Perform distribution math
- Recognize unsafe work conditions
- Select safety equipment
- Use hand tools
- Use power tools

Install Equipment

Sample Questions for Level I

Answers on page 345

1. It is standard practice to install fire hydrants on mains that are at a minimum _____ or larger.

    a. 6 inches
    b. 8 inches
    c. 10 inches
    d. 12 inches

2. A corporation stop is used for a

    a. service line.
    b. pump discharge line.
    c. tank inlet.
    d. tank outlet.

3. When PVC pipe is stacked loose, it should not be stacked more than how high?

    a. 2.0 feet
    b. 3.0 feet
    c. 5.0 feet
    d. 7.5 feet

4. Which is the most common cause for pipe joint failure (leaking) in newly laid pipe?

    a. The use of a cracked gasket
    b. Not pushing the spigot end the full distance into the bell
    c. Not having the joint completely clean
    d. An incorrect trench bedding angle

5. Which should be installed at a dead-end water main?

    a. Vacuum valve
    b. Air valve
    c. Blowoff valve
    d. Water quality sampling station

Sample Questions for Level II

Answers on page 345

1. Compression fittings used with copper or plastic tubing seal by means of a

    a. beveled sleeve.
    b. compression ring.
    c. compressed beveled gasket.
    d. compressed o-rings located at either end of the fitting's beveled neck.

2. The breaking of a buried pipe when it is unevenly supported is called

   a. stress breakage.
   b. shear breakage.
   c. beam breakage.
   d. flexural breakage.

3. Thrust from a water surge almost always acts _____ to the inside surface that it pushes against.

   a. vertically
   b. horizontally
   c. perpendicular
   d. vertically and horizontally

4. Which thrust control is easy to use, especially in locations where existing utilities or structures are numerous?

   a. Restraining fittings
   b. Tie rods
   c. Thrust anchors
   d. Thrust blocks

5. The backfill material for a pipe installation should contain enough _____ to allow for thorough compaction.

   a. moisture
   b. sand
   c. gravel
   d. mixed sizes

Sample Questions for Level III

Answers on page 345

1. A bypass valve would be used with a

   a. gate valve.
   b. butterfly valve.
   c. needle valve.
   d. diaphragm valve.

2. Where ground frost is expected, the riser pipes in a meter pit should be at least how far away from any wall of the meter box to prevent freezing?

   a. 1 to 2 inches
   b. 2 to 4 inches
   c. 4 to 6 inches
   d. 6 to 8 inches

3. How can a distribution operator tell if a new pipe with no obvious cracks or chips is good before it is placed in the trench for installation?

   a. By conducting a pressure test
   b. By gently tapping the length of the pipe with a hammer; the pipe should ring or hum clearly
   c. By attaching a sonic meter at one end and listening at other end with headphones; any defects will give crackling sounds
   d. By placing a sonic meter at one end and an oscilloscope at the other end; defects will show up as sharp peaks and valleys as pitch is varied; no defects if only have smooth curves as pitch is varied

4. Which is the recommended trench width for a 42-inch diameter ductile-iron pipe?

   a. 58 inches
   b. 60 inches
   c. 66 inches
   d. 72 inches

5. The first layer of backfill if compaction is required for a newly installed pipe should come up to

   a. the bottom of the pipe.
   b. one-third up the bottom of the pipe.
   c. the centerline of the pipe.
   d. the top of the pipe.

Sample Questions for Level IV

Answers on page 346

1. If special bedding material is required by the design engineer due to poor local soil conditions, the material should not contain granular material greater than

   a. $\frac{1}{4}$ inch.
   b. $\frac{1}{2}$ inch.
   c. $\frac{3}{4}$ inch.
   d. 1 inch.

2. How much will a flexible disk coupling compensate for in parallel misalignment?

   a. $\frac{1}{32}$ inch
   b. $\frac{1}{16}$ inch
   c. $\frac{1}{8}$ inch
   d. $\frac{1}{4}$ inch

3. Which will a flexible diaphragm coupling compensate for in angular movement?

   a. 1.0 degree
   b. 2.0 degrees
   c. 3.7 degrees
   d. 4.0 degrees

4. When a pressure reducing and a pressure sustaining valve are used in combination, one valve can keep a constant _____ pressure even with fluctuating demand, while the other valve holds the pressure at a minimum predetermined _____.

   a. upstream; flow
   b. upstream; pressure
   c. downstream; flow
   d. downstream; pressure

5. If backsiphonage occurs with a reduced-pressure device and the pressure drops below 2 psi, which process will occur?

   a. A vacuum will occur between the check valves, closing the check valve on the customer's side and fully opening the relief valve
   b. A vacuum will occur on the supply side closing both check valves and fully opening the relief valve
   c. The customer's check valve will close due to atmospheric pressure between the check valves, causing the relief valve to fully open
   d. The relief valve will remain fully open and an air gap (atmospheric pressure) will form between the check valves, closing them both

Operate Equipment

Operate Equipment	Level I	Level II	Level III	Level IV
Blowers and compressors	Comprehension	Comprehension	Comprehension	Comprehension
Chemical feeders	Comprehension	Comprehension	Comprehension	Comprehension
Chlorinators	Comprehension	Comprehension	Application	Application
Computers	Comprehension	Application	Application	Application
Drives	Comprehension	Application	Application	Analysis
Electrical motors	Comprehension	Application	Analysis	Analysis
Electronic testing equipment	Comprehension	Comprehension	Application	Analysis
Engines	Comprehension	Application	Application	Application
Generators	Comprehension	Application	Application	Application
Hand tools	Application	Application	Application	Application
Heavy equipment	Comprehension	Comprehension	Application	Analysis
Hydrants	Application	Application	Application	Application
Hydraulic equipment	Comprehension	Comprehension	Application	Application
Instrumentation	Comprehension	Application	Application	Analysis
Leak correlators/detectors	Comprehension	Application	Application	Analysis
Pipe locators	Application	Application	Application	Analysis
Power tools	Comprehension	Application	Application	Application
Pumps	Application	Application	Application	Analysis
Samplers	Comprehension	Application	Application	Analysis
SCADA	Comprehension	Application	Application	Analysis
Tapping equipment	Comprehension	Application	Application	Analysis
Telemetry system	Comprehension	Application	Analysis	Analysis
Valve locators	Application	Application	Application	Analysis
Valves	Application	Application	Analysis	Analysis

Required Capabilities

Knowledge of:

- Facility operation and maintenance
- Function of tools
- General electrical principles
- General hydraulic principles
- General mechanical principles
- Internal combustion engines
- Lubricant and fluid characteristics
- Operation and maintenance practices
- Pipe fittings and joining methods
- Piping material, type and size
- Pneumatics
- Quality control/quality assurance practices
- Start-up and shut down procedures
- Testing instruments

Ability to:

- Adjust equipment
- Adjust flow patterns
- Adjust system units
- Calibrate equipment
- Calibrate instruments
- Demonstrate safe work habits
- Operate safety equipment
- Perform distribution math
- Perform physical measurements
- Recognize unsafe work conditions
- Select safety equipment
- Use hand tools
- Use power tools

Operate Equipment:

Sample Questions for Level I

Answers on page 347

1. The "heart" of a pump is called the

    a. volute case.
    b. impeller.
    c. motor.
    d. pump.

2. Which device serves the same function as the packing?

    a. Inline suction gland
    b. Packing gland
    c. Mechanical seal
    d. Lantern seal

3. Which is used to stop air leakage into the casing around a pump shaft?

    a. Packing gland
    b. Lantern ring
    c. Seals
    d. Shaft sleeves

4. Which is at the top of a stuffing box?

    a. Packing gland
    b. Lantern ring
    c. Mechanical seal
    d. Seal cage

5. Which assembly holds the lantern ring and packing?

   a. Shaft assembly
   b. Casing ring assembly
   c. Packing gland casing
   d. Stuffing box

Sample Questions for Level II

Answers on page 347

1. Which of the following prevents the impeller of a pump from turning on the shaft?

   a. Lock nut on threaded shaft
   b. Key
   c. Steel pin
   d. Caliper pin

2. Which type of valve is used to isolate a pump on the suction side?

   a. Butterfly valve
   b. Globe valve
   c. Gate valve
   d. Ball valve

3. Water hammer can be described as

   a. particle waves.
   b. acoustic waves.
   c. rogue waves.
   d. longitudinal waves.

4. When fully opened, which valve will have the highest head loss?

   a. Gate valve
   b. Plug valve
   c. Globe valve
   d. Ball valve

5. Which type of pressure sensor uses a wire fastened to a diaphragm?

   a. Bellows sensor
   b. Strain gauge
   c. Helical sensor
   d. Diaphragm element

Sample Questions for Level III

Answers on page 348

1. Which two forces determine the behavior of electricity?

   a. Static and dynamic forces
   b. Dynamic and induced forces
   c. Electric and magnetic forces
   d. Amperage and voltage forces

2. The height a liquid can be raised vertically by a given pressure is called

   a. pressure head.
   b. total head.
   c. velocity head.
   d. pump head.

3. A split case pump has three impellers. Which type of multistage pump is this?

   a. One stage
   b. Two stage
   c. Three stage
   d. Six stage

4. A split case pump has two equal smaller impellers placed on either side of two equally sized large impellers. How many stages does this pump have?

   a. One stage
   b. Two stages
   c. Four stages
   d. Eight stages

5. A pump loses its prime because the suction line has an air pocket. Which is the best solution?

   a. Check pump's amperage and be sure pump's strainer is clean
   b. Clean or repair priming unit
   c. Open suction piping air bleed-off valves
   d. Check external water seal unit

Sample Questions for Level IV

Answers on page 348

1. The lowest pressure point in the pump is the

   a. center of the impeller.
   b. outermost part of the impeller.
   c. suction side of the pump.
   d. discharge side of the pump.

2. The shaft's main function is to transmit _____ from the motor to the impeller.

   a. centrifugal force
   b. torque
   c. kinetic energy
   d. thrust

3. Which is the range of specific speeds for a propeller pump?

   a. 500 to 5,000 rpm
   b. 5,000 to 9,000 rpm
   c. 9,000 to 15,000 rpm
   d. 12,000 to 18,000 rpm

4. A vertical turbine pump would most likely have a specific speed of
    a. 3,000 rpm.
    b. 6,000 rpm.
    c. 9,000 rpm.
    d. 12,000 rpm.
5. Which is the most common cause for surges in the distribution system?
    a. Power failure shutting down a pump suddenly
    b. Valve opening and closing
    c. Failure of flow on pressure regulation
    d. Pump start-up

Perform Maintenance

Perform Maintenance	Level I	Level II	Level III	Level IV
Blowers and compressors	N/A	N/A	N/A	Application
Chemical feeders	Application	Application	Application	Analysis
Chlorinators	Application	Application	Analysis	Analysis
Corrosion control	Application	Application	Analysis	Analysis
Cross-connection control	Application	Application	Analysis	Analysis
Drives	N/A	N/A	N/A	Analysis
Electric motors	Application	Application	Application	Application
Electrical grounding	Application	Application	Application	Application
Engines	Comprehension	Comprehension	Application	Analysis
Evaluate Operation of Equipment	Application	Application	Analssis	Analysis
Facility inspection	Application	Application	Analysis	Analysis
Generators	Comprehension	Comprehension	Application	Application
Hydrants	Application	Application	Analysis	Analysis
Hydraulic equipment	N/A	N/A	Application	Analysis
Hypochlorinators	Application	Analysis	Analysis	Analysis
Instrumentation	Comprehension	Application	Application	Analysis
Leak detection	Application	Application	Analysis	Analysis
Lock-out/tag-out	Application	Application	Application	Application
Meters	Application	Application	Application	Analysis
Pressure sensors	Application	Application	Analysis	Analysis
Pumps	Application	Application	Analysis	Analysis
Service connection	Comprehension	Application	Application	Analysis
Service pipes	Comprehension	Comprehension	Application	Application
Valves	Application	Application	Application	Analysis
Water mains	Application	Application	Analysis	Analysis
Water storage facility	Application	Application	Analysis	Analysis

Required Capabilities

Knowledge of:

- Approved backflow methods and devices
- Blueprint readings
- Building codes
- Corrosion control process (including cathodic protection)
- Different types of cross-connections
- Different types of joints, restraints, and thrust blocks
- Facility operation and maintenance
- Facility security
- Function of tools
- General electrical principles
- General hydraulic principles
- General mechanical principles
- Internal combustion engines
- Laboratory equipment
- Local codes and ordinances
- Lubricant and fluid characteristics
- Measuring instruments
- Operation and maintenance practices
- Personal protective equipment
- Pipe fittings and joining methods
- Piping material, type, and size
- Pneumatics
- Potential causes of disasters in facility
- Potential impact of disasters in facility
- Proper chemical handling and storing
- Proper lifting techniques
- Protective coatings and paints
- Quality control/quality assurance practices
- Record keeping policies
- Safety procedures
- Sanitary survey processes
- Start-up and shut down procedures
- Testing instruments
- Well-head protection

Ability to:

- Adjust equipment
- Adjust flow patterns
- Adjust system units
- Assess likelihood of disaster occurring
- Assign work to proper trade
- Calibrate equipment
- Calibrate instruments
- Demonstrate safe work habits
- Diagnose/troubleshoot equipment
- Diagnose/troubleshoot system units
- Differentiate between preventative/corrective maintenance
- Discriminate between normal/abnormal conditions
- Evaluate facility performance
- Evaluate operation of equipment
- Evaluate system units
- Identify potential safety hazards
- Inspect pumps
- Interpret data
- Interpret Material Safety Data Sheets
- Maintain inventory control system
- Maintain system in normal operating condition
- Monitor electrical equipment
- Monitor mechanical equipment
- Obtain unbiased data
- Operate safety equipment
- Organize information
- Perform distribution math
- Perform general maintenance
- Perform general repairs
- Perform physical measurements
- Recognize normal and abnormal analytical results
- Recognize unsafe work conditions
- Record information
- Review reports
- Select safety equipment
- Translate technical language into common terminology
- Use hand tools
- Use power tools

Perform Maintenance

Sample Questions for Level I

Answers on page 349

1. Why is it so important to monitor the speed of a variable-speed pump?

    a. To prevent excessive temperatures from developing
    b. To prevent vibration from developing
    c. To prevent speed oscillation from occurring
    d. To prevent cavitation from occurring

2. Which basic electrical unit is used to measure a material's opposition to the flow of electricity?

    a. Ampere
    b. Ohm
    c. Volts
    d. Resistance or impedance

3. The first oil change on a new pump should be done

    a. after the first two weeks of operation.
    b. after one month of operation.
    c. after three months of operation.
    d. after six months of operation.

4. How often should the temperature of centrifugal pump motor bearings be checked with a thermometer?

    a. Every day
    b. Once a week
    c. Twice a month
    d. Once a month

5. All sensors that respond to liquid pressure will perform poorly if _____ enter(s) the sensor.

    a. air
    b. corrosive chemicals from water treatment processes
    c. corrosive chemicals from piping
    d. iron bacteria

Sample Questions for Level II

Answers on page 349

1. Packing replacement is usually performed when

    a. water leakage sprays out of the pump housing.
    b. no further tightening can be done on the packing gland.
    c. the packing gland bolts are exposed by more than 2½ inches above the nut.
    d. the packing has completely disintegrated.

2. Which device changes alternating current to direct current by allowing the electric current to flow in one direction but blocking flow in the opposite direction?

   a. Regulator
   b. Converter
   c. Inverter
   d. Rectifier

3. To ease installation of impeller wear rings, they can be

   a. lubricated with a light oil.
   b. greased with lithium.
   c. heated.
   d. cooled.

4. Packing is designed to

   a. add lubricant to the shaft.
   b. expand and deteriorate with normal use.
   c. protect the shaft.
   d. wear and deteriorate with normal use.

5. Bearings on a line shaft turbine can be lubricated with

   a. oil or water.
   b. grease or oil.
   c. lithium or grease.
   d. graphite or grease.

Sample Questions for Level III

Answers on page 349

1. Insulators usually fail because of

   a. abnormally high voltages.
   b. excessive heat.
   c. switching surges.
   d. mechanical failure.

2. Although some phase imbalance will occur in any system, in general, it should not exceed

   a. 5%.
   b. 10%.
   c. 15%.
   d. 20%.

3. Nitrification in the distribution system can best be described as the

   a. breakdown of ammonia by certain bacteria.
   b. utilization of nitrogen in ammonia by certain bacteria.
   c. oxidation of ammonia to nitrites then to nitrates.
   d. reduction of ammonia to nitrites then to nitrates.

4. Which type of rigid coupling is easy to install and remove?

   a. Split coupling
   b. Jaw coupling
   c. Gear coupling
   d. Flexible disc coupling

5. Which is the correct equation for Ohm's Law?

   a. Volts = (amps)(ohms)
   b. Ohms = (volts)(amps)
   c. Ohms = amps/volts
   d. Amps = (volts)(ohms)

Sample Questions for Level IV

Answers on page 350

1. Which are the basic components of every telemetry system?

   a. Transmitter, signal, receiver, and indicator
   b. Transmitter, receiver, and indicator
   c. Transmitter, transmission channel, and receiver
   d. Sensor, transmitter, transmission channel, and receiver

2. The anodes placed in a water storage tank for cathodic protection usually last _____, and should be inspected every _____.

   a. 10 years; 6 months
   b. 10 years; year
   c. 15 years; 6 months
   d. 20 years; year

3. In general, the current drawn by a motor the instant it is connected to the power source, called the locked-motor current, is usually _____ times the normal full-load current of the motor.

   a. 1.5 to 2.0
   b. 2.0 to 3.0
   c. 3.0 to 5.0
   d. 5.0 to 10.0

4. Most telemetry equipment transmit information by

   a. digital signals.
   b. analog signals.
   c. audio signals.
   d. electrical pulses.

5. Control systems consist of the following distinct components:

   a. Signal conditioners and control elements
   b. Signal conditioners, actuators, and control elements
   c. Signal conditioners, actuators, control elements, and indicators
   d. Sensors, signal conditioners, actuators, control elements, and indicators

Perform Security, Safety, and Administrative Procedures

Perform Security, Safety, and Administrative Procedures	Level I	Level II	Level III	Level IV
Manage System				
Administer safety/compliance program	Comprehension	Application	Application	Analysis
Conduct cross-connection surveys	Application	Application	Analysis	Analysis
Develop budget	N/A	N/A	Analysis	Analysis
Develop operation and maintenance plan	Comprehension	Application	Analysis	Analysis
Develop/maintain sample site plan	Application	Application	Analysis	Analysis
Participate in sanitary surveys	Application	Application	Application	Application
Regulatory reporting	Analysis	Analysis	Analysis	Analysis
Promote Public Relations				
Promote customer service program	N/A	Application	Analysis	Analysis
Respond to complaints	Application	Application	Application	Application
Safety Program				
Chemical safety	Application	Application	Application	Analysis
Confined space entry	Application	Application	Application	Application
Excavation, shoring and trenching	Application	Application	Application	Application
General Safety	Application	Application	Application	Application
Personal protective equipment	Application	Application	Application	Application
Public protection	Application	Application	Application	Application
Recordkeeping				
Compliance	Application	Application	Application	Application
Corrective actions to system deficiencies	Application	Application	Application	Application
Equipment repair/replacement	Application	Application	Application	Analysis
Laboratory	Application	Application	Application	Analysis
Maintenance	Application	Application	Application	Application
System operation	Application	Application	Analysis	Analysis

Required Capabilities

Knowledge of:

- Biological science
- Blueprint readings
- Building codes
- Data acquisition techniques
- Disciplinary procedures
- Emergency plans
- Employment laws
- Facility security
- Function of recordkeeping system
- General chemistry
- General electrical principles
- General hydraulic principles
- General mechanical principles
- Human resource practices
- Hydrology
- Local codes and ordinances
- Memorandums of understanding and agreements
- Monitoring requirements
- Potential causes of disasters in facility
- Potential impact of disasters in facility
- Principles of finance
- Principles of general communication
- Principles of management
- Principles of measurement
- Principles of public relations
- Principles of supervision
- Public notification requirements
- Public participation requirements
- Quality control/quality assurance practices
- Record keeping policies
- Regulations
- Reporting requirements
- Risk management
- Safe Drinking Water Act (SDWA)
- Sanitary spring design
- Sanitary survey processes
- Standards
- Water reuse
- Watershed management
- 40 CFR 141 Subpart A: General (definitions, coverage, variances and exemptions, siting requirements, and effective dates)
- 40 CFR 141 Subpart D: Reporting and Recordkeeping Requirements
- 40 CFR 141 Subpart O: Consumer Confidence Reports
- 40 CFR 141 Subpart Q: Public Notification of Drinking Water Violations

Ability to:

- Assess likelihood of disaster occurring
- Assign work to proper trade
- Communicate in writing
- Communicate verbally
- Conduct meetings
- Conduct training programs
- Coordinate emergency response with other water organizations relative to the distribution system
- Determine what information needs to be recorded
- Develop a staffing plan
- Develop a work unit
- Evaluate facility performance
- Evaluate promotional materials
- Evaluate proposals
- Follow written procedures
- Generate a written safety program
- Generate capital plans
- Generate long- and short-term plans
- Identify potential safety hazards
- Interpret data
- Negotiate contracts
- Obtain unbiased data
- Organize information
- Perform distribution math
- Perform impact assessments
- Prepare proposals
- Recognize normal and abnormal analytical results
- Recognize unsafe work conditions
- Record information
- Review reports
- Select safety equipment
- Transcribe data
- Translate technical language into common terminology
- Write policies and procedures

Perform Security, Safety, and Administrative Procedures

Sample Questions for Level I

Answers on page 351

1. Which agency sets legal limits on the concentration levels of harmful contaminants in potable water distributed to customers?

    a. National Primary Drinking Water Regulations
    b. United States Environmental Protection Agency
    c. United States Public Health Service
    d. Occupational Health and Safety Organization

2. Which violations are the most serious?

    a. Tier I
    b. Tier II
    c. Tier III
    d. Tier IV

3. A positive fecal coliform test must be reported to the primacy agency within

    a. 8 hours.
    b. 12 hours.
    c. 24 hours.
    d. 48 hours.

4. The number of monthly distribution system bacteriological samples required is

    a. based on water withdrawal permit limit.
    b. based on system size.
    c. based on population served.
    d. different for each state.

5. Which is the approximate angle of repose for average soils when using the sloping method for the prevention of cave-ins? (Note: horizontal to vertical distance, respectively)

    a. 0.5 : 1.0
    b. 1.0 : 1.0
    c. 1.5 : 1.0
    d. 2.0 : 1.0

Sample Questions for Level II

Answers on page 351

1. Which is the Maximum Contaminant Level for total trihalomethanes (TTHMs)?

    a. 0.040 mg/L
    b. 0.060 mg/L
    c. 0.080 mg/L
    d. 0.100 mg/L

2. A system that fails to collect water samples in their distribution system would fall under which public notification requirement?

   a. Tier I
   b. Tier II
   c. Tier III
   d. Tier IV

3. Under the Surface Water Treatment Rule, disinfection residuals must be collected at the same location in the distribution system as

   a. coliform samples.
   b. total trihalomethanes.
   c. disinfection by-products.
   d. alkalinity, conductivity, and pH for corrosion studies.

4. Iron can cause "red water" and thus customer complaints when its concentration is above its secondary maximum contaminant level of

   a. 0.01 mg/L.
   b. 0.05 mg/L.
   c. 0.10 mg/L.
   d. 0.30 mg/L.

5. Which chemical may encourage the growth of algae and microorganisms?

   a. Lime
   b. Sodium bicarbonate
   c. Sodium hydroxide
   d. Zinc orthophosphate

Sample Questions for Level III

Answers on page 351

1. Water systems that collect samples for coliform testing must not find more than _____ percentage of the samples they collect each month to be coliform positive.

   a. 2%
   b. 4%
   c. 5%
   d. 10%

2. Which is the Maximum Contaminant Level for nitrate as measured as nitrogen?

   a. 1.0 mg/L
   b. 2.0 mg/L
   c. 5.0 mg/L
   d. 10.0 mg/L

3. How much time does a community water system have for public notification if they have any Tier III violation?

   a. 60 days
   b. 3 months
   c. 6 months
   d. 12 months

4. Cohesive soil would contain a lot of

   a. water.
   b. clay.
   c. clay and silt.
   d. clay, silt, and sand.

5. The lead and copper rule requires lead to be less than 0.015 mg/L in which percentage of the total samples collected?

   a. 75%
   b. 90%
   c. 95%
   d. 99%

Sample Questions for Level IV

Answers on page 352

1. Who should inspect the work of a water storage painting contractor?

   a. The water utility supervisor
   b. A water operator or operators responsible for the area the tank is located in
   c. At least two competing painting contractors
   d. A qualified third party

2. Aviation warning lights on water storage tanks have an FAA regulation that requires the bulbs to be replaced before they reach what percentage of their normal life expectancy?

   a. 50%
   b. 60%
   c. 75%
   d. 80%

3. The most expensive part of a pipe installation is/are the

   a. pipe fittings.
   b. valves.
   c. excavation.
   d. engineers' design.

4. Which of the following is a potential biological biotoxin threat?

   a. Sarin
   b. Ricin
   c. Taban
   d. 3-quinucli dinyl benzilate

5. The IDLH (Immediately Dangerous to Life and Health) value for chlorine, set by the National Institution of Occupational Safety and Health, is

   a. 10 ppm.
   b. 15 ppm.
   c. 25 ppm.
   d. 30 ppm.

Additional Distribution Operator Practice Questions

Answers on page 353

1. At which time of day is the age of the water stored in the distribution system the lowest?

    a. Early morning
    b. Late morning
    c. Early afternoon
    d. Late evening

2. Which type of maps should not overlap each other?

    a. Index maps
    b. Comprehensive maps
    c. Construction maps
    d. Sectional maps

3. Which one of the following is a type of joint for ductile iron piping?

    a. Welded joint
    b. Expansion joint
    c. Flexible ball joint
    d. Bell and spigot with rubber o-ring

4. The amount of energy in feet required to overcome resistance to flow in a pipe is called

    a. pump head.
    b. cutoff head.
    c. pressure head.
    d. friction head.

5. The fire insurance underwriters recommend that no main should have a diameter less than

    a. 3 inches.
    b. 4 inches.
    c. 6 inches.
    d. 8 inches.

6. Water velocities through pipe should be limited to 5 ft/sec in order to

    a. minimize friction loss.
    b. reduce the number of pressure reducing stations.
    c. minimize destruction of piping appurtenances.
    d. reduce the amount of corrosion inhibitors applied to piping.

7. Which of the following is a type of joint for polyvinyl chloride piping?

    a. Bell and spigot type
    b. Thermal butt-fusion
    c. Expansion joint
    d. Restrained joint

8. Which of the following is a type of joint for high-density polyethylene piping?

    a. Bell and spigot type
    b. Thermal butt-fusion
    c. Push-on joint
    d. Restrained joint

9. Which of the following is a type of joint for ductile iron piping?

    a. Welded joint
    b. Restrained joint
    c. Expansion joint
    d. Rubber gasket joint

10. Which of the following is a type of joint for ductile iron piping?

    a. Welded joint
    b. Bell and spigot joint
    c. Expansion joint
    d. Flanged joint

11. The height of water in three differently shaped tanks is 22.4 feet. Which tank will have the highest psi at the bottom?

    a. The square tank
    b. The rectangular tank
    c. The cylindrical tank
    d. It will be the same in all three tanks

12. A water tank has a glass tube on its side showing the water level in the tank. What is this surface called?

    a. Hydraulic grade line
    b. Piezometric surface
    c. Potentiometric surface
    d. Potential surface

13. Minor head losses are caused by

    a. slime growths and corrosion or scaling.
    b. corrosion and tuberculation.
    c. type of pipe material and "C" factor.
    d. sudden changes in direction or velocity of flow.

14. Water mains should primarily be sized based on

    a. earthquake size potential.
    b. peak domestic and commercial demands.
    c. peak commercial and industrial demands.
    d. adequate fire flow at an appropriate pressure.

ADDITIONAL DISTRIBUTION OPERATOR PRACTICE QUESTIONS   293

15. The most desirable residential pressure ranges from

    a. 20 to 35 psi.
    b. 35 to 50 psi.
    c. 50 to 75 psi.
    d. 75 to 90 psi.

16. Which is pipe strength expressed in?

    a. Hydrostatic potential
    b. Psi and durability
    c. Tensile and flexural strength
    d. Baud units

17. The resistance of a material to longitudinal pulling forces before it breaks is called

    a. flexural strength.
    b. shear strength.
    c. ductile strength.
    d. tensile strength.

18. Most water systems use hydrants with two nozzles with diameters of _____ and one nozzle with a diameter of _____.

    a. 2.0 inches; 3.0 inches
    b. 2.0 inches; 4.0 inches
    c. 2.5 inches; 3.5 inches
    d. 2.5 inches; 4.5 inches

19. Which type of pipe is durable and strong, has high flexural strength and good corrosion resistance, and is easily tapped?

    a. High-density polyethylene
    b. Steel
    c. Ductile iron
    d. Reinforced concrete

20. Which of the following is a type of joint for concrete piping?

    a. Expansion joint
    b. Push-on joint
    c. Bell and spigot type
    d. Flanged joint

21. Precipitative softening water treatment plants try to end up with distribution system water that is

    a. slightly scale forming.
    b. moderate scale forming.
    c. neutral.
    d. in equilibrium.

22. Which is a closed fire line?

    a. Broken fire hydrant or pipeline that feeds a fire hydrant
    b. Unmetered connection for a fire protection system
    c. Closed valve to a fire hydrant
    d. Closed valve to a fire sprinkler system

23. Comprehensive maps of medium to large systems generally have scales ranging from

    a. 250 to 500 feet to 1 inch.
    b. 500 to 1,000 feet to 1 inch.
    c. 1,000 to 1,500 feet to 1 inch.
    d. 1,500 to 2,000 feet to 1 inch.

24. Sectional maps generally have scales ranging from

    a. 50 to 100 feet to 1 inch.
    b. 100 to 200 feet to 1 inch.
    c. 200 to 250 feet to 1 inch.
    d. 250 to 400 feet to 1 inch.

25. Which type of centrifugal pumps are the most common?

    a. Axial flow impellers
    b. Radial flow impellers
    c. Mixed flow impellers
    d. Vertical flow impellers

26. Which of the following is a type of joint for steel piping?

    a. Flexible ball joint
    b. Restrained joint
    c. Expansion joint
    d. Push-on joint

27. Which of the following is a type of joint for steel piping?

    a. Flexible ball joint
    b. Rubber gasket joint
    c. Restrained joint
    d. Galvanized steel ring

28. The pressure during fire flow conditions should not drop below

    a. 15 psi.
    b. 20 psi.
    c. 25 psi.
    d. 30 psi.

29. During a fire flow event, the residual pressure in a large main should be greater than

    a. 15 psi.
    b. 20 psi.
    c. 25 psi.
    d. 35 psi.

30. Which type of distribution system configuration has smaller mains that generally terminate as dead-ends?

    a. Grid system
    b. Dendritic system
    c. Arterial-loop system
    d. Tree system

31. Which is applied to electrodes placed in the water of a water storage tank?

    a. AC current
    b. DC current
    c. An electrolytic inhibitor around the seals where the wires enter the electrodes
    d. Nothing

32. The frequent starting of distribution system pumps probably indicates

    a. pump is overheating.
    b. inadequate supply to pump or check valve leaking back.
    c. inadequate distribution storage.
    d. pump is in need of maintenance.

33. Which needs to be determined first when designing a pump station?

    a. Discharge requirements
    b. Head requirements
    c. Power requirements
    d. Capacity requirements

34. Which is plotted on the horizontal scale (x-axis) of a pump curve?

    a. Efficiency
    b. Capacity (flow rate)
    c. Total head
    d. Power

35. Hydraulics is the study of

    a. fluid pressure in pipes or conduits.
    b. the force of fluids in motion.
    c. the pressure of fluids in motion.
    d. fluids in motion and at rest.

36. Which type of pipe joint consists of two machined surfaces?

    a. Restrained joint
    b. Push-on joint
    c. Flanged joint
    d. Mechanical joint

37. Which type of pipe joint requires a movable follower ring?

    a. Flanged joint
    b. Grooved joint
    c. Restrained joint
    d. Mechanical joint

38. What is the height limit to which siphoned water can be lifted at sea level?

    a. 22.4 feet
    b. 32.0 feet
    c. 33.9 feet
    d. 34.0 feet

39. How many feet of water will equal the atmospheric pressure at sea level?

    a. 28
    b. 30
    c. 32
    d. 34

40. Which type of impeller is best to use for pumping water with low volumes of solids?

    a. Closed
    b. Open
    c. Semi-open
    d. Radial

41. Which type of pump has the ability to rotate its discharge 360°?

    a. End suction pump
    b. Split case pump
    c. Vertical pump
    d. Jet pump

42. How many times higher than the normal operating pressure should the pressure rating of distribution system piping be?

    a. 1.0 to 2.0 times
    b. 2.5 to 4.0 times
    c. 4.0 to 5.0 times
    d. 5.0 to 7.0 times

43. Which type of pipe has been banned in the water industry?

    a. Ductile iron
    b. Steel
    c. Asbestos cement
    d. High-density polyethylene

44. Which of the following types of pipe comes in diameters as large as 168 inches (14 feet)?

    a. Cast-iron pipe
    b. Reinforced concrete
    c. Ductile iron
    d. Prestressed concrete

45. Which type of pipe has the highest normal maximum working pressure?

    a. Reinforced concrete
    b. Ductile iron
    c. Polyvinyl chloride
    d. High-density polyethylene

46. Which type of pipe has the lowest normal maximum working pressure?

    a. Reinforced concrete
    b. Ductile iron
    c. Polyvinyl chloride
    d. High-density polyethylene

47. Ductile iron piping using flexible ball joints is most appropriate for
    a. low-pressure applications.
    b. high-pressure applications.
    c. river crossings or rugged terrain.
    d. all locations.

48. Ductile iron piping using push-on or mechanical joints is most appropriate
    a. for low-pressure applications.
    b. for high-pressure applications.
    c. for river crossings or rugged terrain.
    d. where flexibility is required.

49. Polyvinyl chloride piping using solvent weld joints is most appropriate
    a. only for small lines.
    b. for high pressure applications.
    c. where flexibility is required.
    d. where valves or fittings are to be attached.

50. Concrete piping using bell and spigot type joints is appropriate
    a. for all locations.
    b. where expansion or contraction will occur.
    c. under low-pressure applications.
    d. under high-pressure applications.

51. Which type of pipe joint is available in both bolted and boltless flexible pipe joint designs?
    a. Ball and socket joint
    b. Push-on joint
    c. Grooved joint
    d. Shouldered joint

52. Which type of pipe joint permits flexible pipe alignment, is inexpensive to manufacture, and is easy to assemble?
    a. Mechanical joint
    b. Push-on joint
    c. Grooved joint
    d. Ball and socket joint

53. Which type of pipe joint has machined ends?
    a. Restrained joint
    b. Flanged joint
    c. Mechanical joint
    d. Shouldered joint

54. New polyvinyl chloride pipe has a C-value of about
    a. 125.
    b. 125 to 128.
    c. 135.
    d. 150+.

55. Which of the following water storage facilities is most likely to contain non-potable water?

    a. Standpipe
    b. Buried storage tanks
    c. Emergency storage tanks
    d. Elevated tanks using a riser; one way in and one way out

56. Which type of valve is used on hydraulic lines connected to valve actuators?

    a. Pinch valve
    b. Needle valve
    c. Pressure-relief valve
    d. Vacuum-relief valve

57. Which type of valve would be particularly useful for throttling the flow of corrosive liquids?

    a. Diaphragm valve
    b. Butterfly valve
    c. Gate valve
    d. Pinch valve

58. Water meter pits are usually used in

    a. areas where flooding will most likely occur.
    b. areas where flooding is very rare.
    c. cold climates.
    d. hot climates.

59. A horizontal load, parallel to a pump shaft is called a

    a. shear load.
    b. thrust load.
    c. radial load.
    d. stress load.

60. Manufacturers typically design pumps such that the first critical speed is at least _____ higher or lower than rated speed.

    a. 10%
    b. 20%
    c. 25%
    d. 30%

61. Regarding fire flow, mains smaller than 6 inches should only be used

    a. in residential areas.
    b. in low-value districts.
    c. in rural areas.
    d. to complete a grid.

62. Which type of pipe requires special external protection in high-chloride soils?

    a. Reinforced concrete
    b. Ductile iron
    c. Steel
    d. High-density polyethylene

63. Which type of pipe would most likely collapse or become distorted under a partial vacuum?

    a. Asbestos-cement
    b. Steel
    c. Cast iron
    d. Plastic

64. Casing pipe used to house a large water main under a railroad crossing should have a diameter that is _____ larger than the outside diameter of the water main bells.

    a. less than 2 inches
    b. 2 to 8 inches
    c. 8 to 6 inches
    d. 6 to 8 inches

65. Data manipulations using the basic elements of an automatic mapping (MS)/Facility mapping (FM)/Geographic Information Systems (GIS) would include

    a. collection and storage.
    b. collection and management.
    c. modeling and analysis.
    d. retrieval, display, and storage.

66. Which data category for water utility data sets includes the following map layers: Control information, planimetric features, and hydrology features?

    a. Land records data
    b. Natural resources data
    c. Watershed resources data
    d. Base data

67. Which of the following chlorine species is called HTH?

    a. Hypochlorous acid
    b. Sodium hypochlorite
    c. Calcium hypochlorite
    d. Hypotrichloro-hydrocarbon

68. Which is the proper detention time for disinfecting a water storage tank that is filled with already chlorinated water such that the free chlorine residual is 10 mg/L after the proper detention time is completed?

    a. 4 hours
    b. 6 hours
    c. 8 hours
    d. 24 hours

69. Which is the proper detention time for disinfecting a water storage tank with water that is mixed with hypochlorite already in the tank such that the free chlorine is 10 mg/L after proper detention time is complete?

    a. 6 hours
    b. 8 hours
    c. 12 hours
    d. 24 hours

70. The disinfectant residual in treated water entering the distribution system must not be less than

    a. 0.2 mg/L at anytime.
    b. 0.2 mg/L for more than 4 hours during periods when the system is serving water to the public.
    c. 0.4 mg/L at anytime.
    d. 0.4 mg/L for more than 5% of the time that the system is serving water to the public.

71. The chemical formula for Hypochlorous acid is

    a. OCl^-.
    b. $NaOCl$.
    c. $HOCl$.
    d. $Ca(OCl)_2$.

72. Usually the best disinfectant to use in large-diameter pipes or very long pipelines is

    a. calcium hypochlorite.
    b. sodium hypochlorite.
    c. chlorine gas.
    d. chloramines.

73. Which method for controlling nitrification problems in a distribution system would result in a temporary increase in trihalomethanes?

    a. Increasing the chlorine to nitrogen weight ratio
    b. Superchlorinating reservoirs and storage tanks
    c. Decreasing chlorine detention time in reservoirs and distribution pipelines
    d. Flushing the distribution pipelines

74. Which is the available chlorine concentration used for disinfecting a water storage tank using a method that involves spraying or painting of all the interior tank surfaces?

    a. 50 mg/L
    b. 100 mg/L
    c. 200 mg/L
    d. 250 mg/L

75. When following a two-step process for disinfecting a water storage tank, the initial dosage of 50 mg/L available chlorine should be held in the partially filled tank for

    a. at least 6 hours.
    b. at least 10 hours.
    c. at least 24 hours.
    d. at least 3 days.

76. Why are altitude valves used on water storage tanks?

    a. To allow water to pass in and out of the tank as pressure fluctuates
    b. To stop the flow of water into the tank when it is full
    c. To allow overflow water to flow out of the tank
    d. To shut the flow of water to the tank off for maintenance and inspection

77. If iron bacteria are causing corrosion problems in the distribution system, which is the most probable solution?

    a. Flushing
    b. Use a corrosion chemical as treatment
    c. Use a surfactant dispersion chemical
    d. Acidify the system, then flush

78. Which type of chlorine gas feeder is most commonly used?

    a. Pressure
    b. Combination water and pressure
    c. Vacuum
    d. Combination pressure and vacuum

79. When concentration cell corrosion occurs in iron pipes the electrons will move through the

    a. water from the anode to the cathode.
    b. water from the cathode to the anode.
    c. iron pipe from the anode to the cathode.
    d. iron pipe from the cathode to the anode.

80. Which type of corrosion occurs when two dissimilar metals are electrically connected and immersed in a common flow of water, where one metal becomes the anode and the other the cathode?

    a. Concentration cell corrosion
    b. Localized corrosion
    c. Uniform corrosion
    d. Galvanic corrosion

81. Which is the main purpose(s) of the meter yoke?

    a. Cushion against stress
    b. Cushion against stress and strain
    c. Provide electrical conductivity
    d. Proper alignment and support for meter

82. The minimum cover over a water main below a road in a warm climate with no possibility of frost is

    a. 2.0 feet.
    b. 2.5 feet.
    c. 3.0 feet.
    d. 3.5 feet.

83. The minimum cover over a water service below a road in a warm climate with no possibility of frost is

    a. 12 inches.
    b. 18 inches.
    c. 24 inches.
    d. 30 inches.

84. How far should the excavation extend ahead of the pipe laying?

    a. 2 pipe lengths
    b. 3 pipe lengths
    c. 4 pipe lengths
    d. Just ahead of the pipe installation

85. Contractors find it easier to lay pipe with the bells facing uphill when the main is being laid on a slope steeper than

    a. 6%.
    b. 9%.
    c. 10%.
    d. 12%.

86. Which is the best mechanical compacting equipment for clays, tills, and silts?

    a. Hand-controlled plate tampers
    b. Vibratory compactors
    c. Irregular drum tampers
    d. Boom-mounted plate tampers

87. Which is the best mechanical compacting equipment for sand in shallow lifts?

    a. Hand-controlled plate tampers
    b. Vibratory compactors
    c. Irregular drum tampers
    d. Boom-mounted plate tampers

88. What is the inlet thread on a standard AWWA corporation stop called?

    a. Tapered Milwaukee thread
    b. Minneapolis thread
    c. Connelly thread
    d. Mueller thread

89. If special bedding material is required by the design engineer due to poor local soil conditions, the material should not contain more than _____ clay or silt that can be sensitive to water.

    a. 5%
    b. 10%
    c. 12%
    d. 18%

90. When pipe cannot be structurally supported by natural trench bedding due to very soft soils, _____ of additional soil should be removed and coarse granular material in well-mixed sizes up to _____ used as backfill and placed on top of the geotextile fabric.

    a. 6 to 12 inches; 2 inches
    b. 12 to 24 inches; 3 inches
    c. 24 to 36 inches; 2 inches
    d. 24 to 36 inches; 4 inches

91. Who should be contacted to ensure proper disposal of groundwater from a pipe trenching project?

    a. US Environmental Protection Agency
    b. Bureau of Land Management, Water Division
    c. Local regulatory agency
    d. Local city government or landowner

92. Which is the bolt torque for a mechanical-joint on a standard 48-inch ductile-iron water main?

    a. 45 to 60 ft-lb
    b. 120 to 150 ft-lb
    c. 225 to 250 ft-lb
    d. 275 to 325 ft-lb

93. Cleaning by the air-purging process is used on water mains that are

    a. 4 inches or smaller.
    b. 6 inches or smaller.
    c. 8 inches or smaller.
    d. 10 inches or smaller.

94. Fuel oil will permeate which type of pipe?

    a. Lead pipe
    b. Polyethylene pipe
    c. Copper pipe
    d. Galvanized pipe

95. Which type of pipe joint is used primarily in exposed locations?

    a. Mechanical joint
    b. Flanged joint
    c. Restrained joint
    d. Shouldered joint

96. Which type of pipe joint lacks flexibility and thus should not normally be buried?

    a. Mechanical joint
    b. Flanged joint
    c. Restrained joint
    d. Shouldered joint

97. Which type of pipe joint is most commonly used for river crossings?

    a. Ball and socket joint
    b. Shouldered joint
    c. Push-on joint
    d. Expansion joint

98. Ball and socket joints can provide deflections of up to

    a. 10.0 degrees.
    b. 13.0 degrees.
    c. 15.0 degrees.
    d. 17.5 degrees.

99. What is the maximum joint deflection for full lengths of standard 48-inch ductile-iron push-joint pipe?

    a. 3 degrees
    b. 6 degrees
    c. 8 degrees
    d. 10 degrees

100. Large meters are usually tested at which percentage of their operating range?

    a. 5 to 10%
    b. 10 to 20%
    c. 20 to 25%
    d. 25 to 33%

101. Which one of the following types of pipe is most prone to sliding out of a push-on joint if not firmly restrained?

    a. Asbestos-cement pipe
    b. Plastic pipe
    c. Ductile-iron pipe
    d. Cast-iron pipe

102. Which is the best mechanical compacting equipment for deep narrow trenches?

    a. Hand-controlled plate tampers
    b. Vibratory compactors
    c. Irregular drum tampers
    d. Boom-mounted plate tampers

103. The lantern ring is also known as the

    a. packing ring.
    b. seal cage.
    c. packing seal.
    d. mechanical seal.

104. How is the capacity of a centrifugal pump related to total head?

    a. Directly
    b. Indirectly
    c. Linear
    d. Disproportionately

105. The bearings that maintain the radial positioning of a shaft are called

    a. radial bearings.
    b. rolling bearings.
    c. sleeve bearings.
    d. thrust bearings.

106. When the discharge of a velocity pump is blocked, head builds up, typically greater than the pressure generated during pumping. When this occurs, water recirculates within the pump impeller and casing. This flow condition is called

    a. run-around.
    b. skirting.
    c. slip.
    d. bypass.

107. Which valves can be used to throttle flow for only a short period of time?

    a. Butterfly valves
    b. Altitude valves
    c. Self-actuating valves
    d. Plug valves

108. Which type of valve operates similar to a diaphragm valve?

    a. Vacuum relief valve
    b. Globe valve
    c. Pressure relief valve
    d. Butterfly valve

109. Which valve can go from fully open to fully closed with ¼ turn?

    a. Plug valve
    b. Needle valve
    c. Globe valve
    d. Pinch valve

110. Foot valves are a special type of

    a. relief valve.
    b. control valve.
    c. check valve.
    d. plug valve.

111. Usually, which type of hydrant is a flush hydrant?

    a. Warm-climate hydrant
    b. Wet-barrel hydrant
    c. Dry-barrel hydrant
    d. Standpipe-barrel hydrant

112. Which type of temperature sensor uses two wires of different materials?

    a. Thermistors
    b. Thermo-resistors
    c. Thermocouples
    d. Thermo-conductors

113. Which type of temperature sensor uses a semi-conductive material?

    a. Thermistors
    b. Thermo-resistors
    c. Thermocouples
    d. Thermo-conductors

114. The amount of energy in feet that a pump supplies to a fluid is called

    a. velocity head.
    b. pump head.
    c. total head.
    d. pressure head.

115. Which is the most frequent cause of hydraulic transients in water distribution systems?

    a. Controlled pump shutdown
    b. Flow demand changes
    c. Pump startup
    d. Valve opening and closing

116. Which type of valve should be used for filling an elevated tank if full system pressure would overflow the tank?

    a. Pressure-relief valve
    b. Altitude valve
    c. Pressure-reducing valve
    d. Globe valve

117. One form of a differential-pressure flow meter is a(n)

    a. propeller.
    b. proportional.
    c. velocity-type.
    d. orifice plate.

118. If a centrifugal pump with a single volute is operated at more than _____ of its design capacity, an excess radial load will result, causing premature bearing failure.

    a. 110%
    b. 120%
    c. 125%
    d. 135%

119. How many inches of mercury is a total vacuum at sea level?

    a. 14.7 inches
    b. 30.0 inches
    c. 34.0 inches
    d. 39.6 inches

120. How many feet of water will equal the atmospheric pressure at sea level?

    a. 28 feet
    b. 30 feet
    c. 32 feet
    d. 34 feet

121. Pressure gauges connected to both suction and discharge sides of a pump should be connected

    a. 2 to 3 feet on either side of a pump, unless there is a change in direction it can go closer.
    b. 1 foot and 2 feet from the pump on the suction side and discharge side, respectively.
    c. next to the pump on the suction side and just before the check valve on the discharge valve.
    d. to the pressure taps supplied on the pump.

122. Which are the two principal types of throttling valves?

    a. Throttling and pressure-reducing valves
    b. Butterfly and relief valves
    c. Butterfly and altitude valves
    d. Pressure-reducing and altitude valves

123. Which valve would be best to use to precisely throttle flow?

    a. Globe valve
    b. Butterfly valve
    c. Rotary valve
    d. Needle valve

124. SCADA systems consist of the following distinct components:

    a. Remote terminal units (RTUs), communications, and human machine interface (HMI)
    b. Sensing instrument, RTUs, communications, and HMI
    c. Sensing instrument, RTUs, communications, master station, and HMI
    d. RTUs, communications, master station, and HMI

125. Stray-current corrosion is caused by

    a. corrosive soils.
    b. acidic soils.
    c. AC currents.
    d. DC currents.

126. When a metal is galvanized, it is coated with

    a. zinc.
    b. aluminum.
    c. aluminum oxide.
    d. aluminum hydroxide.

127. Which system would provide for a "soft start" to a motor?

    a. Wound-rotor induction motor and a controller
    b. Synchronous motor and a variable frequency controller
    c. A variable frequency drive and a squirrel-cage induction motor
    d. Squirrel-cage induction drive and a variable frequency drive

128. Which type of relay is frequently used to detect power loss and initiate a switchover to another power source?

    a. Loss-of-phase relays
    b. Voltage relays
    c. Frequency relays
    d. Circuit breaker relays

129. Which type of relays are placed on each phase of a power supply to open up a control circuit and stop a motor, if the current becomes excessive?

    a. Thermal-overload relays
    b. Frequency relays
    c. Voltage relays
    d. Differential relays

130. Which type of relay is often used where local power generation is involved and on synchronous motor starters to sense when the motor has reached synchronizing speed?

   a. Speed sensor relay
   b. Differential relay
   c. Frequency relay
   d. Voltage relay

131. Which type of relay is frequently used on large equipment to check whether all the current entering a system comes back out of the system?

   a. Voltage relay
   b. Over-current relay
   c. Reverse-current relay
   d. Differential relay

132. Which is the correct sequence of a telemetry signal?

   a. Indicator, sensor, transmission channel, receiver, transmitter, and signal
   b. Sensor, transmitter, transmission channel, receiver, and indicator
   c. Signal, indicator, transmitter, transmission channel, receiver, and sensor
   d. Signal, sensor, transmission channel, transmitter, receiver, and indicator

133. Which type of instrument is used to measure the average power of a load over a specific time interval?

   a. Demand meter
   b. Voltmeter
   c. Ammeter
   d. Wattmeter

134. Sacrificial anodes are also called

   a. corrosion anodes.
   b. corrosion cells.
   c. galvanic anodes.
   d. decaying anodes.

135. Which is becoming the most common, and is one of the least dangerous, methods of thawing a water service line?

   a. Hot water
   b. Electrical
   c. Small torch
   d. Steam vibration

136. When a control switch operated by pressure changes is located too close to a pump station, erratic operation can occur. This problem is known as

   a. pulsing.
   b. seesaw arching.
   c. racking.
   d. roaming.

137. Which is the electronic standard range?

    a. 4 to 20 mA DC
    b. 4 to 20 mA AC
    c. 0 to 100%
    d. 0 to 1 binary

138. Which is the chemical makeup of a tubercle?

    a. $Mn(OH)_2$ on the inside and $Ca(HCO_3)_2$ on the outside
    b. $Mn(OH)_2$ on the outside and $Ca(HCO_3)_2$ on the inside
    c. $Fe(OH)_2$ on the inside and $Fe(OH_3)$ on the outside
    d. $Fe(OH)_2$ on the outside and $Fe(OH_3)$ on the inside

139. Which device consists of a pair of metallic plates that are separated by an insulating material called a dielectric?

    a. Storage battery
    b. Voltage regulator
    c. Rectifier
    d. Capacitor

140. The current rating of a fuse should be _____ the current rating of the circuit or device that it protects.

    a. slightly lower than
    b. 15% greater than
    c. equal to
    d. equal to or slightly larger than

141. Conductors usually fail because of

    a. excessive heat.
    b. mechanical failure.
    c. lightening.
    d. simple aging.

142. Inspection of the shaft and coupling alignment should be performed every _____ or immediately, if problem signs develop.

    a. month
    b. quarter
    c. 6 months
    d. year

143. Bearings should not be lubricated with this type of oil because this oil forms acids as it begins to break down.

    a. Non-detergent mineral oil
    b. Animal oil
    c. SAE 10
    d. SAE 20

144. How much resistance should motor-winding insulation exhibit?

    a. At least 0.5 megohm of resistance
    b. At least 1.0 megohm of resistance
    c. At least 2.0 megohm of resistance
    d. At least 5.0 megohm of resistance

145. Which is the pneumatic standard range?

    a. 0 to 10 psig
    b. 3 to 15 psig
    c. 4 to 20 psig
    d. 0 to 100%

146. Which of the following methods for preventing external corrosion is only a short-term remedy?

    a. Extra thickness for pipe walls
    b. Wrapping pipe in polyethylene plastic sleeves
    c. Applying a coating of bitumastic or coal tar
    d. Installing cathodic protection on the pipe

147. The D'Arsonval meter is

    a. an amperometric meter.
    b. a type of pH meter.
    c. an analog (uses a needle) meter.
    d. a digital (number displays on unit) meter.

148. How does H_2S gas kill a person?

    a. Asphyxiation
    b. Paralyzes the respiratory system
    c. Paralyzes the nervous system
    d. Stops the heart

149. If sodium hypochlorite comes into contact with the skin, it should be immediately flushed with water for at least

    a. 10 minutes
    b. 15 minutes
    c. 20 minutes
    d. 30 minutes

150. Which is the Secondary Maximum Contaminant Level for pH?

    a. 7.0 pH units
    b. 6.5 to 7.5 pH units
    c. 6.5 to 8.5 pH units
    d. 6.0 to 8.5 pH units

151. If an excavation on a road requires that one of the lanes be closed and the speed limit is 25 mph, how many cones are required to divert the traffic?

    a. 6
    b. 9
    c. 13
    d. 15

152. If an excavation on a road requires that one of the lanes be closed and the speed limit is 45 mph, how many cones are required to divert the traffic?

    a. 9
    b. 13
    c. 15
    d. 18

153. Which chemical may cause leaching of lead in stagnant waters?

    a. Sodium zinc phosphate
    b. Lime
    c. Sodium carbonate
    d. Sodium hydroxide

154. During simulated distribution testing, a borate buffer solution maintains a constant pH of

    a. 10.
    b. 8.
    c. 7.
    d. 5.

155. In permit-entry confined space, who is responsible for terminating entries?

    a. Authorized entrant
    b. Standby attendant
    c. Entry supervisor
    d. Entry supervisor and the standby attendant

156. In permit-entry confined space, who is responsible for summoning rescuers?

    a. Standby attendant and authorized attendant
    b. Entry supervisor and authorized attendant
    c. Authorized attendant
    d. Authorized entrant

157. Extremely small chlorine leaks that go for weeks will most likely be found

    a. by using a very strong ammonia solution.
    b. by using soap and looking for bubbles.
    c. by using your nose.
    d. by seeing the discoloration and moisture at the leak point.

158. Which is the Action Level for lead?

    a. 0.015 mg/L
    b. 0.020 mg/L
    c. 0.025 mg/L
    d. 0.040 mg/L

159. Most regulated contaminants

    a. are thought to cause liver or kidney damage.
    b. can cause heart or cardiovascular diseases.
    c. cause health effects only after long exposure.
    d. cause acute effects.

160. High levels of cadmium at the customer's tap would most likely be due to

    a. copper pipe.
    b. galvanized pipe.
    c. iron pipe.
    d. household fixtures.

312   WATER OPERATOR CERTIFICATION STUDY GUIDE

161. Which is the Secondary Maximum Contaminant Level for zinc?

   a. 1 mg/L
   b. 2 mg/L
   c. 5 mg/L
   d. 10 mg/L

162. The prevention of trench cave-ins using the sloping method requires which slope ratio? (Note: horizontal to vertical distance, respectively)

   a. 1.0 : 1.0
   b. 1.5 : 1.0
   c. 2.0 : 1.0
   d. The angle will vary with type of soil, moisture content, and surrounding conditions

163. What is the Maximum Contaminant Level for asbestos fibers greater than 10 micrometers?

   a. 1 million fibers per liter (MFL)
   b. 2.5 MFL
   c. 5.0 MFL
   d. 7.0 MFL

164. Which is the Locational Running Annual Average (LRAA) for trihalomethanes (THM) and haloacetic acid 5 (HAA5) under the Stage 2 Disinfectants/Disinfection By-Products Rule?

   a. 80/60 µg/L
   b. 100/75 µg/L
   c. 100/80 µg/L
   d. 120/100 µg/L

165. Under the Surface Water Treatment Rule, undetected disinfection residuals must not exceed what percentage of samples each month for any two consecutive months that water is served to the public?

   a. Zero tolerance
   b. 1%
   c. 2%
   d. 5%

166. Which would be the most probable solution if Actinomycetes were causing taste and odor problems in the distribution system?

   a. Use copper sulfate and flush the system
   b. Start using chloramines as a disinfectant since it has a longer lasting residual compared to chlorine
   c. Use activated carbon and flush the system
   d. Control nutrients

167. Where is scale most likely to form in a hypochlorinator?

   a. Push rod
   b. Plunger
   c. Pump head
   d. Spring

168. Which type of chlorine gas feeding equipment would be most appropriate for distribution systems where rechlorination is required?

   a. Compound-loop control gas feeder
   b. Automatic proportioning control gas feeder
   c. Automatic residual control gas feeder
   d. Semiautomatic control gas feeder

169. When a sample is collected, its quality begins to change because of

   a. dissolved gases.
   b. chemical activity.
   c. pH.
   d. alkalinity.

170. Chemolithotrophic nitrifying bacteria are

   a. gram-negative and anaerobic.
   b. gram-positive and strictly aerobic.
   c. gram-negative and facultative or aerobic.
   d. gram-negative and strictly aerobic.

171. Which index will indicate scale-forming water with values less than 6 and corrosive water with values greater than 7?

   a. Baylis curve
   b. Langelier saturation index
   c. Marble test
   d. Ryzner index

172. The radiation term corresponding to an energy absorption of 100 ergs per gram of any medium is called a(n)

   a. rad.
   b. rem.
   c. curie.
   d. erg.

173. Community and nontransient community water systems must test for _____ if they use ozone for disinfection or oxidation.

   a. chlorate
   b. bromate
   c. haloacetic acids (five)
   d. total trihalomethanes

174. According to AWWA Standard C652 Method 3, disinfection of water tanks requires exposure to free chlorine at a concentration of _____ for 24-hours with the free chlorine residual ending at a concentration of at least _____.

   a. 10 mg/L; 2 mg/L
   b. 25 mg/L; 10 mg/L
   c. 50 mg/L; 10 mg/L
   d. 100 mg/L; 25 mg/L

175. If a main cannot be installed at an adequate depth in a cold weather climate, Styrofoam insulation board can be used to protect the water in the pipe from freezing. How far above the pipe should this insulation be installed?

   a. 2 to 4 inches
   b. 5 to 7 inches
   c. 8 to 10 inches
   d. 1 foot

176. How much deeper below the grade line of a pipe bottom should a trench be excavated if it is being dug in a rock formation?

   a. 3 to 6 inches
   b. 6 to 9 inches
   c. 9 to 12 inches
   d. 12 to 18 inches

177. Pressure-relief valves are most similar to

   a. globe valves.
   b. vacuum relief valves.
   c. needle valves.
   d. check valves.

178. Current meters are primarily used for measuring flow in lines that are

   a. 5/8 to 1 inch.
   b. 1 to 3 inches.
   c. 3 inches or greater.
   d. 12 inches or greater.

179. Which does beam breakage resemble?

   a. Shear breakage
   b. Stress breakage
   c. Flexural breakage
   d. Tensile breakage

180. Which type of valve should be installed for pressure control and flow control in large pipes?

   a. Butterfly valve
   b. Globe valve
   c. Gate valve
   d. Ball valve

181. Which can be developed by measuring the pressure along a pipeline?

   a. The C factor
   b. The total dynamic head
   c. The pipeline's grade
   d. The hydraulic grade

182. A thrust block would not be required on pipes that are

   a. made of steel.
   b. made of reinforced concrete.
   c. welded at the joints.
   d. of the bell and socket type.

183. Which is the recommended water pressure in commercial districts?

    a. 50 to less than 65 psi
    b. 55 to less than 75 psi
    c. 75 to less than 100 psi
    d. 90 to less than 120 psi

184. Which can be used to avoid excessively high water pressures in distribution mains?

    a. Pressure-reducing valves
    b. Plug valves
    c. Installation of a pneumatic tank
    d. Blowoff valves

185. Which is the commonly accepted water velocity in water mains at maximum flows?

    a. 1 to 3 ft/sec
    b. 2 to 4 ft/sec
    c. 4 to 5 ft/sec
    d. 5 to 6 ft/sec

186. It is recommended that dead-end water mains more than 1,000 feet in length should be constructed of pipe that is at least _____ in diameter.

    a. 6 inches
    b. 8 inches
    c. 10 inches
    d. 12 inches

187. Water mains that are 12 inches or less in diameter should have isolation valves that are located not more than _____ between each other.

    a. 500 feet
    b. 750 feet
    c. 1,000 feet
    d. 1,200 feet

188. The size and capacity of a water distribution system is based largely on

    a. peak hour customer demands.
    b. fire demand.
    c. agricultural demand.
    d. industrial demand.

189. A water storage tank located far from the source of demand would most likely have which type of problem during peak demand, if the area between storage and use is relatively flat?

    a. Low flow due to friction losses
    b. Poor water quality due to transmission pipe being too long
    c. High pumping costs during peak hourly flow
    d. Low pressure

190. Which type of pipe material is prestressed?

    a. Steel cylinder
    b. Ductile iron
    c. Asbestos-cement
    d. C-900 PVC

191. Which type of service pipe material is no longer acceptable for drinking water?

   a. Brass
   b. Lead
   c. Galvanized wrought iron
   d. Copper

192. Which type of joint for connecting water pipes has the advantage of moderate deflection and the gaskets used absorb vibration and pipe movement?

   a. Flanged joint
   b. Victaulic joint
   c. Dressler coupling
   d. Restrained joint

193. Which type of joint for connecting water pipes has the disadvantage of having a grove that weakens the pipe wall?

   a. Flanged joint
   b. Victaulic joint
   c. Dressler coupling
   d. Restrained joint

194. Which type of joint for connecting water pipes is used where the available space to lock a joint in place is lacking and where the soil behind the fitting may be disturbed?

   a. Flanged joint
   b. Victaulic joint
   c. Dressler coupling
   d. Restrained joint

195. Which is normally used on the outside joint space of steel cylinder concrete in order to protect the joint rings from external corrosion?

   a. Bituminous material
   b. Epoxy material
   c. Mastic coating
   d. Cement mortar

196. How much deflection is allowed for asbestos-cement couplings that are 3 to 12 inches in diameter?

   a. 5 degrees
   b. 7 degrees
   c. 10 degrees
   d. 12 degrees

197. Which is the best or first choice corrosion control method(s) to protect pipe?

   a. Use of noncorrosive metals and/or mechanical coatings
   b. Chemical protective coatings
   c. Electrical control by using cathodic protection
   d. Use of metallic coatings such as zinc or aluminum

198. What is an effective way for a distribution system agency to measure financial stability?

    a. Measuring the debt
    b. Calculating the operating and coverage ratios
    c. By dividing the paying customers by the total customers, then multiplying by 100%; anything over 95% is excellent
    d. Simply by how much money in the black that they have each fiscal year

199. Where do good public relations start for a water utility?

    a. With well-maintained fire hydrants
    b. With meaningful advertising
    c. Via public relations campaigns
    d. Through dedicated service-oriented employees

200. One of the most important functions and the major part of the work day for a manager of a water utility or distribution system is involved in

    a. organizing.
    b. planning.
    c. directing.
    d. controlling.

201. Which term defines lead free, when used with respect to solders and flux?

    a. Containing not more than 0.01% lead
    b. Containing not more than 0.02% lead
    c. Containing not more than 0.04% lead
    d. Containing not more than 0.05% lead

202. Which defines lead free when used with respect to pipes and pipe fittings?

    a. Containing not more than 3% lead
    b. Containing not more than 6% lead
    c. Containing not more than 8% lead
    d. Containing not more than 9% lead

203. If a water system collects less than 40 samples per month for the analyses of total coliforms, how many samples can be total coliform positive and the system still remain in compliance with the maximum contaminant level for total coliforms?

    a. Zero
    b. 1
    c. 2
    d. 3

204. Which is the maximum holding time for fecal coliforms using the Fecal Coliform Procedure?

    a. 8 hours
    b. 12 hours
    c. 24 hours
    d. 36 hours

318   WATER OPERATOR CERTIFICATION STUDY GUIDE

205. The peak historical month for a ground water system is the month with the highest total trihalomethanes or haloacetic acids (five) levels or the month with the

 a. warmest water temperature.
 b. lowest pH.
 c. lowest alkalinity.
 d. highest alkalinity.

206. A water system doing a system specific study plan may use the results of samples collected and analyzed for total trihalomethanes and haloacetic acids (five), if the samples were collected within _____ before the study plan submission date.

 a. 1 year
 b. 2 years
 c. 3 years
 d. 5 years

207. A water system is eligible to be certified 40/30, if it had no samples that exceeded 0.040 mg/L for total trihalomethanes and no samples that exceeded 0.030 mg/L for haloacetic acids (five). How many consecutive quarters must a water system be below these numbers to be 40/30 certified?

 a. 6 quarters
 b. 8 quarters
 c. 10 quarters
 d. 12 quarters

208. If a water system is required to conduct quarterly monitoring less frequently then quarterly under the Stage 2 Disinfection By-products Requirements they must begin monitoring in the calendar month

 a. beginning in January 2012.
 b. recommended in the Initial Distribution System Evaluation (IDSE) report.
 c. that includes the compliance date.
 d. that follows the quarter that includes the compliance date.

209. A water system that is required to sample quarterly must make compliance calculations at the end of the

 a. first quarter.
 b. fourth quarter.
 c. sixth quarter.
 d. eighth quarter.

210. All water systems under the Stage 2 Disinfection By-products Requirements must monitor during the month with the highest

 a. disinfection by-products.
 b. total organic carbon.
 c. organic acids.
 d. temperature and lowest pH.

211. A water system monitoring total trihalomethanes and haloacetic acids (five) quarterly must report the number of samples collected, date, and results of each sample during the last quarter. When must this report be submitted to the state?

   a. Within 5 days after quarter ends
   b. Within 10 days after quarter ends
   c. Within 14 days after quarter ends
   d. Within 30 days after quarter ends

212. How are samples for lead and copper analyses collected from services that have a lead service line?

   a. First draw samples that have remained motionless for at least 2 hours
   b. First draw samples that have remained motionless for at least 4 hours
   c. First flush the volume of water between tap and lead service line, then collect the sample
   d. Flush two volumes calculated between tap to end of lead service line, then collect the sample

213. When can a water system that meets the lead and copper action levels for two consecutive six-month periods, reduce the number of samples and reduce the frequency of sampling at the same time such that they only sample every 3 years?

   a. When the 90th percentiles for lead and copper are 0.003 mg/L and 0.50 mg/L, respectively
   b. When the 90th percentiles for lead and copper are 0.004 mg/L and 0.60 mg/L, respectively
   c. When the 90th percentiles for lead and copper are 0.005 mg/L and 0.65 mg/L, respectively
   d. When the 90th percentiles for lead and copper are 0.007 mg/L and 0.70 mg/L, respectively

214. Water systems that are sampling for lead and copper and qualify for a reduction in the number and frequency of sampling, unless the State has approved a different sampling period, shall conduct lead and copper sampling

   a. during October, November, December, and January.
   b. during June, July, August, and September.
   c. in the months January, April, July, and October.
   d. alternate months, collecting 6 sets of samples.

215. A small water system has met certain criteria and has been granted a waiver and thus a reduced frequency for monitoring lead and copper. How often with this waiver does it have to monitor?

   a. Once every 5 years
   b. Once every 9 years
   c. Once every 10 years
   d. Once every 12 years

320   WATER OPERATOR CERTIFICATION STUDY GUIDE

216. When a water system is required to monitor water quality parameters because it exceeded the lead and copper action level, which parameter must be measured every two weeks at each entry point to the distribution system?

   a. Alkalinity
   b. Disinfectant residual
   c. pH
   d. Orthophosphate

217. Each day following a routine sample monitoring result that exceeds the maximum residual disinfectant level for chlorine dioxide, the water system must collect from the distribution system _____ samples for analyses of chlorine dioxide.

   a. 2
   b. 3
   c. 4
   d. 5

218. If a water system monitors for total trihalomethanes (TTHM) and the sum of five haloacetic acids (HAA5), then on a quarterly basis the system must report

   a. the mean of both TTHM and HAA5 results each quarter.
   b. the highest value for both TTHM and HAA5 for each quarter.
   c. the arithmetic average of all samples for TTHM and HAA5.
   d. a running annual average for TTHM and HAA5 once four quarters of data have been collected.

219. How should a distribution operator preserve a sample to be analyzed for asbestos?

   a. Preserve with HNO_3
   b. Preserve with NaOH
   c. Preserve with H_2SO_4
   d. Preserve by keeping it at 4°C

220. How should a distribution operator preserve a sample to be analyzed for nitrate–nitrite?

   a. Preserve with HNO_3
   b. Preserve with NaOH
   c. Preserve with H_2SO_4
   d. Preserve by keeping it at 4°C

221. Which is the sample size for lead and copper analyses?

   a. 100 mL
   b. 250 mL
   c. 500 mL
   d. 1,000 mL

222. Which method can be used to control scaling but is never used to control corrosion?

   a. Controlled $CaCO_3$ scaling
   b. Polyphosphate addition
   c. pH and alkalinity adjustment with lime
   d. Sequestering

223. Samples collected for lead and copper analyses by customers need to avoid problems of residents handling an acid that would otherwise be contained in sample bottles. Thus the samples must be later acidified to

   a. preserve the sample.
   b. resolubilize the metals.
   c. free the lead and copper from possible reactions with other metals in the sample bottles.
   d. free the lead and copper from sodium that may be in the water.

224. Water systems are required to monitor for certain water quality parameters when they exceed the lead and copper action levels. If these systems maintain the range of values for the water quality parameters that reflect optimal corrosion control treatment during each of two consecutive six-month monitoring periods, then although they must still monitor at the entry point(s) to the distribution system, they are allowed to

   a. reduce the number of sites sampled.
   b. reduce the frequency of sampling from 6-month periods to annually.
   c. eliminate all water quality parameters monitored, except pH and alkalinity.
   d. eliminate all water quality parameters monitored, except pH and they must record in mg/L the dosage for any chemical alkalinity adjustment.

225. If the continuous monitoring of the residual disinfectant concentration for water entering the distribution system fails, how often must grab samples be collected in lieu of continuous monitoring until the problem is repaired?

   a. Every hour
   b. Every 4 hours
   c. Every 8 hours
   d. Every 12 hours

Math for Water Distribution Operators Levels I & II

Sample Questions

Answers on Page 371

1. A well yields 2,840 gallons in exactly 20 minutes. What is the well yield in gpm?

    a. 140 gpm
    b. 142 gpm
    c. 145 gpm
    d. 150 gpm

2. Convert 37.4 degrees Fahrenheit to degrees Celsius.

    a. 3.0°C
    b. 5.3°C
    c. 7.9°C
    d. 9.7°C

3. What is the area of a circular tank pad in ft^2, if it has a diameter of 102 ft?

    a. 6,160 ft^2
    b. 6,167 ft^2
    c. 8,170 ft^2
    d. 8,200 ft^2

4. What is the pressure 1.85 feet from the bottom of a water storage tank if the water level is 28.7 feet?

    a. 11.6 psi
    b. 12.4 psi
    c. 62.0 psi
    d. 66.3 psi

5. Calculate the well yield in gpm, given a drawdown of 14.1 ft and a specific yield of 31 gpm/ft.

    a. 2.2 gpm
    b. 7.3 gpm
    c. 45.1 gpm
    d. 440 gpm

6. How many gallons are in a pipe that is 18.0 in. in diameter and 1,165 ft long?

    a. 2,060 gal
    b. 10,300 gal
    c. 15,400 gal
    d. 17,200 gal

324   OPERATOR CERTIFICATION STUDY GUIDE

7. A water tank with a capacity of 5.75 million gallons (mil gal) is being filled at a rate of 2,105 gpm. How many hours will it take to fill the tank?

   a. 31.6 hr
   b. 37.8 hr
   c. 42.9 hr
   d. 45.5 hr

8. Determine the detention time in hours for the following water treatment system:

   - Distribution pipe from water plant to storage tank is 549 ft in length and 14 in. in diameter
   - Storage tank averages 2,310,000 gal of water at any given time
   - Flow through system is 6.72 mgd

   a. 7.2 hr
   b. 7.4 hr
   c. 8.0 hr
   d. 8.3 hr

9. Convert 28.7 cubic feet per second (cfs) to gallons per minute (gpm).

   a. 12,477 gpm
   b. 12,700 gpm
   c. 12,880 gpm
   d. 12,900 gpm

10. Convert 16,912,000 liters to acre-feet.

    a. 13.7 acre-ft
    b. 41.5 acre-ft
    c. 51.9 acre-ft
    d. 767 acre-ft

11. Convert −22.6 °C to degrees Fahrenheit.

    a. −4.6 °F
    b. −8.7 °F
    c. −11.8 °F
    d. −12.8 °F

12. A sodium hypochlorite solution contains 11.3% hypochlorite. Calculate the mg/L hypochlorite in the solution.

    a. 11.3 mg/L sodium hypochlorite
    b. 1,130 mg/L sodium hypochlorite
    c. 11,300 mg/L sodium hypochlorite
    d. 113,000 mg/L sodium hypochlorite

13. Records for a pump show that on June 1st at exactly 9:00 a.m. the number of pumped gallons was 71,576,344 and on July 1st at exactly 9:00 a.m. it was 72,487,008 gallons. Determine the average gallons pumped per day (gal/day) for this month to the nearest gallon.

    a. 18,605 gal/day
    b. 25,875 gal/day
    c. 30,355 gal/day
    d. 34,325 gal/day

14. If chlorine is being fed at a rate of 260 lb/day for a flow rate of 23 cfs, which should be the adjustment on the chlorinator when the flow rate is decreased to 16 cfs, if all other water parameters remain the same?

    a. 160 lb/day
    b. 180 lb/day
    c. 310 lb/day
    d. 370 lb/day

15. Calculate the diameter of a clarifier with a circumference of 215 ft.

    a. 34.8 ft
    b. 56.7 ft
    c. 68.5 ft
    d. 76.2 ft

16. Determine the depth of water in a reservoir, if the psi is 31.9.

    a. 13.8 ft deep
    b. 24.9 ft deep
    c. 45.6 ft deep
    d. 73.7 ft deep

17. Calculate the area of a tank, if the tank's radius is 39.8 ft.

    a. 4,970 ft^2
    b. 5,670 ft^2
    c. 7,820 ft^2
    d. 9,940 ft^2

18. Determine the specific gravity (SG) of an unknown liquid, if the density of the liquid is 70.9 lb/ft^3.

    a. 1.05 SG
    b. 1.14 SG
    c. 1.18 SG
    d. 1.21 SG

19. A water treatment plant is feeding an average of 295 lb/day of chlorine. If the dosage is 2.25 mg/L, which is the number of millions of gallons per day (mgd) being treated?

    a. 15.7 mgd
    b. 35.1 mgd
    c. 58.3 mgd
    d. 79.6 mgd

20. How many gallons of a sodium hypochlorite solution that contains 12.1% available chlorine are needed to disinfect a 1.5-ft diameter pipeline that is 283 ft long, if the dosage required is 50.0 mg/L? Assume the sodium hypochlorite is 9.92 lb/gal.

    a. 0.87 gal sodium hypochlorite
    b. 1.0 gal sodium hypochlorite
    c. 1.3 gal sodium hypochlorite
    d. 1.5 gal sodium hypochlorite

326 OPERATOR CERTIFICATION STUDY GUIDE

21. A water treatment plant is treating 16.4 million gallons per day (mgd). If the chlorine feed rate is 415 lb/day, which is the chlorine dosage in mg/L?

    a. 3.03 mg/L
    b. 3.38 mg/L
    c. 3.43 mg/L
    d. 3.67 mg/L

22. A 1.65-million gallon (mil gal) storage tank needs to be disinfected with a sodium hypochlorite solution that has 11.8% available chlorine. The tank is to be filled at 10% capacity, and the initial chlorine dosage required is 50.0 mg/L. How many gallons of sodium hypochlorite will be needed, if it weighs 9.84 lb/gal?

    a. 50 gal sodium hypochlorite
    b. 53 gal sodium hypochlorite
    c. 59 gal sodium hypochlorite
    d. 63 gal sodium hypochlorite

23. How many pounds of a calcium hypochlorite that contains 64.3% available chlorine are needed to disinfect a water main that is 24 in. in diameter, if the pipeline is 781 ft long and the dosage required is 50.0 mg/L?

    a. 5.95 lb calcium hypochlorite
    b. 8.25 lb calcium hypochlorite
    c. 11.9 lb calcium hypochlorite
    d. 13.8 lb calcium hypochlorite

24. A well is pumping water at a rate of 428 gpm. Which should be the setting on a chlorinator in pounds per day, if the dosage desired is 1.20 mg/L and the chlorine demand is 3.85 mg/L?

    a. 19.8 lb/day of chlorine
    b. 20.6 lb/day of chlorine
    c. 23.7 lb/day of chlorine
    d. 26.0 lb/day of chlorine

25. Convert 48.1 million gallons a day (mgd) to cubic feet per second (cfs).

    a. 68.7 cfs
    b. 74.4 cfs
    c. 79.1 cfs
    d. 82.0 cfs

26. Convert 184 gpm to liters per second (L/s).

    a. 10.3 L/s
    b. 11.1 L/s
    c. 11.6 L/s
    d. 12.3 L/s

27. Which is the average turbidity in ntu at the end of a sedimentation basin given the following data?

1	2	3	4	5	6	7
1.08 ntu	0.98 ntu	0.94 ntu	0.88 ntu	0.96 ntu	1.03 ntu	1.25 ntu

a. 1.00 ntu
b. 1.01 ntu
c. 1.02 ntu
d. 1.03 ntu

28. If 288 is 70.3%, how much is 100%?

a. 410
b. 412
c. 415
d. 418

29. The iron (Fe) content of a water source averages 0.81 mg/L iron. Which is the percent removal, if the treated water averages 0.01 mg/L iron?

a. 96% Fe removal efficiency
b. 97% Fe removal efficiency
c. 98% Fe removal efficiency
d. 99% Fe removal efficiency

30. Which will be the percent of soda ash in the resulting slurry, if 28.2 pounds of soda ash are mixed with exactly 100.0 gallons of water?

a. 3.27% soda ash slurry
b. 3.38% soda ash slurry
c. 3.45% soda ash slurry
d. 3.54% soda ash slurry

31. Which is the exposed exterior surface area of a ground-level storage tank that is 24.0 ft high and has a diameter of 80.1 ft? Assume top is flat.

a. 10,800 ft^2
b. 10,900 ft^2
c. 11,000 ft^2
d. 11,100 ft^2

32. A pipe is 1.43 miles long and has an inner diameter of 18.0 inches. How many gallons are in the pipeline if it is full?

a. 66,500 gal
b. 79,200 gal
c. 99,800 gal
d. 104,000 gal

33. A storage tank has a 60.0-ft radius and averages 25.5 ft in water depth. Calculate the average detention time in hours for this storage tank, if flow through the tank averages 2.91 mgd during a particular month in question.

a. 17.5 hr
b. 17.8 hr
c. 18.6 hr
d. 19.8 hr

34. If the pressure head on a fire hydrant is 134 ft, which is the pressure in psi?

    a. 50 psi
    b. 52 psi
    c. 54 psi
    d. 58 psi

35. A meter indicates the water flow from a fire hydrant is 5.5 ft³/min. How many gallons will flow from the hydrant in 20 minutes?

    a. 820 gal
    b. 850 gal
    c. 880 gal
    d. 920 gal

36. A polymer weighs 8.25 lb and occupies 3.150 liters. Which is the density of the polymer in g/cm³?

    a. 1.18 g/cm³
    b. 1.19 g/cm³
    c. 1.20 g/cm³
    d. 1.21 g/cm³

37. Determine the percent accuracy for a meter being tested, if it reads 245.7 cubic feet and the volumetric tank used to measure the water that flowed through the meter indicates the actual volume as 1,863 gallons.

    a. 97.8% meter efficiency
    b. 98.6% meter efficiency
    c. 99.0% meter efficiency
    d. 99.3% meter efficiency

38. Which is the chlorine dosage at a water treatment plant, if the chlorinator is set on 320 lb/day and the plant is treating 11.6 mgd?

    a. 2.8 mg/L
    b. 3.0 mg/L
    c. 3.3 mg/L
    d. 3.7 mg/L

39. A 1.75-mil gal storage tank needs to be disinfected with a sodium hypochlorite solution that contains 12.0% available chlorine and weighs 8.97 lb/gal. If the chlorine dosage is to be 50.0 mg/L, how many gallons of sodium hypochlorite are required?

    a. 678 gal
    b. 729 gal
    c. 750 gal
    d. 791 gal

40. A 24.0-in. pipeline, 427 ft long, was disinfected with calcium hypochlorite tablets with 65.0% available chlorine. Determine the chlorine dosage in mg/L, if 7.0 lb of calcium hypochlorite was used. Assume that the hypochlorite is so diluted that it weighs 8.34 lb/gal.

    a. 25 mg/L chlorine
    b. 39 mg/L chlorine
    c. 43 mg/L chlorine
    d. 54 mg/L chlorine

41. Water from a well is treated with a sodium hypochlorite solution that contains 10.3% available chlorine and weighs 8.95 lb/gal. The well is pumping water at 260 gpm. Calculate the chlorine dosage, if the chlorinator is pumping at a rate of 95 liters/day.

    a. 5.6 mg/L sodium hypochlorite
    b. 6.3 mg/L sodium hypochlorite
    c. 7.4 mg/L sodium hypochlorite
    d. 8.8 mg/L sodium hypochlorite

42. Which should be the setting on a chlorinator in pounds per day, if the dosage desired is 1.75 mg/L, the chlorine demand averages 2.45 mg/L, and the pumping rate from the well is 208 gpm?

    a. 10.5 lb/day chlorine
    b. 11.2 lb/day chlorine
    c. 12.0 lb/day chlorine
    d. 13.1 lb/day chlorine

43. What is the maximum pumping rate (in gpm) of a pump that is producing 15 water horsepower against a head of 65 ft?

    a. 115 gpm
    b. 910 gpm
    c. 17,000 gpm
    d. 63,000 gpm

44. A water plant serves 23,210 people. If it treats a yearly average of 2.98 mgd, what are the gallons per capita per day (gpcd)? Note: A capita = 1 person.

    a. 115 gpcd
    b. 120 gpcd
    c. 122 gpcd
    d. 128 gpcd

Math for Water Distribution Operators Levels III & IV

Answers on Page 381

1. What is the velocity of flow in feet per second for a 6.0-in. diameter pipe, if it delivers 122 gpm? Assume pipe is full.

    a. 1.3 ft/sec
    b. 1.35 ft/sec
    c. 1.38 ft/sec
    d. 1.4 ft/sec

2. A small cylinder on a hydraulic jack is 10 in. in diameter. A force of 130 lb is applied to the small cylinder. If the diameter of the large cylinder is 2.5 ft, what is the total lifting force?

    a. 1,170 lb
    b. 1,200 lb
    c. 1,250 lb
    d. 1,300 lb

3. A 2.0-ft diameter pipe that is 2.45 miles long was disinfected with chlorine. If 126.9 lb of chlorine were used, what was the initial dosage in mg/L?

    a. 25 mg/L
    b. 40 mg/L
    c. 50 mg/L
    d. 60 mg/L

4. What is the motor horsepower (mhp), if 200 horsepower (hp) is required to run a pump with a motor efficiency (Effic.) of 88% and a pump efficiency of 74%? Note: The 200 hp in this problem is called the water horsepower (whp). The whp is the actual energy (horsepower) available to pump water. Give results to two significant figures.

    a. 130 mhp
    b. 180 mhp
    c. 200 mhp
    d. 310 mhp

5. What is the bowl horsepower (bhp) for a vertical turbine pump given the following parameters?

    - Pumping rate = 385 gpm
    - Bowl head = 215 feet
    - Bowl efficiency = 81%

    a. 17 bhp
    b. 20 bhp
    c. 26 bhp
    d. 33 bhp

6. Water is flowing at a velocity of 1.3 ft/sec in a 4.0-in. diameter pipe. If the pipe changes from the 4.0-inch to a 3.0-in. pipe, what will the velocity be in the 3.0-in. pipe?

    a. 0.73 ft/sec
    b. 1.28 ft/sec
    c. 2.3 ft/sec
    d. 2.6 ft/sec

7. The level in a storage tank drops 2.3 ft in exactly 18 hr. If the tank has a diameter of 120 ft and the plant is producing 4.75 mgd, what is the average discharge rate of the three treated water discharge pumps in gpm?

    a. 3,479 gpm
    b. 3,500 gpm
    c. 4,578 gpm
    d. 4,600 gpm

8. How many gallons of a 12.5% sodium hypochlorite solution (9.34 lb/gal) are required to make exactly 1,000 gal of a 50 mg/L solution?

    a. 0.3 gal
    b. 0.4 gal
    c. 0.43 gal
    d. 0.63 gal

9. What percent hypochlorite solution would result, if 350 gal of an 11% solution were mixed with 225 gal of a 5.8% solution? Assume both solutions have the same density.

    a. 8.9% final solution
    b. 9.0% final solution
    c. 9.1% final solution
    d. 9.12% final solution

10. What water horsepower (whp) is required for a pump that delivers 650 gpm to a total head of 195 feet?

    a. 25 whp
    b. 30 whp
    c. 32 whp
    d. 40 whp

11. Determine the percentage strength of a solution mixture, if 875 lb of a 49.5% strength solution is mixed with 293 lb of a 17.2% strength solution.

    a. 41.4%
    b. 42.4%
    c. 43.0%
    d. 43.1%

12. How many fluid ounces (oz) of sodium hypochlorite (10.5% available chlorine and 9.10 lb/gal) are required to disinfect a well with the following parameters?

    - Depth of well is 287 ft
    - 12-in. diameter well casing extends down to 100.0 ft
    - The remainder is a 10.0-in. diameter casing
    - The residual desired dose is 50.0 mg/L
    - The depth to water is 168.4 ft
    - The chlorine demand is 4.7 mg/L

    a. 21 oz
    b. 25 oz
    c. 27 oz
    d. 30 oz

13. Determine the cost to the nearest cent to operate a 300 Hp motor for one month (assume 30 days), if it runs an average of 4.2 hr/day, is 82% efficient, and the electrical costs are $0.041 per kW.

    a. $948.04
    b. $970.92
    c. $1,156.15
    d. $1,184.05

14. A storage tank has a level capacity of 24.50 ft. Currently the water level is 16.55 ft in the tank. Calculate the SCADA reading on the board in mA for a 4 mA to 20 mA signal.

    a. 13.5 mA
    b. 13.51 mA
    c. 14.8 mA
    d. 14.81 mA

15. A pipe that is 3,270 ft long has a diameter of 14.0 in. for two-thirds of its length and 10.0 in. for the remaining one-third. How many gallons will it take to completely fill this pipe?

    a. 17,598 gal
    b. 21,900 gal
    c. 52,670 gal
    d. 87,500 gal

16. How many pounds of lime must be added to exactly 200 gal of water to produce a lime slurry of 15%?

    a. 220 lb
    b. 290 lb
    c. 340 lb
    d. 420 lb

17. Determine the volume in gallons of a trapezoid-shaped canal that has the following dimensions:

    - Length = 6,091 ft
    - Height = 4.10 ft
    - Bottom width = 5.85 ft (b_1)
    - Top width = 10.6 ft (b_2)

    a. 1,270,000 gal
    b. 1,330,000 gal
    c. 1,480,000 gal
    d. 1,540,000 gal

18. Calculate the detention time to the nearest 100 hr for the following system:

    - The clear well is 308 ft long, 118 ft wide, and has an average water depth of 12.85 ft
    - Distribution pipe from clear well to storage tank is 1.34 miles long and has a diameter of 2.00 ft
    - The storage tank has a diameter of 99.8 ft and averages a height of 26.48 ft of water
    - The water production for the year averaged 30.02 mgd

    a. 4.02 hr
    b. 4.16 hr
    c. 4.22 hr
    d. 4.29 hr

19. A well has a depth of 276.5 ft. If the depth to water is 153.8 ft, which is the pressure in psi 5.0 ft above the bottom? Disregard the additional atmospheric pressure in the well.

    a. 42 psi
    b. 46 psi
    c. 48 psi
    d. 51 psi

20. How many gallons per minute should a flowmeter register, if a 10.0-in. diameter main is to be flushed at 5.10 ft/sec?

    a. 1,050 gpm
    b. 1,100 gpm
    c. 1,250 gpm
    d. 1,350 gpm

21. Water is flowing at a velocity of 2.0 ft/sec in an 8.0-in. diameter pipe. If the pipe changes from the 8.0-in. to a 10.0-in. pipe, the velocity in the 10.0-in. pipe will be

    a. 1.3 ft/sec
    b. 1.5 ft/sec
    c. 1.7 ft/sec
    d. 1.8 ft/sec

22. An 18-in. diameter distribution pipe delivers 988,000 gallons in 24 hr. Which is the average flow during the 24 hr in ft/sec?

    a. 0.60 ft/sec
    b. 0.73 ft/sec
    c. 0.87 ft/sec
    d. 0.94 ft/sec

23. A 64.5% calcium hypochlorite solution was used to treat 10.6 mil gal. The tank containing the hypochlorite solution is 6.0 ft in diameter. If the tank dropped 8.03 in. during the time the 10.6 mil gal were treated, which must have been the chlorine dosage in mg/L?

    a. 5.78 mg/L
    b. 6.33 mg/L
    c. 7.25 mg/L
    d. 8.61 mg/L

24. A well that is 227 ft deep and 12 in. in diameter requires disinfection. Depth to water from the casing top is 143 ft. If the desired dose is 50.0 mg/L, how many gallons of sodium hypochlorite (12.5% available chlorine) are required? Note: The specific gravity of the sodium hypochlorite is 1.15 or 9.59 lb/gal.

    a. 0.17 gal
    b. 0.19 gal
    c. 0.21 gal
    d. 0.25 gal

25. A pipe that is 2.50 ft in diameter and 1,058 ft long is to be disinfected with 64.5% calcium hypochlorite tablets. If the desired dose is 25.0 mg/L, how many pounds of calcium hypochlorite are required?

    a. 10.1 lb
    b. 11.7 lb
    c. 12.6 lb
    d. 13.2 lb

26. A tank 84.0 ft in diameter and 24.25 ft high at the overflow requires disinfection. How much 12.5% sodium hypochlorite that is 9.59 lb/gal will be required for a dosage of 50.0 mg/L?

    a. 310 gal
    b. 350 gal
    c. 380 gal
    d. 410 gal

27. How many calcium hypochlorite tablets, each weighing 0.45 lb, are needed to disinfect a water main, given the following information:

    - Length of pipe = 513 ft
    - Pipe diameter = 2.50 ft
    - Calcium hypochlorite = 64.0% available chlorine
    - Dosage required = 25.0 mg/L

    a. 10 tablets
    b. 12 tablets
    c. 14 tablets
    d. 16 tablets

28. A well that is 210 ft in depth and 14.0 in. in diameter requires disinfection. The depth to water from top of casing is 91 ft. If the desired dose is 50.0 mg/L, which is the number of pounds and ounces of sodium hypochlorite (12.5% available chlorine) required? Assume the sodium hypochlorite solution is 9.59 lb/gal.

    a. 43 oz of NaOCl
    b. 45 oz of NaOCl
    c. 49 oz of NaOCl
    d. 54 oz of NaOCl

29. Soda ash slurry is being added to water being released from a clear well to the distribution system to raise the pH. If the amount of soda ash being added averages 124.5 grams per minute for that day and the water leaving the distribution system averages 3,075 gpm for that day, which must have been the soda ash dosage in mg/L?

    a. 8.18 mg/L
    b. 9.70 mg/L
    c. 10.69 mg/L
    d. 12.47 mg/L

30. The level in a clearwell tank drops 7.08 ft in exactly 12.0 hr. If the tank has a diameter of 149.8 ft and the plant is producing 4.75 mgd, calculate the average discharge rate for each pump of the four same capacity treated water discharge pumps in gallons per minute.

    a. 1,150 gpm
    b. 1200 gpm
    c. 1,250 gpm
    d. 1,680 gpm

31. Determine the horsepower (hp) required for a clear well water pump that needs to pump water to a storage tank given the following parameters:

    - Elevation of clear well water pump = 170.84 ft
    - Elevation of water storage tank = 478.16 ft
    - Length of pipeline from clear well water pump to storage tank = 2,107 ft
    - Pump above clear well (suction lift) = 2.5 ft
    - Friction loss in pipeline = 1.57 ft per 1,000 ft
    - Assume velocity head = 2.38 ft
    - Required flow per day (maximum) = 4,000 gpm
    - Pump efficiency = 85%
    - Motor efficiency = 89%

    a. 375 hp
    b. 400 hp
    c. 420 hp
    d. 450 hp

32. Which is the net positive suction head available (NPSHA) given the following data? Will the pump cavitate, if the net positive suction head required (NPSHR) is 18.4 ft? Note: There are 1.11 ft/in. of Hg.

    - Atmospheric pressure (AP) = 29.8 in. Hg
    - Static suction lift (SSL) = 15.1 ft
    - Friction headloss (Hf) = 0.61 ft
    - Vapor pressure at 12°F (VP) = 0.50 ft

    a. 14 ft, therefore NPSHA < NPSHR so cavitation should occur
    b. 17 ft, therefore NPSHA < NPSHR so cavitation should occur
    c. 20 ft, therefore NPSHA > NPSHR so cavitation should not occur
    d. 22 ft, therefore NPSHA > NPSHR so cavitation should not occur

33. Determine the approximate C factor for a pipe that is 1.0 ft in diameter and has a flow of 1,225 gpm given the following data:

    - Upstream pressure gauge = 120 ft
    - Downstream pressure gauge = 105 ft
    - Distance between gauges = 2,274 ft

    a. 95
    b. 100
    c. 110
    d. 120

34. A storage tank has a capacity of 34.0 ft. Currently there are 22.89 ft of water in the tank. Which would the SCADA reading be on the board in milliamps (mA) for a 4-mA to 20-mA signal?

    a. 13.9 mA
    b. 14.1 mA
    c. 14.3 mA
    d. 14.8 mA

Answers to Water Distribution Operator Questions

System Information/Components

Sample Questions for Level I—Answers

1. Answer: **c.** feet.

    Reference: *Basic Science Concepts and Applications,* 4th edition. 2010. Nicholas G. Pizzi, Editor. American Water Works Association. Hydraulics 3.

2. Answer: **a.** a map.

    Reference: *Water Transmission and Distribution*, 4th edition. 2010. Larry Mays, Editor. American Water Works Association. Glossary.

3. Answer: **d.** Air-and-vacuum relief valve

    Reference: *Water Transmission and Distribution*, 4th edition. 2010. Larry Mays, Editor. American Water Works Association. Chapter 6.

4. Answer: **a.** cathodic protection.

    Reference: *Water Transmission and Distribution*, 4th edition. 2010. Larry Mays, Editor. American Water Works Association. Chapter 2.

5. Answer: **a.** Early morning

    Reference: *Water Transmission and Distribution,* 4th edition. 2010. Larry Mays, Editor. American Water Works Association. Chapter 5.

Sample Questions for Level II—Answers

1. Answer: **a.** pressure head.

    Reference: *Pumping: Fundamentals for the Water & Wastewater Maintenance Operator Series.* 2001. Frank R. Spellman and Joanne Drinan. Technomic Publishing Company.

2. Answer: **b.** pump head.

    Reference: *Pumping: Fundamentals for the Water & Wastewater Maintenance Operator Series.* 2001. Frank R. Spellman and Joanne Drinan. Technomic Publishing Company.

3. Answer: **d.** roughness that retards flow due to friction.

    Reference: *Water Transmission and Distribution*, 4th edition. 2010. Larry Mays, Editor. American Water Works Association. Chapter 2.

4. Answer: **b.** Push-on joint

    Reference: *Water Transmission and Distribution*, 4th edition. 2010. Larry Mays, Editor. American Water Works Association. Chapter 2.

5. Answer: **d.** air gap, RPZ, DCVA, and VB.

   Reference: *Water Transmission and Distribution*, 4th edition. 2010. Larry Mays, Editor. American Water Works Association. Chapter 11.

Sample Questions for Level III—Answers

1. Answer: **c.** proportionately.

   Reference: *Pumping: Fundamentals for the Water & Wastewater Maintenance Operator Series*. 2001. Frank R. Spellman and Joanne Drinan. Technomic Publishing Company. Chapter 3.

2. Answer: **c.** Arterial-loop system

   Reference: *Water Transmission and Distribution*, 4th edition. 2010. Larry Mays, Editor. American Water Works Association. Chapter 2.

3. Answer: **a.** Grid system

   Reference: *Water Transmission and Distribution*, 4th edition. 2010. Larry Mays, Editor. American Water Works Association. Chapter 2.

4. Answer: **c.** 35 psi or greater.

   Reference: *Water Transmission and Distribution*, 4th edition. 2010. Larry Mays, Editor. American Water Works Association. Chapter 2.

5. Answer: **b.** Restrained joint

   Reference: *Water Transmission and Distribution*, 4th edition. 2010. Larry Mays, Editor. American Water Works Association. Chapter 2.

Sample Questions for Level IV—Answers

1. Answer: **c.** to increase flow.

   Reference: *Pumping: Fundamentals for the Water & Wastewater Maintenance Operator Series*. 2001. Frank R. Spellman and Joanne Drinan. Technomic Publishing Company. Chapter 2.

2. Answer: **a.** It has the potential for both internal and external corrosion

   Reference: *Water Transmission and Distribution*, 4th edition. 2010. Larry Mays, Editor. American Water Works Association. Chapter 2.

3. Answer: **c.** the Hazen-Williams formula.

   Reference: *Basic Science Concepts and Applications*, 4th edition. 2010. Nicholas G. Pizzi, Editor. American Water Works Association. Hydraulics 5.

4. Answer: **c.** potential energy.

   Reference: *Pumps & Pumping*, 9th edition. 2010. E.E. "Skeet" Arasmith, Mitch Scheele, & Kimon Zentz. ACR Publications. Lesson 1.

5. Answer: **a.** 1; 2

   Reference: *Basic Science Concepts and Applications*, 4th edition. 2010. Nicholas G. Pizzi, Editor. American Water Works Association. Hydraulics 4.

Monitor, Evaluate, and Adjust Disinfection

Sample Questions for Level I—Answers

1. Answer: **c.** has a persistent residual.

   Reference: *Water Treatment*, 4th edition. 2010. Nicholas G. Pizzi, Editor. American Water Works Association. Chapter 7.

2. Answer: **b.** 2.5

   Reference: *Water Treatment Operator Handbook*, Revised edition. 2005. Nicholas G. Pizzi. American Water Works Association. Chapter 8.

3. Answer: **c.** Coliform test

   Reference: *Water Transmission and Distribution*, 4th edition. 2010. Larry Mays, Editor. American Water Works Association. Chapter 3.

4. Answer: **d.** Use a chlorine gas detector and place it next to any area suspected of leaking

   Reference: M20, *Water Chlorination/Chlorination Practices and Principles*, 2nd edition. 2006. American Water Works Association. Chapter 4.

5. Answer: **a.** a detectable level

   Reference: 40 CFR 141 Subpart G: National Primary Drinking Water Regulations: Maximum Contaminant Levels and Maximum Residual Disinfectant Levels.

Sample Questions for Level II—Answers

1. Answer: **b.** 5 to 20%

   Reference: M20, *Water Chlorination/Chlorination Practices and Principles*, 2nd edition. 2006. American Water Works Association. Chapter 2.

2. Answer: **d.** somewhere in the distribution system.

   Reference: M20, *Water Chlorination/Chlorination Practices and Principles*, 2nd edition. 2006. American Water Works Association. Chapter 6.

3. Answer: **b.** 25 mg/L

   Reference: M20, *Water Chlorination/Chlorination Practices and Principles,* 2nd edition. 2006. American Water Works Association. Appendix C.

4. Answer: **c.** Residual disinfectant concentration

   Reference: National Primary Drinking Water Regulations, Subpart H – Filtration and Disinfection, §141.74 – Analytical and monitoring requirements (b)(6)(i)

5. Answer: **a.** a weak acid.

   Reference: *Water Treatment Operator Handbook*, Revised edition. 2005. Nicholas G. Pizzi. American Water Works Association. Chapter 8.

Sample Questions for Level III—Answers

1. Answer: **b.** Chlorine

   Reference: *Water Treatment*, 4th edition. 2010. Nicholas G. Pizzi, Editor. American Water Works Association. Chapter 7.

2. Answer: **a.** 4.0 mg/L

   Reference: M20, *Water Chlorination/Chlorination Practices and Principles*, 2nd edition. 2006. American Water Works Association. Chapter 3.

3. Answer: **a.** Clogged equipment

   Reference: *Water Treatment*, 4th edition. 2010. Nicholas G. Pizzi, Editor. American Water Works Association. Chapter 7.

4. Answer: **c.** Organic material

   Reference: *Water Treatment Operator Handbook*, Revised edition. 2005. Nicholas G. Pizzi. American Water Works Association. Chapter 8.

5. Answer: **b.** Every 4 hours

   Reference: National Primary Drinking Water Regulations, Subpart H – Filtration and Disinfection, §141.74 – Analytical and monitoring requirements (b)(5)

Sample Questions for Level IV—Answers

1. Answer: **c.** Nematodes

   Reference: M7, *Problem Organisms in Water: Identification and Treatment*, 3rd edition. 2003. American Water Works Association. Chapter 5.

2. Answer: **a.** 0.4 to 0.5 mg/L

   Reference: *Water Treatment*, 4th edition. 2010. Nicholas G. Pizzi, Editor. American Water Works Association. Chapter 7.

3. Answer: **d.** 4.5 : 1 to 5.0 : 1

   Reference: M20, *Water Chlorination/Chlorination Practices and Principles*, 2nd edition. 2006. American Water Works Association. Chapter 6.

4. Answer: **c.** sodium chlorite + chlorine at low pH values.

   Reference: *Water Treatment Operator Handbook*, Revised edition. 2005. Nicholas G. Pizzi. American Water Works Association. Chapter 8.

5. Answer: **d.** Routine use of disinfectant and penetrant

   Reference: M7, *Problem Organisms in Water: Identification and Treatment*, 3rd edition. 2003. American Water Works Association. Chapter 11.

Laboratory Analysis

Sample Questions for Level I—Answers

1. Answer: **b.** 6 hours.

    Reference: *Water Treatment Operator Handbook*, Revised edition. 2005. Nicholas G. Pizzi. American Water Works Association. Chapter 10.

2. Answer: **c.** sodium thiosulfate.

    Reference: *Basic Microbiology for Drinking Water Personnel*, 2nd edition. 2006. Dennis R. Hill. American Water Works Association. Chapter 4.

3. Answer: **a.** 100 mL.

    Reference: *Water Treatment Operator Handbook*, Revised edition. 2005. Nicholas G. Pizzi. American Water Works Association. Chapter 12.

4. Answer: **b.** 15 minutes.

    Reference: *Water Quality*, 4th edition. 2010. Joseph A. Ritter, Editor. American Water Works Association. Chapter 6.

5. Answer: **a.** Heterotrophic plate count (HPC)

    Reference: National Primary Drinking Water Regulations, Subpart H – Filtration and Disinfection, §141.74 – Analytical and monitoring requirements (b)(6)(i)

Sample Questions for Level II—Answers

1. Answer: **a.** pH

    Reference: *Water Treatment Operator Handbook*, Revised edition. 2005. Nicholas G. Pizzi. American Water Works Association. Chapter 12.

2. Answer: **a.** accept an electron pair.

    Reference: *The Water Dictionary: A Comprehensive Reference of Water Terminology*, 2nd edition. 2010. Nancy McTigue, Editor, and James M. Symons, Editor Emeritus. American Water Works Association.

3. Answer: **b.** Lime

    Reference: *Water Treatment*, 4th edition. 2010. Nicholas G. Pizzi, Editor. American Water Works Association. 4th edition. Chapter 4.

4. Answer: **a.** faucet without threads.

    Reference: *Water Quality*, 4th edition. 2010. Joseph A. Ritter, Editor. American Water Works Association. Chapter 2.

5. Answer: **a.** DPD color comparater

    Reference: *Water Quality*, 4th edition. 2010. Joseph A. Ritter, Editor. American Water Works Association. Chapter 6.

Sample Questions for Level III—Answers

1. Answer: **d.** concentrated nitric acid.

    Reference: *Water Quality*, 4th edition. 2010. Joseph A. Ritter, Editor. American Water Works Association. Chapter 6.

2. Answer: **b.** Every 4 hours

   Reference: National Primary Drinking Water Regulations, Subpart H – Filtration and Disinfection, §141.74 – Analytical and monitoring requirements (c)(1)

3. Answer: **b.** Lowest value for chlorine residual and contact time, lowest temperature, and highest pH

   Reference: National Primary Drinking Water Regulations, Subpart H – Filtration and Disinfection, §141.75 – Reporting and record keeping requirements (a)(2) (i, iii, iv, v)

4. Answer: **c.** 0.015 mg/L.

   Reference: National Primary Drinking Water Regulations, Subpart I – Control of Lead and Copper, §141.80 – General requirements (c)(1)

5. Answer: **c.** TOC < 2.0 mg/L for 2 years or < 1.0 mg/L for 1 year

   Reference: National Primary Drinking Water Regulations Subpart L – Disinfectant Residuals, Disinfectant By-products and Disinfectant By-product precursors, §141.132 – Monitoring requirements (d)(2)

Sample Questions for Level IV—Answers

1. Answer: **a.** By taking the average of the highest and second highest concentrations

   Reference: National Primary Drinking Water Regulations, Subpart I – Control of Lead and Copper, §141.80 – General requirements (c)(3)(iv)

2. Answer: **c.** TTHM ≤ 0.040 mg/L and HAA5 ≤ 0.030 mg/L.

   Reference: National Primary Drinking Water Regulations, Subpart L – Disinfectant Residuals, Disinfectant By-products and Disinfectant By-product Precursors, §141.132 – Monitoring requirements (b)(1)(ii), Table – Reduced monitoring frequency for TTHM and HAA5

3. Answer: **a.** one three-sample set per quarter

   Reference: National Primary Drinking Water Regulations, Subpart L – Disinfectant Residuals, Disinfectant By-products and Disinfectant By-product Precursors, §141.132 – Monitoring requirements (b)(2)(iii)(B)

4. Answer: **d.** No more than 5%

   Reference: National Primary Drinking Water Regulations, Subpart G – Maximum Contaminant Levels and Maximum Residual Disinfectant Levels, §141.63 – Maximum contaminant levels (MCLs) for microbiological contaminants (a)(1)

5. Answer: **b.** develop a disinfection benchmark.

   Reference: National Primary Drinking Water Regulations Subpart P – Enhanced Filtration and Disinfection—Systems Serving 10,000 or More People, §141.172 – Disinfection profiling and benchmarking (c)(1)

Install Equipment

Sample Questions for Level I—Answers

1. Answer: **a.** 6 inches

   Reference: *Water Transmission and Distribution,* 4th edition. 2010. Larry Mays, Editor. American Water Works Association. Chapter 7.

2. Answer: **a.** service line.

   Reference: *Water Transmission and Distribution*, 4th edition. 2010. Larry Mays, Editor. American Water Works Association. Chapter 6.

3. Answer: **b.** 3.0 feet

   Reference: *Water Transmission and Distribution*, 4th edition. 2010. Larry Mays, Editor. American Water Works Association. Chapter 12.

4. Answer: **c.** Not having the joint completely clean

   Reference: *Water Transmission and Distribution*, 4th edition. 2010. Larry Mays, Editor. American Water Works Association. Chapter 12.

5. Answer: **c.** Blowoff valve

   Reference: *Water Distribution System Operation and Maintenance*, 5th edition. 2005. Ken Kerri. California State University – Sacramento. Page 71.

Sample Questions for Level II—Answers

1. Answer: **c.** compressed beveled gasket.

   Reference: *Water Transmission and Distribution*, 4th edition. 2010. Larry Mays, Editor. American Water Works Association. Chapter 15.

2. Answer: **c.** beam breakage.

   Reference: *Water Transmission and Distribution*, 4th edition. 2010. Larry Mays, Editor. American Water Works Association. Chapter 2.

3. Answer: **c.** perpendicular

   Reference: *Water Transmission and Distribution*, 4th edition. 2010. Larry Mays, Editor. American Water Works Association. Chapter 12.

4. Answer: **a.** Restraining fittings

   Reference: *Water Transmission and Distribution*, 4th edition. 2010. Larry Mays, Editor. American Water Works Association. Chapter 12.

5. Answer: **a.** moisture.

   Reference: *Water Transmission and Distribution*, 4th edition. 2010. Larry Mays, Editor. American Water Works Association. Chapter 13.

Sample Questions for Level III—Answers

1. Answer: **a.** gate valve.

   Reference: *Water Transmission and Distribution*, 4th edition. 2010. Larry Mays, Editor. American Water Works Association. Chapter 6.

2. Answer: **a.** 1 to 2 inches

   Reference: *Water Transmission and Distribution*, 4th edition. 2010. Larry Mays, Editor. American Water Works Association. Chapter 10.

3. Answer: **b.** By gently tapping the length of the pipe with a hammer; the pipe should ring or hum clearly

   Reference: *Water Distribution System Operation and Maintenance*, 5th edition. 2005. Ken Kerri. California State University – Sacramento. Page 90.

4. Answer: **c.** 66 inches

   Reference: *Water Transmission and Distribution*, 4th edition. 2010. Larry Mays, Editor. American Water Works Association. Chapter 12.

5. Answer: **c.** the centerline of the pipe.

   Reference: *Water Distribution System Operation and Maintenance*, 5th edition. 2005. Ken Kerri. California State University – Sacramento. Page 105.

Sample Questions for Level IV—Answers

1. Answer: **d.** 1 inch.

   Reference: *Water Transmission and Distribution*, 4th edition. 2010. Larry Mays, Editor. American Water Works Association. Chapter 12.

2. Answer: **a.** $\frac{1}{32}$ inch

   Reference: *Pumps & Pumping*, 9th edition. 2010. E.E. "Skeet" Arasmith, Mitch Scheele, & Kimon Zentz. ARC Publications. Chapter 4.

3. Answer: **d.** 4.0 degrees

   Reference: *Pumps & Pumping*, 9th edition. 2010. E.E. "Skeet" Arasmith, Mitch Scheele, & Kimon Zentz. ARC Publications. Chapter 4.

4. Answer: **d.** downstream; pressure

   Reference: *Water Distribution System Operation and Maintenance*, 5th edition. 2005. Ken Kerri. California State University – Sacramento. Page 121.

5. Answer: **d.** The relief valve will remain fully open and an air gap (atmospheric pressure) will form between the check valves, closing them both

   Reference: *Water Distribution System Operation and Maintenance*, 5th edition. 2005. Ken Kerri. California State University – Sacramento. Page 146.

Operate Equipment

Sample Questions for Level I—Answers

1. Answer: **b.** impeller.

   Reference: *Pumps & Pumping,* 9th edition. 2010. E.E. "Skeet" Arasmith, Mitch Scheele, & Kimon Zentz. ARC Publications. Lesson 3.

2. Answer: **c.** Mechanical seal

   Reference: *Pumps & Pumping,* 9th edition. 2010. E.E. "Skeet" Arasmith, Mitch Scheele, & Kimon Zentz. ARC Publications. Lesson 3.

3. Answer: **c.** Seals

   Reference: *Pumping: Fundamentals for the Water & Wastewater Maintenance Operator Series.* 2001. Frank R. Spellman and Joanne Drinan. Technomic Publishing Company. Chapter 3.

4. Answer: **a.** Packing gland

   Reference: *Pumping: Fundamentals for the Water & Wastewater Maintenance Operator Series.* 2001. Frank R. Spellman and Joanne Drinan. Technomic Publishing Company. Chapter 4.

5. Answer: **d.** Stuffing box

   Reference: *Pumping: Fundamentals for the Water & Wastewater Maintenance Operator Series.* 2001. Frank R. Spellman and Joanne Drinan. Technomic Publishing Company. Chapter 3.

Sample Questions for Level II—Answers

1. Answer: **b.** Key

   Reference: *Pumps & Pumping,* 9th edition. 2010. E.E. "Skeet" Arasmith, Mitch Scheele, & Kimon Zentz. ARC Publications. Lesson 4.

2. Answer: **c.** Gate valve

   Reference: *Water Transmission and Distribution,* 4th edition. 2010. Larry Mays, Editor. American Water Works Association. Chapter 4.

3. Answer: **b.** acoustic waves.

   Reference: *Water Transmission and Distribution,* 4th edition. 2010. Larry Mays, Editor. American Water Works Association. Chapter 5.

4. Answer: **c.** Globe valve

   Reference: *Water Transmission and Distribution,* 4th edition. 2010. Larry Mays, Editor. American Water Works Association. Chapter 6.

5. Answer: **b.** Strain gauge

   Reference: *Water Transmission and Distribution,* 4th edition. 2010. Larry Mays, Editor. American Water Works Association. Chapter 9.

Sample Questions for Level III—Answers

1. Answer: **c.** Electric and magnetic forces

    Reference: *Basic Science Concepts and Applications*, 4th edition. 2010. Nicholas G. Pizzi, Editor. American Water Works Association. Electricity 1.

2. Answer: **a.** pressure head.

    Reference: *Pumping: Fundamentals for the Water & Wastewater Maintenance Operator Series*. 2001. Frank R. Spellman and Joanne Drinan. Technomic Publishing Company. Chapter 2.

3. Answer: **b.** Two stage

    Reference: *Pumps & Pumping,* 9th edition. 2010. E.E. "Skeet" Arasmith, Mitch Scheele, & Kimon Zentz. ARC Publications. Lesson 2.

4. Answer: **b.** Two stages

    Reference: *Pumps & Pumping,* 9th edition. 2010. E.E. "Skeet" Arasmith, Mitch Scheele, & Kimon Zentz. ARC Publications. Lesson 2.

5. Answer: **c.** Open suction piping air bleed-off valves

    Reference: *Pumping: Fundamentals for the Water & Wastewater Maintenance Operator Series*. 2001. Frank R. Spellman and Joanne Drinan. Technomic Publishing Company. Chapter 9.

Sample Questions for Level IV—Answers

1. Answer: **a.** center of the impeller.

    Reference: *Pumps & Pumping,* 9th edition. 2010. E.E. "Skeet" Arasmith, Mitch Scheele, & Kimon Zentz. ARC Publications. Lesson 3.

2. Answer: **b.** torque

    Reference: *Pumps & Pumping,* 9th edition. 2010. E.E. "Skeet" Arasmith, Mitch Scheele, & Kimon Zentz. ARC Publications. Lesson 3.

3. Answer: **c.** 9,000 to 15,000 rpm

    Reference: *Pumping: Fundamentals for the Water & Wastewater Maintenance Operator Series*. 2001. Frank R. Spellman and Joanne Drinan. Technomic Publishing Company.

4. Answer: **a.** 3,000 rpm.

    Reference: *Pumping: Fundamentals for the Water & Wastewater Maintenance Operator Series*. 2001. Frank R. Spellman and Joanne Drinan. Technomic Publishing Company.

5. Answer: **a.** Power failure shutting down a pump suddenly

    Reference: *Water Transmission and Distribution,* 4th edition. 2010. Larry Mays, Editor. American Water Works Association. Chapter 5.

Perform Maintenance

Sample Questions for Level I—Answers

1. Answer: **d.** To prevent cavitation from occurring

    Reference: *Water Transmission and Distribution*, 4th edition. 2010. Larry Mays, Editor. American Water Works Association. Chapter 4.

2. Answer: **b.** Ohm

    Reference: *Basic Science Concepts and Applications*, 4th edition. 2010. Nicholas G. Pizzi, Editor. American Water Works Association. Electricity 2.

3. Answer: **b.** after one month of operation.

    Reference: *Water Transmission and Distribution*, 4th edition. 2010. Larry Mays, Editor. American Water Works Association. Chapter 4.

4. Answer: **d.** Once a month

    Reference: *Pumping: Fundamentals for the Water & Wastewater Maintenance Operator Series*. 2001. Frank R. Spellman and Joanne Drinan. Technomic Publishing. Chapter 6.

5. Answer: **a.** air

    Reference: *Water Transmission and Distribution*, 4th edition. 2010. Larry Mays, Editor. American Water Works Association.

Sample Questions for Level II—Answers

1. Answer: **b.** no further tightening can be done on the packing gland.

    Reference: *Pumps & Pumping,* 9th edition. 2010. E.E. "Skeet" Arasmith, Mitch Scheele, & Kimon Zentz. ARC Publications. Lesson 5.

2. Answer: **d.** Rectifier

    Reference: *Basic Science Concepts and Applications*, 4th edition. 2010. Nicholas G. Pizzi, Editor. American Water Works Association. Electricity 3.

3. Answer: **c.** heated.

    Reference: *Pumps & Pumping,* 9th edition. 2010. E.E. "Skeet" Arasmith, Mitch Scheele, & Kimon Zentz. ARC Publications. Lesson 4.

4. Answer: **d.** wear and deteriorate with normal use.

    Reference: *Pumps & Pumping,* 9th edition. 2010. E.E. "Skeet" Arasmith, Mitch Scheele, & Kimon Zentz. ARC Publications. Lesson 5.

5. Answer: **a.** oil or water.

    Reference: *Pumps & Pumping,* 9th edition. 2010. E.E. "Skeet" Arasmith, Mitch Scheele, & Kimon Zentz. ARC Publications. Lesson 10.

Sample Questions for Level III—Answers

1. Answer: **b.** excessive heat.

    Reference: *Basic Science Concepts and Applications*, 4th edition. 2010. Nicholas G. Pizzi, Editor. American Water Works Association. Electricity 3.

2. Answer: **a.** 5%.

   Reference: *Water Transmission and Distribution*, 4th edition. 2010. Larry Mays, Editor. American Water Works Association. Chapter 8.

3. Answer: **c.** oxidation of ammonia to nitrites then to nitrates.

   Reference: *Water Treatment Operator Handbook*, Revised edition. 2005. Nicholas G. Pizzi. American Water Works Association. Chapter 8.

4. Answer: **a.** Split coupling

   Reference: *Pumps & Pumping,* 9th edition. 2010. E.E. "Skeet" Arasmith, Mitch Scheele, & Kimon Zentz. ARC Publications. Lesson 9.

5. Answer: **a.** Volts = (amps)(ohms)

   Reference: *Water Transmission and Distribution*, 4th edition. 2010. Larry Mays, Editor. American Water Works Association. Chapter 8.

Sample Questions for Level IV—Answers

1. Answer: **c.** Transmitter, transmission channel, and receiver

   Reference: *Water Transmission and Distribution*, 4th edition. 2010. Larry Mays, Editor. American Water Works Association. Chapter 9.

2. Answer: **b.** 10 years; year

   Reference: *Water Transmission and Distribution*, 4th edition. 2010. Larry Mays, Editor. American Water Works Association. Chapter 3.

3. Answer: **d.** 5.0 to 10.0

   Reference: *Water Transmission and Distribution*, 4th edition. 2010. Larry Mays, Editor. American Water Works Association. Chapter 8.

4. Answer: **a.** digital signals.

   Reference: *Water Transmission and Distribution*, 4th edition. 2010. Larry Mays, Editor. American Water Works Association. Chapter 9.

5. Answer: **b.** Signal conditioners, actuators, and control elements

   Reference: *Water Transmission and Distribution*, 4th edition. 2010. Larry Mays, Editor. American Water Works Association. Chapter 9.

Perform Security, Safety, and Administrative Procedures

Sample Questions for Level I—Answers

1. Answer: **b.** United States Environmental Protection Agency

   Reference: *Water Quality*, 4th edition. 2010. Joseph A. Ritter, Editor. American Water Works Association. Chapter 1.

2. Answer: **a.** Tier I

   Reference: *Water Quality*, 4th edition. 2010. Joseph A. Ritter, Editor. American Water Works Association. Chapter 1.

3. Answer: **c.** 24 hours.

   Reference: *Water Quality*, 4th edition. 2010. Joseph A. Ritter, Editor. American Water Works Association. Chapter 4.

4. Answer: **c.** based on population served.

   Reference: *Water Treatment Operator Handbook*, Revised edition. 2005. Nicholas G. Pizzi. American Water Works Association. Chapter 12.

5. Answer: **b.** 1.0 : 1.0

   Reference: *Water Transmission and Distribution*, 4th edition. 2010. Larry Mays, Editor. American Water Works Association. Chapter 12.

Sample Questions for Level II—Answers

1. Answer: **c.** 0.080 mg/L

   Reference: *Water Quality*, 4th edition. 2010. Joseph A. Ritter, Editor. American Water Works Association. Chapter 1.

2. Answer: **a.** Tier I

   Reference: *Water Quality*, 4th edition. 2010. Joseph A. Ritter, Editor. American Water Works Association. Chapter 1.

3. Answer: **a.** coliform samples.

   Reference: *Water Quality*, 4th edition. 2010. Joseph A. Ritter, Editor. American Water Works Association. Chapter 1.

4. Answer: **d.** 0.30 mg/L.

   Reference: *Water Quality*, 4th edition. 2010. Joseph A. Ritter, Editor. American Water Works Association. Chapter 6.

5. Answer: **d.** Zinc orthophosphate

   Reference: *Water Treatment,* Nicholas G. Pizzi, Editor, 4th edition. Chapter 9.

Sample Questions for Level III—Answers

1. Answer: **c.** 5%

   Reference: *Water Quality*, 4th edition. 2010. Joseph A. Ritter, Editor. American Water Works Association. Chapter 1.

2. Answer: **d.** 10.0 mg/L

   Reference: *Water Quality*, 4th edition. 2010. Joseph A. Ritter, Editor. American Water Works Association. Chapter 1.

3. Answer: **d.** 12 months

   Reference: *Water Quality*, 4th edition. 2010. Joseph A. Ritter, Editor. American Water Works Association. Chapter 1.

4. Answer: **b.** clay.

   Reference: *The Water Dictionary: A Comprehensive Reference of Water Terminology,* 2nd edition. 2010. Nancy McTigue, Editor, and James M. Symons, Editor Emeritus. American Water Works Association.

5. Answer: **b.** 90%

   Reference: *Water Treatment Operator Handbook,* Revised edition. 2005. Nicholas G. Pizzi. American Water Works Association. Chapter 10.

Sample Questions for Level IV—Answers

1. Answer: **d.** A qualified third party

   Reference: *Water Transmission and Distribution*, 4th edition. 2010. Larry Mays, Editor. American Water Works Association. Chapter 3.

2. Answer: **c.** 75%

   Reference: *Water Transmission and Distribution*, 4th edition. 2010. Larry Mays, Editor. American Water Works Association. Chapter 3.

3. Answer: **c.** excavation.

   Reference: *Water Transmission and Distribution*, 4th edition. 2010. Larry Mays, Editor. American Water Works Association. Chapter 12.

4. Answer: **b.** Ricin

   Reference: *Water Transmission and Distribution*, 4th edition. 2010. Larry Mays, Editor. American Water Works Association. Chapter 17.

5. Answer: **a.** 10 ppm.

   Reference: *Water Treatment Operator Handbook,* Revised edition. 2005. Nicholas G. Pizzi. American Water Works Association. Chapter 13.

Answers to Additional Distribution Operator Practice Questions

1. Answer: **d.** Late evening

   Reference: *Water Transmission and Distribution*, 4th edition. 2010. Larry Mays, Editor. American Water Works Association. Chapter 5.

2. Answer: **d.** Sectional maps

   Reference: *Water Transmission and Distribution*, 4th edition. 2010. Larry Mays, Editor. American Water Works Association. Chapter 16.

3. Answer: **c.** Flexible ball joint

   Reference: *Water Transmission and Distribution*, 4th edition. 2010. Larry Mays, Editor. American Water Works Association. Chapter 2.

4. Answer: **d.** friction head.

   Reference: *Pumping: Fundamentals for the Water and Wastewater Maintenance Operator Series*. 2001. Frank R. Spellman and Joanne Drinan. Technomic Publishing Company.

5. Answer: **c.** 6 inches.

   Reference: *Water Transmission and Distribution*, 4th edition. 2010. Larry Mays, Editor. American Water Works Association. Chapter 2.

6. Answer: **a.** minimize friction loss.

   Reference: *Water Transmission and Distribution*, 4th edition. 2010. Larry Mays, Editor. American Water Works Association. Chapter 2.

7. Answer: **a.** Bell and spigot type

   Reference: *Water Transmission and Distribution*, 4th edition. 2010. Larry Mays, Editor. American Water Works Association. Chapter 2.

8. Answer: **b.** Thermal butt-fusion

   Reference: *Water Transmission and Distribution*, 4th edition. 2010. Larry Mays, Editor. American Water Works Association. Chapter 2.

9. Answer: **b.** Restrained joint

   Reference: *Water Transmission and Distribution*, 4th edition. 2010. Larry Mays, Editor. American Water Works Association. Chapter 2.

10. Answer: **d.** Flanged joint

    Reference: *Water Transmission and Distribution*, 4th edition. 2010. Larry Mays, Editor. American Water Works Association. Chapter 2.

11. Answer: **d.** It will be the same in all three tanks

    Reference: *Basic Science Concepts and Applications*, 4th edition. 2010. Nicholas G. Pizzi, Editor. American Water Works Association. Hydraulics 2.

12. Answer: **b.** Piezometric surface

    Reference: *Basic Science Concepts and Applications*, 4th edition. 2010. Nicholas G. Pizzi, Editor. American Water Works Association. Hydraulics 3.

13. Answer: **d.** sudden changes in direction or velocity of flow.

    Reference: *Basic Science Concepts and Applications*, 4th edition. 2010. Nicholas G. Pizzi, Editor. American Water Works Association. Hydraulics 5.

14. Answer: **d.** adequate fire flow at an appropriate pressure.

    Reference: *Water Transmission and Distribution*, 4th edition. 2010. Larry Mays, Editor. American Water Works Association. Chapter 2.

15. Answer: **c.** 50 to 75 psi.

    Reference: *Water Transmission and Distribution*, 4th edition. 2010. Larry Mays, Editor. American Water Works Association. Chapter 2.

16. Answer: **c.** Tensile and flexural strength

    Reference: *Water Transmission and Distribution*, 4th edition. 2010. Larry Mays, Editor. American Water Works Association. Chapter 2.

17. Answer: **d.** tensile strength.

    Reference: *Water Transmission and Distribution*, 4th edition. 2010. Larry Mays, Editor. American Water Works Association. Chapter 2.

18. Answer: **d.** 2.5 inches; 4.5 inches

    Reference: *Water Transmission and Distribution*, 4th edition. 2010. Larry Mays, Editor. American Water Works Association. Chapter 7.

19. Answer: **c.** Ductile iron

    Reference: *Water Transmission and Distribution*, 4th edition. 2010. Larry Mays, Editor. American Water Works Association. Chapter 3.

20. Answer: **c.** Bell and spigot type

    Reference: *Water Transmission and Distribution*, 4th edition. 2010. Larry Mays, Editor. American Water Works Association. Chapter 2.

21. Answer: **a.** slightly scale forming.

    Reference: *Water Treatment Operator Handbook*, Revised edition. 2005. Nicholas G. Pizzi. American Water Works Association. Chapter 9.

22. Answer: **b.** Unmetered connection for a fire protection system

    Reference: *The Water Dictionary: A Comprehensive Reference of Water Terminology*, 2nd edition. 2010. Nancy McTigue, Editor, and James M. Symons, Editor Emeritus. American Water Works Association.

23. Answer: **b.** 500 to 1,000 feet to 1 inch.

    Reference: *Water Transmission and Distribution*, 4th edition. 2010. Larry Mays, Editor. American Water Works Association. Chapter 16.

24. Answer: **a.** 50 to 100 feet to 1 inch.

    Reference: *Water Transmission and Distribution*, 4th edition. 2010. Larry Mays, Editor. American Water Works Association. Chapter 16.

ANSWERS TO ADDITIONAL DISTRIBUTION OPERATOR PRACTICE QUESTIONS 355

25. Answer: **b.** Radial flow impellers

    Reference: *Pumping: Fundamentals for the Water and Wastewater Maintenance Operator Series*. 2001. Frank R. Spellman and Joanne Drinan. Technomic Publishing Company. Chapter 3.

26. Answer: **c.** Expansion joint

    Reference: *Water Transmission and Distribution*, 4th edition. 2010. Larry Mays, Editor. American Water Works Association. Chapter 2.

27. Answer: **b.** Rubber gasket joint

    Reference: *Water Transmission and Distribution*, 4th edition. 2010. Larry Mays, Editor. American Water Works Association. Chapter 2.

28. Answer: **b.** 20 psi.

    Reference: *Water Transmission and Distribution*, 4th edition. 2010. Larry Mays, Editor. American Water Works Association. Chapter 2.

29. Answer: **b.** 20 psi.

    Reference: *Water Transmission and Distribution*, 4th edition. 2010. Larry Mays, Editor. American Water Works Association. Chapter 7.

30. Answer: **d.** Tree system

    Reference: *Water Transmission and Distribution*, 4th edition. 2010. Larry Mays, Editor. American Water Works Association. Chapter 2.

31. Answer: **b.** DC current

    Reference: *Water Transmission and Distribution*, 4th edition. 2010. Larry Mays, Editor. American Water Works Association. Chapter 3.

32. Answer: **c.** inadequate distribution storage.

    Reference: *Water Transmission and Distribution*, 4th edition. 2010. Larry Mays, Editor. American Water Works Association. Chapter 3.

33. Answer: **b.** Head requirements

    Reference: *Water Transmission and Distribution*, 4th edition. 2010. Larry Mays, Editor. American Water Works Association. Chapter 4.

34. Answer: **b.** Capacity (flow rate)

    Reference: *Basic Science Concepts and Applications*, 4th edition. 2010. Nicholas G. Pizzi, Editor. American Water Works Association. Hydraulics 6.

35. Answer: **d.** fluids in motion and at rest.

    Reference: *Pumping: Fundamentals for the Water and Wastewater Maintenance Operator Series*. 2001. Frank R. Spellman and Joanne Drinan. Technomic Publishing Company. Chapter 2.

36. Answer: **c.** Flanged joint

    Reference: *Water Transmission and Distribution*, 4th edition. 2010. Larry Mays, Editor. American Water Works Association. Chapter 2.

37. Answer: **d.** Mechanical joint

    Reference: *Water Transmission and Distribution*, 4th edition. 2010. Larry Mays, Editor. American Water Works Association. Chapter 2.

38. Answer: **c.** 33.9 feet

    Reference: *Water Transmission and Distribution*, 4th edition. 2010. Larry Mays, Editor. American Water Works Association. Chapter 11.

39. Answer: **d.** 34

    Reference: *Pumping: Fundamentals for the Water and Wastewater Maintenance Operator Series*. 2001. Frank R. Spellman and Joanne Drinan. Technomic Publishing Company.

40. Answer: **a.** Closed

    Reference: *Pumps & Pumping,* 9th edition. E.E. "Skeet" Arasmith, Mitch Scheele, and Kimon Zentz. ACR Publications. Lesson 1.

41. Answer: **a.** End suction pump

    Reference: *Pumps & Pumping,* 9th edition. E.E. "Skeet" Arasmith, Mitch Scheele, and Kimon Zentz. ACR Publications. Lesson 2.

42. Answer: **b.** 2.5 to 4.0 times

    Reference: *Water Transmission and Distribution*, 4th edition. 2010. Larry Mays, Editor. American Water Works Association. Chapter 2.

43. Answer: **c.** Asbestos cement

    Reference: *Water Transmission and Distribution*, 4th edition. 2010. Larry Mays, Editor. American Water Works Association. Chapter 2.

44. Answer: **b.** Reinforced concrete

    Reference: *Water Transmission and Distribution*, 4th edition. 2010. Larry Mays, Editor. American Water Works Association. Chapter 2.

45. Answer: **b.** Ductile iron

    Reference: *Water Transmission and Distribution*, 4th edition. 2010. Larry Mays, Editor. American Water Works Association. Chapter 2.

46. Answer: **c.** Polyvinyl chloride

    Reference: *Water Transmission and Distribution*, 4th edition. 2010. Larry Mays, Editor. American Water Works Association. Chapter 2.

47. Answer: **c.** river crossings or rugged terrain.

    Reference: *Water Transmission and Distribution*, 4th edition. 2010. Larry Mays, Editor. American Water Works Association. Chapter 2.

48. Answer: **d.** where flexibility is required.

    Reference: *Water Transmission and Distribution*, 4th edition. 2010. Larry Mays, Editor. American Water Works Association. Chapter 2.

49. Answer: **a.** only for small lines.

    Reference: *Water Transmission and Distribution*, 4th edition. 2010. Larry Mays, Editor. American Water Works Association. Chapter 2.

50. Answer: **a.** for all locations.

    Reference: *Water Transmission and Distribution*, 4th edition. 2010. Larry Mays, Editor. American Water Works Association. Chapter 2.

ANSWERS TO ADDITIONAL DISTRIBUTION OPERATOR PRACTICE QUESTIONS

51. Answer: **a.** Ball and socket joint

    Reference: *Water Transmission and Distribution*, 4th edition. 2010. Larry Mays, Editor. American Water Works Association. Chapter 2.

52. Answer: **b.** Push-on joint

    Reference: *Water Transmission and Distribution*, 4th edition. 2010. Larry Mays, Editor. American Water Works Association. Chapter 2.

53. Answer: **d.** Shouldered joint

    Reference: *Water Transmission and Distribution*, 4th edition. 2010. Larry Mays, Editor. American Water Works Association. Chapter 2.

54. Answer: **d.** 150+.

    Reference: *Water Transmission and Distribution*, 4th edition. 2010. Larry Mays, Editor. American Water Works Association. Chapter 2.

55. Answer: **c.** Emergency storage tanks

    Reference: *Water Transmission and Distribution*, 4th edition. 2010. Larry Mays, Editor. American Water Works Association. Chapter 3.

56. Answer: **b.** Needle valve

    Reference: *Water Transmission and Distribution*, 4th edition. 2010. Larry Mays, Editor. American Water Works Association. Chapter 6.

57. Answer: **d.** Pinch valve

    Reference: *Water Transmission and Distribution*, 4th edition. 2010. Larry Mays, Editor. American Water Works Association. Chapter 6.

58. Answer: **c.** cold climates.

    Reference: *Water Transmission and Distribution*, 4th edition. 2010. Larry Mays, Editor. American Water Works Association. Chapter 15.

59. Answer: **b.** thrust load.

    Reference: *Pumps & Pumping,* 9th edition. 2010. E.E. "Skeet" Arasmith, Mitch Scheele, and Kimon Zentz. ACR Publications. Lesson 3.

60. Answer: **b.** 20%

    Reference: *Pumping: Fundamentals for the Water and Wastewater Maintenance Operator Series.* 2001. Frank R. Spellman and Joanne Drinan. Technomic Publishing Company.

61. Answer: **d.** to complete a grid.

    Reference: *Water Transmission and Distribution*, 4th edition. 2010. Larry Mays, Editor. American Water Works Association. Chapter 2.

62. Answer: **a.** Reinforced concrete

    Reference: *Water Transmission and Distribution*, 4th edition. 2010. Larry Mays, Editor. American Water Works Association. Chapter 2.

63. Answer: **b.** Steel

    Reference: *Water Transmission and Distribution*, 4th edition. 2010. Larry Mays, Editor. American Water Works Association. Chapter 2.

64. Answer: **b.** 2 to 8 inches

    Reference: *Water Transmission and Distribution*, 4th edition. 2010. Larry Mays, Editor. American Water Works Association. Chapter 12.

65. Answer: **c.** modeling and analysis.

    Reference: *Water Transmission and Distribution*, 4th edition. 2010. Larry Mays, Editor. American Water Works Association. Chapter 16.

66. Answer: **d.** Base data

    Reference: *Water Transmission and Distribution*, 4th edition. 2010. Larry Mays, Editor. American Water Works Association. Chapter 16.

67. Answer: **c.** Calcium hypochlorite

    Reference: *Water Treatment Operator Handbook*, Revised edition. 2005. Nicholas G. Pizzi. American Water Works Association. Chapter 8.

68. Answer: **b.** 6 hours

    Reference: *Water Transmission and Distribution*, 4th edition. 2010. Larry Mays, Editor. American Water Works Association. Chapter 3.

69. Answer: **d.** 24 hours

    Reference: *Water Transmission and Distribution*, 4th edition. 2010. Larry Mays, Editor. American Water Works Association. Chapter 3.

70. Answer: **b.** 0.2 mg/L for more than 4 hours during periods when the system is serving water to the public.

    Reference: M20, *Water Chlorination/Chlorination Practices and Principles*, 2nd edition. 2006. American Water Works Association. Chapter 6.

71. Answer: **c.** HOCl.

    Reference: *Water Treatment Operator Handbook*, Revised edition. 2005. Nicholas G. Pizzi. American Water Works Association. Chapter 8.

72. Answer: **c.** chlorine gas.

    Reference: *Water Transmission and Distribution*, 4th edition. 2010. Larry Mays, Editor. American Water Works Association. Chapter 13.

73. Answer: **b.** Superchlorinating reservoirs and storage tanks

    Reference: M7, *Problem Organisms in Water: Identification and Treatment*, 3rd edition. 2003. American Water Works Association. Chapter 4.

74. Answer: **c.** 200 mg/L

    Reference: *Water Transmission and Distribution*, 4th edition. 2010. Larry Mays, Editor. American Water Works Association. Chapter 3.

75. Answer: **a.** at least 6 hours.

    Reference: *Water Transmission and Distribution*, 4th edition. 2010. Larry Mays, Editor. American Water Works Association. Chapter 3.

76. Answer: **b.** To stop the flow of water into the tank when it is full

    Reference: *Water Transmission and Distribution*, 4th edition. 2010. Larry Mays, Editor. American Water Works Association. Chapter 3.

ANSWERS TO ADDITIONAL DISTRIBUTION OPERATOR PRACTICE QUESTIONS 359

77. Answer: **b.** Use a corrosion chemical as treatment

Reference: M7, *Problem Organisms in Water: Identification and Treatment,* 3rd edition. 2003. American Water Works Association. Chapter 11.

78. Answer: **c.** Vacuum

Reference: *Water Treatment Operator Handbook*, Revised edition. 2005. Nicholas G. Pizzi. American Water Works Association. Chapter 8.

79. Answer: **c.** iron pipe from the anode to the cathode.

Reference: *Water Treatment*, 4th edition. 2010. Nicholas G. Pizzi, Editor. American Water Works Association. Chapter 9.

80. Answer: **d.** Galvanic corrosion

Reference: *Water Treatment*, 4th edition. 2010. Nicholas G. Pizzi, Editor. American Water Works Association. Chapter 9.

81. Answer: **d.** Proper alignment and support for meter

Reference: *Water Transmission and Distribution*, 4th edition. 2010. Larry Mays, Editor. American Water Works Association. Chapter 10.

82. Answer: **b.** 2.5 feet.

Reference: *Water Transmission and Distribution*, 4th edition. 2010. Larry Mays, Editor. American Water Works Association. Chapter 12.

83. Answer: **b.** 18 inches.

Reference: *Water Transmission and Distribution*, 4th edition. 2010. Larry Mays, Editor. American Water Works Association. Chapter 12.

84. Answer: **d.** Just ahead of the pipe installation

Reference: *Water Transmission and Distribution*, 4th edition. 2010. Larry Mays, Editor. American Water Works Association. Chapter 12.

85. Answer: **a.** 6%.

Reference: *Water Transmission and Distribution*, 4th edition. 2010. Larry Mays, Editor. American Water Works Association. Chapter 12.

86. Answer: **c.** Irregular drum tampers

Reference: *Water Transmission and Distribution*, 4th edition. 2010. Larry Mays, Editor. American Water Works Association. Chapter 13.

87. Answer: **a.** Hand-controlled plate tampers

Reference: *Water Transmission and Distribution*, 4th edition. 2010. Larry Mays, Editor. American Water Works Association. Chapter 13.

88. Answer: **d.** Mueller thread

Reference: *Water Transmission and Distribution*, 4th edition. 2010. Larry Mays, Editor. American Water Works Association. Chapter 15.

89. Answer: **c.** 12%

Reference: *Water Transmission and Distribution*, 4th edition. 2010. Larry Mays, Editor. American Water Works Association. Chapter 12.

90. Answer: **b.** 12 to 24 inches; 3 inches

    Reference: *Water Transmission and Distribution*, 4th edition. 2010. Larry Mays, Editor. American Water Works Association. Chapter 12.

91. Answer: **c.** Local regulatory agency

    Reference: *Water Transmission and Distribution*, 4th edition. 2010. Larry Mays, Editor. American Water Works Association. Chapter 12.

92. Answer: **b.** 120 to 150 ft-lb

    Reference: *Water Transmission and Distribution*, 4th edition. 2010. Larry Mays, Editor. American Water Works Association. Chapter 12.

93. Answer: **a.** 4 inches or smaller.

    Reference: *Water Transmission and Distribution*, 4th edition. 2010. Larry Mays, Editor. American Water Works Association. Chapter 14.

94. Answer: **b.** Polyethylene pipe

    Reference: *Water Transmission and Distribution*, 4th edition. 2010. Larry Mays, Editor. American Water Works Association. Chapter 15.

95. Answer: **b.** Flanged joint

    Reference: *Water Transmission and Distribution*, 4th edition. 2010. Larry Mays, Editor. American Water Works Association. Chapter 2.

96. Answer: **b.** Flanged joint

    Reference: *Water Transmission and Distribution*, 4th edition. 2010. Larry Mays, Editor. American Water Works Association. Chapter 2.

97. Answer: **a.** Ball and socket joint

    Reference: *Water Transmission and Distribution*, 4th edition. 2010. Larry Mays, Editor. American Water Works Association. Chapter 2.

98. Answer: **c.** 15.0 degrees.

    Reference: *Water Transmission and Distribution*, 4th edition. 2010. Larry Mays, Editor. American Water Works Association. Chapter 2.

99. Answer: **a.** 3 degrees

    Reference: *Water Transmission and Distribution*, 4th edition. 2010. Larry Mays, Editor. American Water Works Association. Chapter 12.

100. Answer: **a.** 5 to 10%

    Reference: *Water Transmission and Distribution*, 4th edition. 2010. Larry Mays, Editor. American Water Works Association. Chapter 10.

101. Answer: **b.** Plastic pipe

    Reference: *Water Transmission and Distribution*, 4th edition. 2010. Larry Mays, Editor. American Water Works Association. Chapter 12.

102. Answer: **d.** Boom-mounted plate tampers

    Reference: *Water Transmission and Distribution*, 4th edition. 2010. Larry Mays, Editor. American Water Works Association. Chapter 13.

ANSWERS TO ADDITIONAL DISTRIBUTION OPERATOR PRACTICE QUESTIONS

103. Answer: **b.** seal cage.

    Reference: *Pumping: Fundamentals for the Water and Wastewater Maintenance Operator Series.* 2001. Frank R. Spellman and Joanne Drinan. Technomic Publishing Company. Chapter 3.

104. Answer: **a.** Directly

    Reference: *Pumping: Fundamentals for the Water and Wastewater Maintenance Operator Series.* 2001. Frank R. Spellman and Joanne Drinan. Technomic Publishing Company. Chapter 3.

105. Answer: **a.** radial bearings.

    Reference: *Pumping: Fundamentals for the Water and Wastewater Maintenance Operator Series.* 2001. Frank R. Spellman and Joanne Drinan. Technomic Publishing Company. Chapter 4.

106. Answer: **c.** slip.

    Reference: *Water Transmission and Distribution*, 4th edition. 2010. Larry Mays, Editor. American Water Works Association. Chapter 4.

107. Answer: **a.** Butterfly valves

    Reference: *Water Transmission and Distribution*, 4th edition. 2010. Larry Mays, Editor. American Water Works Association. Chapter 4.

108. Answer: **b.** Globe valve

    Reference: *Water Transmission and Distribution*, 4th edition. 2010. Larry Mays, Editor. American Water Works Association. Chapter 6.

109. Answer: **a.** Plug valve

    Reference: *Water Transmission and Distribution*, 4th edition. 2010. Larry Mays, Editor. American Water Works Association. Chapter 6.

110. Answer: **c.** check valve.

    Reference: *Water Transmission and Distribution*, 4th edition. 2010. Larry Mays, Editor. American Water Works Association. Chapter 6.

111. Answer: **c.** Dry-barrel hydrant

    Reference: *Water Transmission and Distribution*, 4th edition. 2010. Larry Mays, Editor. American Water Works Association. Chapter 7.

112. Answer: **c.** Thermocouples

    Reference: *Water Transmission and Distribution*, 4th edition. 2010. Larry Mays, Editor. American Water Works Association. Chapter 9.

113. Answer: **a.** Thermistors

    Reference: *Water Transmission and Distribution*, 4th edition. 2010. Larry Mays, Editor. American Water Works Association. Chapter 9.

114. Answer: **b.** pump head.

    Reference: *Pumping: Fundamentals for the Water and Wastewater Maintenance Operator Series.* 2001. Frank R. Spellman and Joanne Drinan. Technomic Publishing Company. Chapter 2.

115. Answer: **d.** Valve opening and closing

    Reference: *Water Transmission and Distribution*, 4th edition. 2010. Larry Mays, Editor. American Water Works Association. Chapter 5.

116. Answer: **b.** Altitude valve

    Reference: *Water Transmission and Distribution*, 4th edition. 2010. Larry Mays, Editor. American Water Works Association. Chapter 6.

117. Answer: **d.** orifice plate.

    Reference: *Water Transmission and Distribution*, 4th edition. 2010. Larry Mays, Editor. American Water Works Association. Chapter 9.

118. Answer: **b.** 120%

    Reference: *Pumping: Fundamentals for the Water and Wastewater Maintenance Operator Series.* 2001. Frank R. Spellman and Joanne Drinan. Technomic Publishing Company. Chapter 4.

119. Answer: **b.** 30.0 inches

    Reference: *Pumping: Fundamentals for the Water and Wastewater Maintenance Operator Series.* 2001. Frank R. Spellman and Joanne Drinan. Chapter 2.

120. Answer: **d.** 34 feet

    Reference: *Pumping: Fundamentals for the Water and Wastewater Maintenance Operator Series.* 2001. Frank R. Spellman and Joanne Drinan. Technomic Publishing Company. Chapter 2.

121. Answer: **d.** to the pressure taps supplied on the pump.

    Reference: *Water Transmission and Distribution*, 4th edition. 2010. Larry Mays, Editor. American Water Works Association. Chapter 4.

122. Answer: **d.** Pressure-reducing and altitude valves

    Reference: *Water Transmission and Distribution*, 4th edition. 2010. Larry Mays, Editor. American Water Works Association. Chapter 6.

123. Answer: **d.** Needle valve

    Reference: *Water Transmission and Distribution*, 4th edition. 2010. Larry Mays, Editor. American Water Works Association. Chapter 6.

124. Answer: **d.** RTU's, communications, master station, and HMI

    Reference: *Water Transmission and Distribution*, 4th edition. 2010. Larry Mays, Editor. American Water Works Association. Chapter 9.

125. Answer: **d.** DC currents.

    Reference: *Water Transmission and Distribution*, 4th edition. 2010. Larry Mays, Editor. American Water Works Association. Chapter 14.

126. Answer: **a.** zinc.

    Reference: *The Water Dictionary: A Comprehensive Reference of Water Terminology,* 2nd edition. 2010. Nancy McTigue, Editor, and James M. Symons, Editor Emeritus. American Water Works Association.

127. Answer: **a.** Wound-rotor induction motor and a controller

    Reference: *Water Transmission and Distribution*, 4th edition. 2010. Larry Mays, Editor. American Water Works Association. Chapter 8.

ANSWERS TO ADDITIONAL DISTRIBUTION OPERATOR PRACTICE QUESTIONS 363

128. Answer: **b.** Voltage relays

Reference: *Water Transmission and Distribution*, 4th edition. 2010. Larry Mays, Editor. American Water Works Association. Chapter 8.

129. Answer: **a.** Thermal-overload relays

Reference: *Water Transmission and Distribution*, 4th edition. 2010. Larry Mays, Editor. American Water Works Association. Chapter 8.

130. Answer: **c.** Frequency relay

Reference: *Water Transmission and Distribution,* 4th edition. 2010. Larry Mays, Editor. American Water Works Association. Chapter 8.

131. Answer: **d.** Differential relay

Reference: *Water Transmission and Distribution,* 4th edition. 2010. Larry Mays, Editor. American Water Works Association. Chapter 8.

132. Answer: **b.** Sensor, transmitter, transmission channel, receiver, and indicator

Reference: *Water Transmission and Distribution*, 4th edition. 2010. Larry Mays, Editor. American Water Works Association. Chapter 9.

133. Answer: **a.** Demand meter

Reference: *Basic Science Concepts and Applications,* 4th edition. 2010. Nicholas G. Pizzi, Editor. American Water Works Association. Electricity 3.

134. Answer: **c.** galvanic anodes.

Reference: *Water Transmission and Distribution*, 4th edition. 2010. Larry Mays, Editor. American Water Works Association. Chapter 14.

135. Answer: **a.** Hot water

Reference: *Water Transmission and Distribution*, 4th edition. 2010. Larry Mays, Editor. American Water Works Association. Chapter 15.

136. Answer: **c.** racking.

Reference: *Water Transmission and Distribution*, 4th edition. 2010. Larry Mays, Editor. American Water Works Association. Chapter 8.

137. Answer: **a.** 4 to 20 mA DC

Reference: *Water Transmission and Distribution*, 4th edition. 2010. Larry Mays, Editor. American Water Works Association. Chapter 9.

138. Answer: **c.** $Fe(OH)_2$ on the inside and $Fe(OH)_3$ on the outside

Reference: *Basic Chemistry for Water and Wastewater Operators,* Revised edition. 2005. Darshan Singh Sarai. American Water Works Association. Chapter 14.

139. Answer: **d.** Capacitor

Reference: *Basic Science Concepts and Applications,* 4th edition. 2010. Nicholas G. Pizzi, Editor. American Water Works Association. Electricity 3.

140. Answer: **d.** equal to or slightly larger than

Reference: *Basic Science Concepts and Applications*, 4th edition. 2010. Nicholas G. Pizzi, Editor. American Water Works Association. Electricity 3.

141. Answer: **b.** mechanical failure.

    Reference: *Basic Science Concepts and Applications*, 4th edition. 2010. Nicholas G. Pizzi, Editor. American Water Works Association. Electricity 3.

142. Answer: **c.** 6 months

    Reference: *Pumping: Fundamentals for the Water and Wastewater Maintenance Operator Series.* 2001. Frank R. Spellman and Joanne Drinan. Technomic Publishing Company. Chapter 6.

143. Answer: **b.** Animal oil

    Reference: *Pumping: Fundamentals for the Water and Wastewater Maintenance Operator Series.* 2001. Frank R. Spellman and Joanne Drinan. Technomic Publishing Company. Chapter 8.

144. Answer: **b.** At least 1.0 megohm of resistance

    Reference: *Water Transmission and Distribution*, 4th edition. 2010. Larry Mays, Editor. American Water Works Association. Chapter 8.

145. Answer: **b.** 3 to 15 psig

    Reference: *Water Transmission and Distribution*, 4th edition. 2010. Larry Mays, Editor. American Water Works Association. Chapter 9.

146. Answer: **a.** Extra thickness for pipe walls

    Reference: *Water Transmission and Distribution*, 4th edition. 2010. Larry Mays, Editor. American Water Works Association. Chapter 14.

147. Answer: **c.** an analog (uses a needle) meter.

    Reference: *Water Transmission and Distribution*, 4th edition. 2010. Larry Mays, Editor. American Water Works Association. Glossary.

148. Answer: **b.** Paralyzes the respiratory system

    Reference: *Water Treatment*, 4th edition. 2010. Nicholas G. Pizzi, Editor. American Water Works Association. Chapter 14.

149. Answer: **b.** 15 minutes

    Reference: M20, *Water Chlorination/Chlorination Practices and Principles,* 2nd edition. 2006. American Water Works Association. Chapter 5.

150. Answer: **c.** 6.5 to 8.5 pH units

    Reference: *Water Quality*, 4th edition. 2010. Joseph A. Ritter, Editor. American Water Works Association. Chapter 1.

151. Answer: **a.** 6

    Reference: *Water Transmission and Distribution*, 4th edition. 2010. Larry Mays, Editor. American Water Works Association. Chapter 13.

152. Answer: **b.** 13

    Reference: *Water Transmission and Distribution*, 4th edition. 2010. Larry Mays, Editor. American Water Works Association. Chapter 13.

153. Answer: **a.** Sodium zinc phosphate

    Reference: *Water Treatment*, 4th edition. 2010. Nicholas G. Pizzi, Editor. American Water Works Association. Chapter 9.

ANSWERS TO ADDITIONAL DISTRIBUTION OPERATOR PRACTICE QUESTIONS 365

154. Answer: **b.** 8.

    Reference: *The Water Dictionary: A Comprehensive Reference of Water Terminology,* 2nd edition. 2010. Nancy McTigue, Editor, and James M. Symons, Editor Emeritus. American Water Works Association.

155. Answer: **c.** Entry supervisor

    Reference: *Water Treatment Operator Handbook,* Revised edition. 2005. Nicholas G. Pizzi. American Water Works Association. Chapter 13.

156. Answer: **c.** Authorized attendant

    Reference: *Water Treatment Operator Handbook,* Revised edition. 2005. Nicholas G. Pizzi. American Water Works Association. Chapter 13.

157. Answer: **d.** by seeing the discoloration and moisture at the leak point.

    Reference: *Water Treatment*, 4th edition. 2010. Nicholas G. Pizzi, Editor. American Water Works Association. Chapter 7.

158. Answer: **a.** 0.015 mg/L

    Reference: *Water Quality*, 4th edition. 2010. Joseph A. Ritter, Editor. American Water Works Association. Chapter 1.

159. Answer: **c.** cause health effects only after long exposure.

    Reference: *Water Treatment Operator Handbook*, Revised edition. 2005. Nicholas G. Pizzi. American Water Works Association. Chapter 1.

160. Answer: **b.** galvanized pipe.

    Reference: *Water Treatment*, 4th edition. 2010. Nicholas G. Pizzi, Editor. American Water Works Association. Chapter 9.

161. Answer: **c.** 5 mg/L

    Reference: *Water Quality*, 4th edition. 2010. Joseph A Ritter, Editor. American Water Works Association. Chapter 1.

162. Answer: **d.** The angle will vary with type of soil, moisture content, and surrounding conditions

    Reference: *Water Transmission and Distribution,* 4th edition. 2010. Larry Mays, Editor. American Water Works Association. Chapter 12.

163. Answer: **d.** 7.0 MFL

    Reference: *Water Quality*, 4th edition. 2010. Joseph A. Ritter, Editor. American Water Works Association. Chapter 1.

164. Answer: **a.** 80/60 µg/L

    Reference: *Water Quality*, 4th edition. 2010. Joseph A. Ritter, Editor. American Water Works Association. Chapter 1.

165. Answer: **d.** 5%

    Reference: *Water Quality*, 4th edition. 2010. Joseph A. Ritter, Editor. American Water Works Association. Chapter 1.

166. Answer: **c.** Use activated carbon and flush the system

    Reference: M7, *Problem Organisms in Water: Identification and Treatment,* 3rd edition. 2003. American Water Works Association. Chapter 11.

167. Answer: **c.** Pump head

    Reference: *Water Treatment*, 4th edition. 2010. Nicholas G. Pizzi, Editor. American Water Works Association. Chapter 7.

168. Answer: **a.** Compound-loop control gas feeder

    Reference: M20, *Water Chlorination/Chlorination Practices and Principles*, 2nd edition. 2006. American Water Works Association. Chapter 4.

169. Answer: **b.** chemical activity.

    Reference: *Water Quality*, 4th edition. 2010. Joseph A. Ritter, Editor. American Water Works Association. Chapter 2.

170. Answer: **d.** gram-negative and strictly aerobic.

    Reference: M7, *Problem Organisms in Water: Identification and Treatment*, 3rd edition. 2003. American Water Works Association, Chapter 4.

171. Answer: **d.** Ryzner index

    Reference: *Basic Chemistry for Water & Wastewater Operators*, Revised edition. 2005. Darshan Singh Sarai. American Water Works Association. Chapter 14.

172. Answer: **a.** rad.

    Reference: *Water Quality*, 4th edition. 2010. Joseph A. Ritter, Editor. American Water Works Association. Chapter 7.

173. Answer: **b.** bromate

    Reference: National Primary Drinking Water Regulations, Subpart L – Disinfectant Residuals, Disinfectant By-products and Disinfectant By-product Precursors, §141.132 – Monitoring requirements (b)(3)

174. Answer: **a.** 10 mg/L; 2 mg/L

    Reference: M20, *Water Chlorination/Chlorination Practices and Principles*, 2nd edition. 2006. American Water Works Association. Appendix C.

175. Answer: **a.** 2 to 4 inches

    Reference: *Water Transmission and Distribution*, 4th edition. 2010. Larry Mays, Editor. American Water Works Association. Chapter 12.

176. Answer: **b.** 6 to 9 inches

    Reference: *Water Transmission and Distribution*, 4th edition. 2010. Larry Mays, Editor. American Water Works Association. Chapter 12.

177. Answer: **a.** globe valves.

    Reference: *Water Transmission and Distribution*, 4th edition. 2010. Larry Mays, Editor. American Water Works Association. Chapter 6.

178. Answer: **c.** 3 inches or greater.

    Reference: *Water Transmission and Distribution*, 4th edition. 2010. Larry Mays, Editor. American Water Works Association. Chapter 12.

179. Answer: **a.** Shear breakage

    Reference: *Water Transmission and Distribution*, 4th edition. 2010. Larry Mays, Editor. American Water Works Association. Chapter 2.

180. Answer: **b.** Globe valve

Reference: *Water Distribution System Operation and Maintenance*, 5th edition. 2005. Ken Kerri. California State University – Sacramento. Page 120.

181. Answer: **d.** The hydraulic grade

Reference: *Water Distribution System Operation and Maintenance*, 5th edition. 2005. Ken Kerri. California State University – Sacramento. Page 63.

182. Answer: **c.** welded at the joints.

Reference: *Water Distribution System Operation and Maintenance*, 5th edition. 2005. Ken Kerri. California State University – Sacramento. Page 63.

183. Answer: **c.** 75 to less than 100 psi

Reference: *Water Distribution System Operation and Maintenance*, 5th edition. 2005. Ken Kerri. California State University – Sacramento. Page 69.

184. Answer: **a.** Pressure-reducing valves

Reference: *Water Distribution System Operation and Maintenance*, 5th edition. 2005. Ken Kerri. California State University – Sacramento. Page 69.

185. Answer: **b.** 2 to 4 ft/sec

Reference: *Water Distribution System Operation and Maintenance*, 5th edition. 2005. Ken Kerri. California State University – Sacramento. Page 71.

186. Answer: **a.** 6 inches

Reference: *Water Distribution System Operation and Maintenance*, 5th edition. 2005. Ken Kerri. California State University – Sacramento. Page 71.

187. Answer: **c.** 1,000 feet

Reference: *Water Distribution System Operation and Maintenance*, 5th edition. 2005. Ken Kerri. California State University – Sacramento. Page 71.

188. Answer: **b.** fire demand.

Reference: *Water Distribution System Operation and Maintenance*, 5th edition. 2005. Ken Kerri. California State University – Sacramento. Page 71.

189. Answer: **d.** Low pressure

Reference: *Water Distribution System Operation and Maintenance*, 5th edition. 2005. Ken Kerri. California State University – Sacramento. Page 72.

190. Answer: **a.** Steel cylinder

Reference: *Water Distribution System Operation and Maintenance*, 5th edition. 2005. Ken Kerri. California State University – Sacramento. Page 78.

191. Answer: **b.** Lead

Reference: *Water Distribution System Operation and Maintenance*, 5th edition. 2005. Ken Kerri. California State University – Sacramento. Page 81.

192. Answer: **c.** Dressler coupling

Reference: *Water Distribution System Operation and Maintenance*, 5th edition. 2005. Ken Kerri. California State University – Sacramento. Page 81.

193. Answer: **b.** Victaulic joint

    Reference: *Water Distribution System Operation and Maintenance*, 5th edition. 2005. Ken Kerri. California State University – Sacramento. Page 81.

194. Answer: **d.** Restrained joint

    Reference: *Water Distribution System Operation and Maintenance*, 5th edition. 2005. Ken Kerri. California State University – Sacramento. Page 81.

195. Answer: **d.** Cement mortar

    Reference: *Water Distribution System Operation and Maintenance*, 5th edition. 2005. Ken Kerri. California State University – Sacramento. Page 86.

196. Answer: **a.** 5 degrees

    Reference: *Water Distribution System Operation and Maintenance*, 5th edition. 2005. Ken Kerri. California State University – Sacramento. Page 86.

197. Answer: **a.** Use of noncorrosive metals and/or mechanical coatings

    Reference: *Water Distribution System Operation and Maintenance*, 5th edition. 2005. Ken Kerri. California State University – Sacramento. Page 88.

198. Answer: **b.** Calculating the operating and coverage ratios

    Reference: *Water Distribution System Operation and Maintenance*, 5th edition. 2005. Ken Kerri. California State University – Sacramento. Page 469.

199. Answer: **d.** Through dedicated service-oriented employees

    Reference: *Water Distribution System Operation and Maintenance*, 5th edition. 2005. Ken Kerri. California State University – Sacramento. Page 466.

200. Answer: **b.** planning.

    Reference: *Water Distribution System Operation and Maintenance*, 5th edition. 2005. Ken Kerri. California State University – Sacramento. Page 444.

201. Answer: **b.** Containing not more than 0.02% lead

    Reference: National Primary Drinking Water Regulations, Subpart E – Special Regulations, Including Monitoring Regulations And Prohibition On Lead Use, §141.43 – Prohibition on use of lead pipes, solder and flux (d)(1)

202. Answer: **c.** Containing not more than 8% lead

    Reference: National Primary Drinking Water Regulations, Subpart E – Special Regulations, Including Monitoring Regulations And Prohibition On Lead Use, §141.43 – Prohibition on use of lead pipes, solder and flux (d)(2)

203. Answer: **b.** 1

    Reference: National Primary Drinking Water Regulations, Subpart G – Maximum Contaminant Levels and Maximum Residual Disinfectant Levels, §141.63 – Maximum contaminant levels (MCLs) for microbiological contaminants (a)(2)

204. Answer: **a.** 8 hours

    Reference: National Primary Drinking Water Regulations, Subpart H – Filtration and Disinfection, §141.74 – Analytical and monitoring requirements (a)(1) Table explanation 2

ANSWERS TO ADDITIONAL DISTRIBUTION OPERATOR PRACTICE QUESTIONS 369

205. Answer: **a.** warmest water temperature.

Reference: National Primary Drinking Water Regulations, Subpart U – Initial Distribution System Evaluations, §141.601 – Standard monitoring (b)(1) Table Endnote 2

206. Answer: **d.** 5 years

Reference: National Primary Drinking Water Regulations, Subpart U – Initial Distribution System Evaluations, §141.601 – System specific studies (a)(1)(i)

207. Answer: **b.** 8 quarters

Reference: National Primary Drinking Water Regulations, Subpart U – Initial Distribution System Evaluations, §141.603 – 40/30 certification (a)

208. Answer: **b.** recommended in the Initial Distribution System Evaluation (IDSE) report.

Reference: National Primary Drinking Water Regulations, Subpart V – Stage 2 Disinfection By-products Requirements, §141.620 – General requirements (c)(6)(ii)

209. Answer: **b.** fourth quarter.

Reference: National Primary Drinking Water Regulations, Subpart V – Stage 2 Disinfection By-products Requirements, §141.620 – General requirements (c)(7)

210. Answer: **a.** disinfection by-products.

Reference: National Primary Drinking Water Regulations, Subpart V – Stage 2 Disinfection By-products Requirements, §141.621 – Routine monitoring (a)(2) Table Endnote 1

211. Answer: **b.** Within 10 days after the quarter ends

Reference: National Primary Drinking Water Regulations, Subpart V – Stage 2 Disinfection By-products Requirements, §141.629 – Reporting and recordkeeping requirements (a)(1)(i, ii)

212. Answer: **c.** First flush the volume of water between tap and lead service line, then collect the sample

Reference: National Primary Drinking Water Regulations, Subpart I – Control of Lead and Copper, §141.86 – Monitoring requirements for lead and copper in tap water (b)(3)(i, ii, iii)

213. Answer: **c.** When the 90th percentiles for lead and copper are 0.005 mg/L and 0.65 mg/L, respectively

Reference: National Primary Drinking Water Regulations, Subpart I – Control of Lead and Copper, §141.86 – Monitoring requirements for lead and copper in tap water (d)(4)(iv)

214. Answer: **b.** during June, July, August, and September.

Reference: National Primary Drinking Water Regulations, Subpart I – Control of Lead and Copper, §141.86 – Monitoring requirements for lead and copper in tap water (d)(4)(iv)

215. Answer: **b.** Once every 9 years

Reference: National Primary Drinking Water Regulations, Subpart I – Control of Lead and Copper, §141.86 – Monitoring requirements for lead and copper in tap water (g)

216. Answer: **c.** pH

Reference: National Primary Drinking Water Regulations, Subpart I – Control of Lead and Copper, §141.87 – Monitoring requirements for water quality parameters (c)(2)(i)

217. Answer: **b.** 3

Reference: National Primary Drinking Water Regulations, Subpart L – Disinfectant Residuals, Disinfectant By-products and Disinfectant By-product Precursors, §141.132 – Monitoring requirements (c)(2)(ii)

218. Answer: **c.** the arithmetic average of all samples for TTHM and HAA5.

Reference: National Primary Drinking Water Regulations, Subpart L – Disinfectant Residuals, Disinfectant By-products and Disinfectant By-product precursors, §141.134 – Reporting and recordkeeping requirements (b) Table – (1)(iii)

219. Answer: **d.** Preserve by keeping it at 4°C

Reference: National Primary Drinking Water Regulations, Subpart C – Monitoring and Analytical Requirements, §141.23 – Inorganic Chemical Sampling and Analytical Requirements (k)(2)

220. Answer: **c.** Preserve with H_2SO_4

Reference: National Primary Drinking Water Regulations, Subpart C – Monitoring and Analytical Requirements, §141.23 – Inorganic Chemical Sampling and Analytical Requirements (k)(2)

221. Answer: **d.** 1,000 mL

Reference: National Primary Drinking Water Regulations, Subpart I – Control of Lead and Copper, §141.86 – Monitoring requirements for lead and copper in tap water (b)(2)

222. Answer: **d.** Sequestering

Reference: *Basic Science Concepts and Applications*, 4th edition. 2010. Nicholas G. Pizzi, Editor. Chemistry 6.

223. Answer: **b.** resolubilize the metals.

Reference: National Primary Drinking Water Regulations, Subpart I – Control of Lead and Copper, §141.86 – Monitoring requirements for lead and copper in tap water (b)(2)

224. Answer: **a.** reduce the number of sites sampled.

Reference: National Primary Drinking Water Regulations, Subpart I – Control of Lead and Copper, §141.87 – Monitoring requirements for water quality parameters (e)(1)

225. Answer: **b.** Every 4 hours

Reference: National Primary Drinking Water Regulations, Subpart H – Filtration and Disinfection, §141.74 – Analytical and monitoring requirements (b)(5)

Answers to Math Questions for Water Distribution Operators Levels I & II

1. Answer: **b.** 142 gpm

   Equation: Well yield, gpm = $\dfrac{\text{Gallons produced}}{\text{Test duration, min}}$

   Well yield, gpm = $\dfrac{2{,}840 \text{ gpm}}{20 \text{ min}}$ = **142 gpm**

2. Answer: **a.** 3.0°C

   Equation: °C = (°F − 32) × 5/9

   First: 37.4 − 32 = 5.4

   Then: 5.4 × 5 = 270 ÷ 9 = **3.0 °C**

3. Answer: **c.** 8,170 ft²

   Equation: Area = πr^2, where π = 3.14

   First find the radius: Radius = Diameter/2 = 102/2 = 51 ft

   Area of tank = (3.14)(51 ft)(51 ft) = 8,167.14 ft², round to **8,170 ft²**

4. Answer: **a.** 11.6 psi

   Equation: First subtract water level from the level in question, 1.85 feet.

   Number of feet in question = 28.7 ft − 1.85 ft = 26.85 ft

   Pressure, psi = (26.85 ft)(0.433 psi/ft) = 11.626 psi, round to **11.6 psi**

5. Answer: **d.** 440 gpm

   Equation: Well yield, gpm = (Specific yield, gpm/ft)(Drawdown, ft)

   Substitution: Well yield, gpm = (31 gpm/ft)(14.1 ft) = 437.1 gpm, round to **440 gpm**

6. Answer: **c.** 15,400 gal

   First convert the diameter from inches to feet.

   Number of feet = $\dfrac{18.0 \text{ in.}}{12 \text{ in./ft}}$ = 1.50 ft

   Next, calculate the volume.

   Equation: Pipe volume, gal = (0.785)(Diameter, ft)²(Length, ft)(7.48 gal/ft³)

   Pipe volume, gal = (0.785)(1.50 ft)(1.50 ft)(1,165 ft)(7.48 gal/ft³)
   = 15,391 gal, round to **15,400 gal**

7. Answer: **d.** 45.5 hr

   First, convert gpm to gal per hour (gph): (2,105 gpm)(60 min/hr) = 126,300 gph

   Next, convert million gallons (mil gal) to gallons.

   Water tank, gal = (5.75 mil gal)(1,000,000) = 5,750,000 gal

   Equation: Time, hr = Number of gallons ÷ gph

   Time, hr = 5,750,000 gal ÷ 126,300 gph = 45.527 hr, round to **45.5 hr**

8. Answer: **d.** 8.3 hr

First, convert pipe diameter from inches to feet.

Pipe diameter, ft = (14 in.)(1 ft/12 in.) = 1.167 ft

Next, find the number of gallons in the pipeline.

Number of gal = (0.785)(Diameter, ft)^2(Length, ft)(7.48 gal/ft^3)

Number of gal = (0.785)(1.167 ft)(1.167 ft)(549 ft)(7.48 gal/ft^3) = 4,390 gal

Add the pipe and tank volume to get the total number of gallons.

Pipe and tank volume, gal = 4,390 gal + 2,310,000 gal = 2,314,390 gal

Then convert mgd to gallons per day.

Number of gal = (6.72 mgd)(1,000,000) = 6,720,000 gal/day

Using the following equation, solve for the detention time.

Equation: Detention time, hr = $\dfrac{\text{(Total Volume)(24 hr/day)}}{\text{Flow, gal/day}}$

Substitute known values and solve:

$\dfrac{(2{,}314{,}390 \text{ gal})(24 \text{ hr/day})}{6{,}720{,}000 \text{ gal/day}}$ = 8.266 hr, round to **8.3 hr**

9. Answer: **d.** 12,900 gpm

Number of gpm = (28.7 cfs)(60 sec/min)(7.48 gal/ft^3)
= 12,880.56 gpm, round to **12,900 gpm**

10. Answer: **a.** 13.7 acre-ft

First, convert the number of liters to gallons.

Number of gal = $\dfrac{16{,}912{,}000 \text{ liters}}{3.785 \text{ gal/liter}}$ = 4,468,164 gal

Next convert gallons to acre-feet.

Number of acre-ft = $\dfrac{\text{(Number of gal)}}{(43{,}560 \text{ ft}^3/\text{acre-ft})(7.48 \text{ gal/ft}^3)}$

Number of acre-ft = $\dfrac{(4{,}468{,}164 \text{ gal})}{(43{,}560 \text{ ft}^3/\text{acre-ft})(7.48 \text{ gal/ft}^3)}$ = 13.713 acre-ft,

round to **13.7 acre-ft**

11. Answer: **b.** −8.7 °F

Equation: °F = (9/5 × °C) + 32

°F = (9/5 × −22.6 °C) + 32 = [(9 × −22.6) ÷ 5] + 32 = (−203.4 ÷ 5) + 32
= −40.68 + 32 = −8.68 °F, round to **−8.7 °F**

12. Answer: **d.** 113,000 mg/L sodium hypochlorite

Know: 1% solution = 10,000 mg/L

(11.3%)(10,000 mg/L) = **113,000 mg/L**

13. Answer: **c.** 30,355 gal/day

Equation: Pumped, gal/day = $\dfrac{\text{(Last read of gal pumped − First read of gallons pumped)}}{\text{Number of days}}$

$$\text{Pumped, gal/day} = \frac{(72{,}487{,}008 \text{ gal} - 71{,}576{,}344 \text{ gal})}{30 \text{ days}} = \frac{910{,}664 \text{ gal}}{30 \text{ days}}$$

$$= 30{,}355.467, \text{ round to } \mathbf{30{,}355 \text{ gal/day}}$$

14. Answer: **b.** 180 lb/day

    Set up a ratio and solve for the unknown, x

    $$\frac{x \text{ lb/day}}{16 \text{ cfs}} = \frac{260 \text{ lb/day}}{23 \text{ cfs}}$$

    $$x = \frac{(260 \text{ lb/day})(16 \text{ cfs})}{23 \text{ cfs}} = \mathbf{180 \text{ lb/day}}$$

15. Answer: **c.** 68.5 ft

    Equation: Circumference = (π)(Diameter)

    Rearrange the equation to solve for the diameter.

    $$\text{Diameter} = \frac{\text{Circumference}}{\pi}$$

    $$\text{Diameter} = \frac{215 \text{ ft}}{3.14} = \mathbf{68.5 \text{ ft}}$$

16. Answer: **d.** 73.7 ft deep

    $$\text{Equation: psi} = \frac{\text{Depth, ft}}{2.31 \text{ ft/psi}}$$

    Rearrange and solve:

    Depth, ft = (31.9 psi)(2.31 ft/psi) = 73.689 ft, round to **73.7 ft**

17. Answer: **a.** 4,970 ft^2

    Equation: Area = $(\pi)r^2$

    Area = (3.14)(39.8 ft)(39.8 ft) = 4,973.8856 ft^2, round to **4,970 ft^2**

18. Answer: **b.** 1.14 SG

    Know: Water has a density of 62.4 lb/ft^3.

    Divide the density of the unknown by the density of water.

    Equation: Specific Gravity (SG) = Density of substance ÷ Density of water

    $$\text{SG of the unknown liquid} = \frac{70.9 \text{ lb/ft}^3}{62.4 \text{ lb/ft}^3} = \mathbf{1.14 \text{ SG}}$$

19. Answer: **a.** 15.7 mgd

    Equation: lb/day = (mgd)(Dosage, mg/L)(8.34 lb/day)

    Rearrange to solve for mgd.

    Treated amount, mgd = Chlorine, lb/day ÷ [(Dosage, mg/L)(8.34 lb/gal)]

    Treated amount, mgd = 295 lb/day ÷ [(2.25 mg/L)(8.34 lb/gal)] = **15.7 mgd**

20. Answer: **c.** 1.3 gal sodium hypochlorite

    First, find the volume of the pipe using.

    Equation: Volume, gal = (0.785)(Diameter, ft)^2(Length, ft)(7.48 gal/ft^3)

Volume = (0.785)(1.5)(1.5)(283 ft)(7.48 gal/ft^3) = 3,739 gal

Next, find the number of million gallons (mil gal).

mil gal = (3,739 gal)(1 M/1,000,000) = 0.003739 mil gal

Then use the "pound" equation:

$$\text{Sodium hypochlorite solution, lb} = \frac{(\text{mil gal})(\text{Dosage, mg/L})(8.34 \text{ lb/gal})}{\% \text{ Available chlorine} \div 100\%}$$

$$\text{Sodium hypochlorite solution, lb} = \frac{(0.003739 \text{ mil gal})(50.0 \text{ mg/L})(8.34 \text{ lb/gal})}{12.1\% \text{ Available chlorine} \div 100\%}$$

$$= 12.89 \text{ lb}$$

Lastly, calculate the number of gallons of sodium hypochlorite.

Sodium hypochlorite, gal = 12.89 lb ÷ 9.92 lb/gal = 1.299 gal, round to **1.3 gal**

21. Answer: **a.** 3.03 mg/L

    Equation: Chlorine, lb/day = (mgd)(Dosage, mg/L)(8.34 lb/day)

    Rearrange to solve for dosage.

    $$\text{Dosage, mg/L} = \frac{\text{Chlorine, lb/day}}{(\text{mgd})(8.34 \text{ lb/gal})} = \frac{415 \text{ lb/day}}{(16.4 \text{ mgd})(8.34 \text{ lb/gal})} = \textbf{3.03 mg/L}$$

22. Answer: **c.** 59 gal sodium hypochlorite

    First, find the initial amount of water to be disinfected, 10% capacity
    = (1.65 mil gal)(10% ÷ 100%) = 0.165 mil gal

    Next, determine the number of pounds of chlorine needed by using the "pounds" equation.

    $$\text{Sodium hypochlorite, lb} = \frac{(\text{mil gal})(\text{Dosage, mg/L})(8.34 \text{ lb/gal})}{(\% \text{ Available chlorine})(100\%)}$$

    $$\text{Sodium hypochlorite, lb} = \frac{(0.165 \text{ mil gal})(50.0 \text{ mg/L})(8.34 \text{ lb/gal})(100\%)}{(11.8\%)}$$

    Lastly, convert the pounds of sodium hypochlorite to gallons by dividing by 9.84 lb/gal.

    Sodium hypochlorite, gal = 583.1 lb ÷ 9.84 lb/gal = 59.26, round to **59 gal**

23. Answer: **c.** 11.9 lb calcium hypochlorite

    First, convert 24 in. to feet: 24 in. ÷ 12 in. per foot = 2 ft

    Next, find the volume of the pipe in gallons using the following formula:

    Equation: Pipe volume, gal = (0.785)(Diameter, ft)^2(Length, ft)(7.48 gal/ft^3)

    Pipe volume, gal = (0.785)(2.0 ft)(2.0 ft)(781 ft)(7.48 gal/ft^3) = 18,344 gal

    Next, find the number of million gallons (mil gal).

    mil gal = (18,344 gal)(1 M ÷ 1,000,000) = 0.018344 mil gal

    Then use the "pounds" equation:

    $$\text{Calcium hypochlorite, lb} = \frac{(\text{mil gal})(\text{Dosage, mg/L})(8.34 \text{ lb/gal})}{(\% \text{ Available chlorine})(100\%)}$$

    $$\text{Calcium hypochlorite, lb} = \frac{(0.018344 \text{ mil gal})(50.0 \text{ mg/L})(8.34 \text{ lb/gal})}{(64.3\%)(100\%)} = \textbf{11.9 lb}$$

24. Answer: **d.** 26.0 lb/day of chlorine

First, convert the pumping rate to million gallons per day (mgd).

Equation: $\text{mgd} = \dfrac{(\text{Pumping rate, gpm})(1{,}440 \text{ min/day})}{1{,}000{,}000}$

Substitute known values and solve: $\text{mgd} = \dfrac{(428 \text{ gpm})(1{,}440 \text{ min/day})}{1{,}000{,}000} = 0.61632 \text{ mgd}$

Next, find the total chlorine dose required.
Total chlorine dose, mg/L = 1.20 mg/L required + 3.85 mg/L demand = 5.05 mg/L
Next, use the "pounds" equation to solve the problem.
Chlorine, lb/day = (mgd)(Dosage, mg/L)(8.34 lb/gal)
= (0.61632 mgd)(5.05 mg/L)(8.34 lb/gal)
= 25.958 lb/day, round to **26.0 lb/day**

25. Answer: **b.** 74.4 cfs

Equation: Number of cfs = $\dfrac{(\text{mgd})(1{,}000{,}000 \text{ gal})(1 \text{ ft}^3)(1 \text{ day})(1 \text{ min})}{(1 \text{ mil gal})(7.48 \text{ gal})(1{,}440 \text{ min})(60 \text{ sec})}$

or: Number of cfs = $\dfrac{(\text{mgd})(1{,}000{,}000 \text{ gal})(1 \text{ ft}^3)(1 \text{ day})}{(1 \text{ mil gal})(7.48 \text{ gal})(86{,}400 \text{ sec})}$

Substitute known values and solve.

Number of cfs = $\dfrac{(48.1 \text{ mgd})(1{,}000{,}000 \text{ gal})(1 \text{ ft}^3)(1 \text{ day})}{(1 \text{ mil gal})(7.48 \text{ gal})(86{,}400 \text{ sec})}$ = **74.4 cfs**

26. Answer: **c.** 11.6 L/s

Equation: Flow, L/s = $\dfrac{(\text{Flow, gpm})(3.785 \text{ L/gal})}{60 \text{ sec/min}}$

Flow, L/s = $\dfrac{(184 \text{ gpm})(3.785 \text{ Liters/gal})}{60 \text{ sec/min}}$ = **11.6 L/s**

27. Answer: **c.** 1.02 ntu

1	2	3	4	5	6	7
1.08 ntu	0.98 ntu	0.94 ntu	0.88 ntu	0.96 ntu	1.03 ntu	1.25 ntu

First add all seven measurements:
1.08 + 0.98 + 0.94 + 0.88 + 0.96 + 1.03 + 1.25 = 7.12 ntu

Equation: Average = $\dfrac{\text{Sum of measurements}}{\text{Number of measurements}}$

Average sedimentation basin ntu = $\dfrac{7.12 \text{ mg/L}}{7}$ = 1.017 ntu, round to **1.02 ntu**

28. Answer: **a.** 410

Equation: 100% Number = $\dfrac{(\text{Number given})(100\%)}{\text{Percent of given number}}$

100% Number = $\dfrac{(288)(100\%)}{70.3\%}$ = 409.67, round to **410**

29. Answer: **d.** 99% Fe removal efficiency

Equation: Percent Fe removal efficiency = $\dfrac{(\text{In} - \text{Out})(100\%)}{\text{In}}$

Percent Fe removal efficiency = $\dfrac{(0.81 - 0.01)(100\%)}{0.81}$ = **99% Fe removal efficiency**

30. Answer: **a.** 3.27% soda ash slurry

Equation: Percent soda ash slurry = $\dfrac{(\text{Soda ash, lb})(100\%)}{\text{Soda ash, lb} + (8.34 \text{ lb/gal})(\text{Water, gal})}$

Substitute known values and solve.

Percent soda ash slurry = $\dfrac{(28.2 \text{ lb})(100\%)}{28.2 \text{ lb} + (8.34 \text{ lb/gal})(100.0 \text{ gal})}$

= $\dfrac{(28.2 \text{ lb})(100\%)}{28.2 \text{ lb} + 834 \text{ lb}}$ = $\dfrac{(28.2 \text{ lb})(100\%)}{862.2 \text{ lb}}$

= **3.27% soda ash slurry**

31. Answer: **d.** 11,100 ft²

First, find the square footage of the wall area.

Equation: Wall area, ft² = (Diameter, ft)(π)(Height, ft); where π equals 3.14

Wall area, ft² = (80.1 ft)(3.14)(24.0 ft) = 6,036 ft²

Next, find the area of the top. Note: There is no bottom exterior area.

Top area, ft² = (0.785)(Diameter, ft)² = (0.785)(80.1 ft)(80.1 ft) = 5,037 ft²

Total exterior surface area of tank, ft² = 6,036 ft² + 5,037 ft²
= 11,073 ft², round to **11,100 ft²**

32. Answer: **c.** 99,800 gal

First, convert miles to feet:

Number of miles = (1.43 miles)(5,280 ft/mile) = 7,550.4 ft

Next, convert 18.0 inches to feet:

Number of ft = 18.0 inches/12 inches per foot = 1.50 ft

Substitute known values and solve.

Equation: Volume, gal = (0.785)(Diameter, ft)²(Length, ft)(7.48 gal/ft³)

Volume, gal = (0.785)(1.50 ft)(1.50 ft)(7,550.4 ft)(7.48 gal/ft³)
= 99,752 gal, round to **99,800 gal**

33. Answer: **b.** 17.8 hr

First, find the diameter: Diameter, ft = 2(radius) = 2(60.0 ft) = 120 ft

Then, determine the volume of water in the storage tank.

Equation: Volume, gal = (0.785)(Diameter, ft)²(Depth, ft)(7.48 gal/ft³)

Average Tank Volume, gal = (0.785)(120 ft)(120 ft)(25.5 ft)(7.48 gal/ft³)
= 2,156,125 gal

Next, convert mgd to gallons per day.

Flow through tank, gal/day = (2.91 mgd)(1,000,000 gal) = 2,910,000 gal/day

Next, solve for the detention.

Equation: Detention time, hr = [(Tank Volume)(24 hr/day)] ÷ (Flow, gal/day)

ANSWERS TO MATH QUESTIONS FOR WATER DISTRIBUTION OPERATORS LEVELS I & II 377

Substitute known values and solve:

Detention time, hr = [(2,156,125 gal)(24 hr/day)] ÷ (2,910,000 gal/day) = **17.8 hr**

34. Answer: **d.** 58 psi

    Equation: Pressure Head, ft = (Pressure, psi)(2.31 ft/psi)

    Rearrange to solve for pressure in psi:

    Pressure, psi = (Pressure head, ft) ÷ (2.31 ft/psi)

    Pressure, psi = 134 ft ÷ 2.31 ft/psi = **58 psi**

35. Answer: **a.** 820 gal

    First, convert the flow in ft^3/min to gallons per minute (gpm).

    gpm = (5.5 ft^3/min)(7.48 gal/ft^3) = 41.14 gpm

    Then determine the number of gallons that flowed through the fire hydrant.

    Gallons = (41.14 gpm)(20 min) = 822.8 gal, round to **820 gal**

36. Answer: **b.** 1.19 g/cm^3

    First, convert the number of pounds to grams.

    Know from conversion tables that 1 pound = 454 grams and 1 liter = 1000.027 cm^3

    Number of grams = (Number of lb)(454 g/lb)

    Number of grams = (8.25 lb)(454 g/lb) = 3,745.5 g

    Number of cm^3 = (1000.027 cm^3/1 L)(3.150 L) = 3,150.085 cm^3

    Equation: Density = Mass/Volume

    Density = 3,745.5 g/3,150.085 cm^3 = **1.19 g/cm^3**

37. Answer: **b.** 98.6% meter efficiency

    First, convert cubic feet to gallons.

    Number of gal = (245.7 ft^3)(7.48 gal/ft^3) = 1,837.836 gal

    Equation: Meter accuracy, % = $\dfrac{\text{(Meter reading, gal)(100\%)}}{\text{Actual volume, gal}}$

    Meter accuracy, % = $\dfrac{(1{,}837.836 \text{ gal})(100\%)}{1{,}863 \text{ gal}}$ = **98.6% meter efficiency**

38. Answer: **c.** 3.3 mg/L

    Equation: lb/day = (mgd)(Dosage, mg/L)(8.34 lb/day)

    Rearrange the equation and solve for dosage.

    Dosage, mg/L = $\dfrac{\text{lb/day}}{\text{(mgd)(8.34 lb/gal)}}$ = $\dfrac{320 \text{ lb/day}}{(11.6 \text{ mgd})(8.34 \text{ lb/gal})}$

    = 3.308 mg/L, round to **3.3 mg/L**

39. Answer: **a.** 678 gal of sodium hypochlorite

    First, determine the number of pounds of chlorine needed by using the "pounds" equation.

    Equation: Sodium hypochlorite, lb = (mil gal)(Dosage, mg/L)(8.34 lb/gal)

Chlorine, lb = (1.75 mil gal)(50.0 mg/L)(8.34 lb/gal) = 729.75 lb

Next, find the number of gallons of sodium hypochlorite.

Equation: Sodium hypochlorite, gal = $\dfrac{(\text{Chlorine, lb})(100\%)}{(\text{lb/gal})(\%\text{ Solution})}$

Substitute known values and solve.

Sodium hypochlorite, gal = $\dfrac{(729.75 \text{ lb of Cl}_2)(100\%)}{(8.97 \text{ lb/gal})(12.0\%)}$ = 677.95 gal, round to **678 gal**

40. Answer: **d.**   54 mg/L chlorine

First, convert the diameter of the pipeline from inches to feet.

Number of feet = 24.0 in. ÷ 12 in./ft = 2.0 ft

Next, find the number of gallons by determining the volume of the pipeline.

Equation: Volume of pipe, gal = (0.785)(Diameter, ft)^2(Length, ft)

Volume of pipe, gal = (0.785)(2.0 ft)(2.0 ft)(427 ft)(7.48 gal/ft^3) = 10,029 gal

Then convert number of gallons to mil gal.

Number of mil gal = 10,029 gal ÷ 1,000,000 = 0.010029 mil gal

Finally, calculate the dosage by rearranging the "pounds" equation.

Calcium hypochlorite, lb/day = $\dfrac{(\text{mgd})(\text{Dosage, mg/L})(8.34 \text{ lb/gal})(100\%)}{\text{Percent available chlorine}}$

Rearrange the equation and drop the day on each side of the equation as it is not needed.

Dosage, mg/L = $\dfrac{(\text{Calcium hypochlorite, lb})(65.0\% \text{ Available chlorine})}{(\text{mil gal})(8.34 \text{ lb/gal})(100\% \text{ calcium hypochlorite})}$

= $\dfrac{(7.0)(65.0\%)}{(0.010029)(8.34)(100\%)}$

= 54.4 mg/L, round to **54 mg/L**

41. Answer: **c.**   7.4 mg/L sodium hypochlorite

First, convert the production rate of the well pump to mgd.

Equation: mgd = $\dfrac{(\text{Pumping rate})(1,440 \text{ min/day})}{1,000,000}$

mgd = $\dfrac{(260 \text{ gpm})(1,440 \text{ min/day})}{1,000,000 \text{ mil gal}}$ = 0.3744 mgd

Second, convert liters/day to gallons/day.

Number of gal = $\dfrac{95 \text{ Liters/day}}{3.785 \text{ Liters/gal}}$ = 25.1 gal/day

Next, calculate the chlorine usage in pounds per day. Equation:

Chlorine usage, lb/day

= $\dfrac{(\text{Hypochlorinator flow, gal/day})(\%\text{Cl}_2 \text{ in hypochlorite solution})(8.95 \text{ lb/gal})}{100\%}$

= $\dfrac{(25.1 \text{ gal/day})(10.3\%)(8.95 \text{ lb/gal})}{100\%}$ = 23.14 lb/day

Lastly, calculate the chlorine dosage using the "pounds" equation.

Sodium hypochlorite, lb/day = [(mgd)(Dosage, mg/L)(8.34 lb/gal)]

Rearrange the equation to solve for dosage.

Equation: Dosage, mg/L = $\dfrac{\text{Sodium hypochlorite, lb/day}}{(\text{mgd})(8.34 \text{ lb/gal})}$

Dosage, mg/L = $\dfrac{23.14}{(0.3744)(8.34)}$ = **7.4 mg/L sodium hypochlorite**

42. Answer: **a.** 10.5 lb/day of chlorine

    First, convert the pumping rate to mgd.

    Equation: mgd = $\dfrac{(\text{pumping rate, gpm})(1{,}440 \text{ min/day})}{1{,}000{,}000}$

    Substitute known values and solve:

    mgd = $\dfrac{(208 \text{ gpm})(1{,}440 \text{ min/day})}{1{,}000{,}000}$ = 0.29952 mgd

    Next determine the total chlorine dosage required.

    Total chlorine, mg/L = Chlorine demand, mg/L + chlorine residual, mg/L

    Total chlorine, mg/L = 2.45 mg/L + 1.75 mg/L = 4.20 mg/L

    Next, use the "pounds" equation to solve the problem.

    Equation: Chlorine, lb/day = (mgd)(Dosage, mg/L)(8.34 lb/gal)

    Chlorine, lb/day = (0.29952 mgd)(4.20 mg/L)(8.34 lb/gal) = **10.5 lb/day chlorine**

43. Answer: **b.** 910 gpm

    Know: 1 hp = 33,000 ft-lb/min

    Convert 15 hp to ft-lb/min: 15 × 33,000 = 495,000 ft-lb/min

    Solve for the unknown value, x:

    (65)(x lb/min) = 495,000 ft-lb/min

    x lb/min = 495,000 ft-lb/min ÷ 65 = 7,615 lb/min

    Now express this maximum pumping rate in gallons per minute

    7,615 lb/min ÷ 8.34 lb/gal = 913.70 gpm, round to **910 gpm**

44. Answer: 128 gpcd

    First, convert 2.98 mgd to mil gal.

    Number of gallons = 2.98 mgd × 1,000,000 mil gal = 2,980,000 gal

    Equation:
    Gallons per capita per day (gpcd) = Volume gal/day ÷ Population served/day

    gpcd = 2,980,000 gal/day ÷ 23,210 capita/day = **128 gpcd**

Answers to Math Questions for Water Distribution Operators Levels III & IV

1. Answer: **d.** 1.4 ft/sec

   First, convert gpm to cubic feet per second (cfs).

   Number of cfs = $\dfrac{122 \text{ gpm}}{(7.48 \text{ gal/ft}^2)(60 \text{ sec/min})}$ = 0.2718 cfs

   Next, convert the diameter from inches to feet.

   Number of ft = (6.0-in)(1 ft/12-in) = 0.50 ft

   Equation: Flow, cfs = (Area, ft²)(Velocity, ft/sec); where the Area = (0.785)(Diameter)²

   0.2718 cfs = (0.785)(0.50 ft)(0.50 ft)(Flow, ft/sec)

   Rearrange and solve for the flow in ft/sec.

   Flow, ft/sec = $\dfrac{0.2718 \text{ cfs}}{(0.785)(0.50 \text{ ft})(0.50 \text{ ft})}$ = 1.38, round to **1.4 ft/sec**

2. Answer: **b.** 1,200 lb

   Equation: Pressure = $\dfrac{\text{Force, lb}}{\text{Area, ft}^2}$ for pressure on the small cylinder.

   First convert 10-inches to ft: (10 in)(1 ft/12 in) = 0.833 ft

   Pressure = $\dfrac{130 \text{ lb}}{(0.785)(0.833 \text{ ft})(0.833 \text{ ft})}$ = 238.66 lb/ft²

   Next, calculate the total force on the large cylinder.

   Equation: Total force = (Pressure)(Area)

   Total force = (238.66 lb/ft²)(0.785)(2.5 ft)(2.5 ft) = 1,170.926 lb, round to **1,200 lb**

3. Answer: **c.** 50 mg/L

   First, find the number of feet in 2.45 miles

   Length in ft = (5,280 ft/mile)(2.45 miles) = 12,936 ft

   Next, find the volume in cubic feet (ft³) for the pipe.

   Equation: Volume, ft³ = (0.785)(Diameter, ft)²(Length, ft)

   Volume, ft³ = (0.785)(2.0 ft)(2.0 ft)(12,936 ft) = 40,619.04 ft³

   Then, determine the number of gallons.

   Number of gal = (40,619.04 ft³)(7.48 gal/ft³) = 303,830.42 gal

   Convert the number of gallons to million gallons (mil gal).

   mil gal = $\dfrac{303,830.42 \text{ gal}}{1,000,000}$ = 0.30383 mil gal

   Lastly, find the dosage in mg/L.

   Dosage, mg/L = $\dfrac{\text{lb of Chlorine}}{(\text{mil gal})(8.34 \text{ lb/gal})}$ = $\dfrac{126.9}{(0.30383)(8.34)}$

   = 50.08 mg/L, round to **50 mg/L**

4. Answer: **d.** 310 mhp

Equation: Motor hp = $\dfrac{\text{whp}}{(\text{Motor effic.})(\text{Pump effic.})}$

Motor hp = $\dfrac{200 \text{ whp}}{(88\%/100\% \text{ Motor effic.})(74\%/100\% \text{ Pump effic.})}$

= $\dfrac{(200 \text{ whp})}{(0.88 \text{ Motor effic.})(0.74 \text{ Pump effic.})}$ = 307 mhp, round to **310 mhp**

5. Answer: **c.** 26 bhp

Solution: bhp = $\dfrac{(\text{Bowl head, ft})(\text{Capacity, gpm})}{(3960)(\text{Bowl efficiency, \%/100\%})}$

bhp = $\dfrac{(215 \text{ ft})(385 \text{ gpm})}{(3960)(81\%/100\%)}$ = 25.8 bhp, round to **26 bhp**

6. Answer: **c.** 2.3 ft/sec

Flow in 4.0-inch pipe equals the flow in the 3.0-inch pipe as the flow must remain constant: $Q_1 = Q_2$

Equation: (Area 1)(Velocity 1) = (Area 2)(Velocity 2)

First, find the diameter for the 3.0-inch and 4.0-inch pipes:

Diameter for 3.0-inch = (3.0-inch)(1 ft/12-in) = 0.25 ft

Diameter for 4.0-inch = (4.0-inch)(1 ft/12-in) = 0.333 ft

Then determine the areas of each size pipe: Area = $(0.785)(\text{Diameter, ft})^2$

Area 1 (3.0-in) = (0.785)(0.25 ft)(0.25 ft) = 0.049 ft^2

Area 2 (4.0-in) = (0.785)(0.333 ft)(0.333 ft) = 0.087 ft^2

Lastly, substitute areas calculated and known velocity in 4.0-inch pipe.

(0.049 ft^2)(x, ft/sec) = (0.087 ft^2)(1.3 ft/sec)

Solve for x:

x, ft/sec = $\dfrac{(0.087 \text{ ft}^2)(1.3 \text{ ft/sec})}{(0.049 \text{ ft}^2)}$ = 2.308 ft/sec, round to **2.3 ft/sec**

7. Answer: **b.** 3,500 gpm

First, find the water production during the 18-hour interval.

Gallons of water treated in 18-hr interval =

[(4.75 mgd)(1,000,000)(18 hrs)] ÷ 24 hrs = 3,562,500 gal

Next, find the gallons contained in the 2.3 ft drop in water level.

Volume, tank = $(0.785)(\text{Diameter, ft})^2(\text{Height})$

Volume of 2.3 ft, 120 ft diameter tank = (0.785)(120 ft)(120 ft)(2.3 ft)(7.48 gal/ft^3)
= 194,474 gal

Production plus the loss in level is the amount the discharge pump had to send into the distribution system, but first find the number of minutes in 18 hrs.

(18 hrs)(60 min/hr) = 1,080 min

Then determine total gallons discharge pump moved.

Total gal discharge pump moved in 18 hrs = 3,562,500 gal + 194,474 gal
= 3,756,974 gal

Lastly, divide the number of gallons the discharge pump moved by the time in minutes.

Discharge pump, gpm = 3,756,974 gal ÷ 1,080 min
= 3,479 gpm, round to **3,500 gpm**

8. Answer: **b.** 0.4 gal of sodium hypochlorite

First, convert 1,000 gallons to million gallons:

Number of mil gal = 1,000 gal ÷ 1,000,000 = 0.001 mil gal

Next determine the number of pounds of hypochlorite that are required using the "pounds" formula.

Equation: Sodium hypochlorite, lb = $\dfrac{(\text{Dosage, mg/L})(\text{mil gal})(8.34 \text{ lb/gal})(100\%)}{\text{Percent sodium hypochlorite}}$

Substitute known values and solve:

Sodium hypochlorite, lb = $\dfrac{(50 \text{ mg/L})(0.001 \text{ mil gal})(8.34 \text{ lb/gal})(100\%)}{12.5\% \text{ Sodium hypochlorite}}$ = 3.736 lb

Lastly, determine the number of gallons of sodium hypochlorite by dividing by 9.34 lb/gal.

Sodium hypochlorite, gal = 3.736 lb ÷ 9.34 lb/gal = **0.4 gal**

9. Answer: **b.** 9.0% final solution

First, find the total volume that would result from mixing these two solutions:

Total Volume = 350 gal + 225 gal = 575 gal

Then write the equation:

(Concentration_1)(Volume_1) + (Concentration_2)(Volume_2) = (Concentration_3)(Volume_3), or condensed as

$C_1 V_1 + C_2 V_2 = C_3 V_3$,

where C_1 and C_2 = % Concentration of the two solutions before being mixed,

V_1 and V_2 = Volume of the two solutions before being mixed,

and C_3 and V_3 = the resulting % Concentration and Volume, respectively.

Substituting:

$\dfrac{(11\%)(350 \text{ gal})}{100\%} + \dfrac{(5.8\%)(225 \text{ gal})}{100\%} = \dfrac{C_3(575 \text{ gal})}{100\%}$

$38.5 \text{ gal} + 13.05 \text{ gal} = \dfrac{C_3(575 \text{ gal})}{100\%}$

Solving for C_3:

$C_3 = \dfrac{(38.5 \text{ gal} + 13.05 \text{ gal})(100\%)}{575 \text{ gal}} = \dfrac{(51.55 \text{ gal})(100\%)}{575 \text{ gal}}$ = 8.965, round to **9.0%**

10. Answer: **c.** 32 whp

Equation: whp = $\dfrac{(\text{Pump rate, gpm})(\text{Total head, ft})(8.34 \text{ lb/gal})}{33,000 \text{ ft-lb/min/hp}}$

whp = $\dfrac{(650 \text{ gpm})(195 \text{ ft})(8.34 \text{ lb/gal})}{33,000 \text{ ft-lb/min/hp}}$ = **32 whp**

11. Answer: **a.** 41.4%

    Equation: Percent mixture

    $$= \frac{\frac{(\text{Sol. 1-lb})(\text{Percent Strength of Sol. 1})}{100\%} + \frac{(\text{Sol. 2-lb})(\text{Percent Strength of Sol. 2})}{100\%}(100\%)}{\text{Sol. 1-lb} + \text{Sol. 2-lb}}$$

    Where Sol. = solution

    Substitute known values and solve:

    $$\text{Percent mixture} = \frac{\frac{(875 \text{ lb})(49.5\%)}{100\%} + \frac{(293 \text{ lb})(17.2\%)}{100\%}(100\%)}{875 \text{ lb} + 293 \text{ lb}}$$

    Reduce:

    $$\text{Percent mixture} = \frac{(433.125 \text{ lb} + 50.396 \text{ lb})(100\%)}{1,168 \text{ lb}} = \frac{(483.521 \text{ lb})(100\%)}{1,168 \text{ lb}}$$

    $$= 41.397\%, \text{ round to } \mathbf{41.4\%}$$

12. Answer: **d.** 30 oz

    First, find the diameter in feet for both well casings.

    Diameter for 12-in. casing = 12 in. = 1.00 ft

    Diameter for 10-in. casing = 10 in. ÷ 12 in. = 0.833 ft

    Then, find the length (in feet) of water in the casing.

    Length of water-filled casing = Depth of well − Depth of water to top of casing

    Length of water-filled casing for 12-in. diameter = 100 ft − 168.4 ft = −68.4 ft; since it is negative there is no water in this section, and

    Length of water-filled casing for 10-in. diameter = 287 ft − 168.4 ft = 118.6 ft;

    Next, determine the volume in gallons of water in the 10-in. casing using the equation:

    Volume, in gal = $(0.785)(\text{Diameter, ft})^2(\text{Length, ft})(7.48 \text{ gal/ft}^3)$

    Vol., in gal for 10.0-inch casing = $(0.785)(0.833 \text{ ft})(0.833 \text{ ft})(118.6 \text{ ft})(7.48 \text{ gal/ft}^3)$

    $$= 483 \text{ gal}$$

    Next, determine the number of mil gal.

    mil gal = (483 gal)(1 M/1,000,000) = 0.000483 mil gal.

    Then, find the chlorine required.

    Chlorine required = Chlorine demand + Chlorine residual.

    Chlorine required = 4.7 mg/L + 50.0 mg/L = 54.7 mg/L

    Then, using the "pounds" equation, calculate the number of lb of sodium hypochlorite.

    $$\text{Sodium Hypochlorite, lb} = \frac{(0.000483 \text{ mil gal})(54.7 \text{ mg/L})(8.34 \text{ lb/gal})}{10.5\%/100\% \text{ available chlorine}}$$

    Sodium Hypochlorite, lb = 2.0985 lb

    Next, calculate the number of gallons of sodium hypochlorite required.

    2.0985 lb ÷ 9.10 lb/gal = 0.2306 gal

    Lastly, convert to fluid ounces (oz):

    Sodium hypochlorite, oz = (0.2306 gal)(128 oz/gal) = 29.5 oz, round to **30 oz**

ANSWERS TO MATH QUESTIONS FOR WATER DISTRIBUTION OPERATORS LEVELS III & IV   385

13. Answer: **c.** $1,156.15

   Equation: Cost, month = (Hp)(hrs/day)(# days)(0.746 kW/Hp)(Cost/kW-hr)

   Substitute known values:

   Cost, month = (300 Hp)(4.2 hrs/day)(30 days)(0.746 kW/Hp)($0.041/kW)

   Cost, month = **$1,156.15**

14. Answer: **d.** 14.81 mA

   Equation:
   $$\text{Current process reading} = \frac{(\text{Live signal, mA} - 4\text{ mA offset})(\text{Maximum capacity})}{16 \text{ milliamp span}}$$

   Substitute known values and solve:

   $$16.55 \text{ ft} = \frac{(\text{Live signal, mA} - 4\text{ mA offset})(24.50 \text{ ft})}{16 \text{ mA span}}$$

   Rearrange formula to solve for the current number of milliamps.

   $$\text{Live signal, mA} - 4 \text{ mA offset} = \frac{(16.55 \text{ ft})(16 \text{ mA})}{24.50 \text{ ft}}$$

   $$\text{Live signal, mA} = \frac{(16.55 \text{ ft})(16 \text{ mA})}{24.5 \text{ ft}} + 4 \text{ mA}$$

   Live signal, mA = 10.808 mA + 4 mA = 14.808 mA, round to **14.81 mA**

15. Answer: **b.** 21,900 gal

   First, convert the pipe diameters from inches to feet.

   Large pipe, ft = 14-in. ÷ 12-in. = 1.167 ft

   Small pipe, ft = 10-in. ÷ 12 in. = 0.833 ft

   Next, convert the diameter in feet to the radius in feet by dividing by 2.

   Large pipe, ft = 1.167 ft ÷ 2 = 0.5835 ft

   Small pipe, ft = 0.833 ft ÷ 2 = 0.4165 ft

   Next determine the gallons in the 14-inch pipe and then the 10-inch pipe.

   Equation: Number, gal = ⅔ (Length, ft)$(\pi)(r)^2$(7.48 gal/ft^3); for large pipe

   where r = radius and π = 3.14

   Number, gal = 2/3(3,270 ft)(3.14)(0.5835)^2(7.48 gal/ft^3) = 17,432.89 gal

   Number, gal = 1/3(Length, ft)$(\pi)(r)^2$(7.48 gal/ft^3)

   Number, gal = 1/3(3,270 ft)(3.14)(0.4165)^2(7.48 gal/ft^3) = 4,441.07 gal

   Lastly add the number of gallons for each pipe together.

   Total number of gal = 17,432.89 + 4,441.07 = 21,873.96, round to **21,900 gal**

16. Answer: **b.** 290 lb

   $$\text{Equation: \% Lime} = \frac{(x \text{ lb})(100\%)}{x \text{ lb} + (\text{Water, gal})(8.34 \text{ lb/gal})}$$

   $$15\% \text{ Lime} = \frac{(x \text{ lb})(100\%)}{x \text{ lb} + (200 \text{ gal})(8.34 \text{ lb/gal})}$$

   Multiply both sides of the equation by {x lb + (200 gal)(8.34 lb/gal)}. This leaves the following:

15% Lime {x lb + (200 gal)(8.34 lb/gal)} = (x lb)(100%)

Complete the multiplication.

15x%, lb + 25,020%, lb = 100x%, lb

Subtract 15x%, lb from both sides of the equation.

25,020%, lb = 85x%, lb

Lastly, divide both sides of the equation by 85%.

x = 294.35 lb, round to **290 lb**

17. Answer: **d.** 1,540,000 gal

Equation: Volume, gal = $\dfrac{(b_1 + b_2)(\text{Height, ft})(\text{Length, ft})(7.48 \text{ gal/ft}^3)}{2}$

Volume, gal = $\dfrac{(5.85 \text{ ft} + 10.6 \text{ ft})(4.10 \text{ ft})(6,091 \text{ ft})(7.48 \text{ gal/ft}^3)}{2}$

Volume, gal = $\dfrac{(16.45 \text{ ft})(4.10 \text{ ft})(6,091 \text{ ft})(7.48 \text{ gal/ft}^3)}{2}$

Reduce problem by dividing 2 into 4.10 ft

Volume, gal = (16.45 ft)(2.05 ft)(6,091 ft)(7.48 gal/ft^3)

Volume, gal = 1,536,420 gal, round to **1,540,000 gal**

18. Answer: **b.** 4.16 hr

First, determine the number of gallons in the clearwell, distribution pipe, and storage tank.

Equation: Volume, gal = (Length, ft)(Width, ft)(Depth, ft)(7.48 gal/ft^3)

Clearwell volume, gal = (308 ft)(118 ft)(12.85 ft)(7.48 gal/ft^3) = 3,493,313 gal

Next, convert number of miles to feet.

Number of ft = (5,280 ft/mi)(1.34 miles) = 7,075.2 ft

Vol., gal in pipeline = (0.785 ft)(2.00 ft)(2.00 ft)(7,075.2 ft)(7.48 gal/ft^3)
= 166,177 gal

Volume, gal = (0.785)(Diameter, ft)^2(Height, ft)(7.48 gal/ft^3)

Vol., gal in tank = (0.785)(99.8 ft)(99.8 ft)(26.48 ft)(7.48 gal/ft^3) = 1,548,639 gal

Equation: Clearwell gal + Pipe gal + Tank gal = Total volume of gallons

3,493,313 gal + 166,177 gal + 1,548,639 gal = 5,208,129 gal

Next, convert average mgd to gallons per hour (gph).

(30.02 mgd ÷ 24 hours)(1,000,000) = 1,250,833 gph

Now solve for detention time.

Equation: Detention time, hr = Volume, gal ÷ Flow rate, gph

Detention time, hr = 5,208,129 gal ÷ 1,250,833 gph = **4.16 hr**

19. Answer: **d.** 51 psi

First, calculate the column of water in feet.

Water column = Total depth of well − Depth to water − Number of ft above bottom

Water column in ft = 276.5 − 153.8 ft − 5.0 ft = 117.7 ft

Next, find the pressure in psi.

Pressure, psi = (Water column in ft)(0.433 psi/ft)

Pressure, psi = (117.7 ft)(0.433 psi/ft) = 50.96 psi, round to **51 psi**

20. Answer: **c.** 1,250 gpm

First, convert the diameter of the pipe from inches to feet.

Diameter, ft = (10.0 in.)(1 ft/12 in.) = 0.833 ft

Next, calculate the pipe's cross-sectional area in square feet.

Area, ft² = (0.785)(Diameter, ft)²

Area, ft² = (0.785)(0.833 ft)(0.833 ft) = 0.5447 ft²

Next, find the flow in the pipe in cfs.

Flow, cfs = (Area, ft²)(Velocity, ft/sec)

Flow, cfs = (0.5447 ft²)(5.10 ft/sec) = 2.778 cfs

Last, determine the reading on the flowmeter in gpm.

Flow, gpm = (2.778 cfs)(7.48 gal/ft³)(60 sec/min) = 1,247 gpm, round to **1,250 gpm**

21. Answer: **a.** 1.3 ft/sec

Flow in 8.0-in. pipe equals the flow in the 10.0-in. pipe as the flow must remain constant: $Q_1 = Q_2$

Since Q_1 flow = (Area)(Velocity), it follows that:
(Area 1)(Velocity 1) = (Area 2)(Velocity 2)

First, find the diameters in feet for the 8.0-in. and 10.0-in. pipes.

Diameter for 8.0-in. = 8.0 in. ÷ 12 in. = 0.667 ft

Diameter for 10.0-in. = 10.0 in. ÷ 12 in. = 0.833 ft

Then determine the areas of each size pipe: Area = (0.785)(Diameter, ft)²

Area 1 (8.0-in.) = (0.785)(0.667 ft)(0.667 ft) = 0.349 ft²

Area 2 (10.0-in.) = (0.785)(0.833 ft)(0.833 ft) = 0.5447 ft²

Lastly, substitute areas calculated and known velocity in 8.0-in. pipe.

(0.349 ft²)(2.0 ft/sec) = (0.5447 ft²)(x, ft/sec)

Solve for x, ft/sec.

$$x, \text{ft/sec} = \frac{(0.349 \text{ ft}^2)(2.0 \text{ ft/sec})}{(0.5447 \text{ ft}^2)} = 1.28 \text{ fps, round to } \mathbf{1.3 \text{ ft/sec}}$$

22. Answer: **c.** 0.87 ft/sec

First, convert the diameter in inches to feet.

Diameter, ft = 18 in. ÷ 12 in. = 1.5 ft

Next, determine the cross-sectional area of the pipe.

Equation: Area, ft² = (0.785)(Diameter, ft)²

Area, ft² = (0.785)(1.5 ft)(1.5 ft) = 1.76625 ft²

Next, determine the cfs flowing in the 18.0-in. pipe.

Number of cfs = (988,000 gal/24 hr)(1 ft³/7.48 gal)(1 hr/3,600 sec) = 1.5288 cfs

Lastly, determine the flow in ft/sec.

Equation: Flow, cfs = (Area, ft^2)(Velocity, ft/sec)

Rearrange the formula to solve for velocity.

Average velocity, ft/sec = Flow, cfs ÷ Area, ft^2

Substitute known values and solve.

Average velocity, ft/sec = 1.5288 cfs ÷ 1.76625 ft^2

= 0.8656 ft/sec, round to **0.87 ft/sec**

23. Answer: **d.** 8.61 mg/L

    First, convert the level drop in the tank from inches to feet.

    Level drop, ft = 8.03 in. ÷ 12 in. = 0.6692 ft

    Next, determine the amount of hypochlorite used.

    Equation: Volume, gal = (0.785)(Tank diameter, ft)^2(Level drop, ft)(7.48 gal/ft^3)

    Volume, gal = (0.785)(6.0 ft)(6.0 ft)(0.6692 ft)(7.48 gal/ft^3) = 141.46 gal

    Next, find how many pounds of available chlorine this is.

    Chlorine, lb = [(141.46 gal)(8.34 lb/gal)(64.5%)] ÷ 100% = 760.96 lb of chlorine

    Lastly, calculate the dosage.

    Equation: Chlorine, lb = (mil gal)(Dosage, mg/L)(8.34 lb/gal)

    Solve for dosage:

    Dosage, mg/L = Chlorine, lb ÷ [(mil gal)(8.34 lb/gal)]

    Dosage, mg/L = 760.96 lb ÷ [(10.6 mil gal)(8.34 lb/gal)] = **8.61 mg/L**

24. Answer: **a.** 0.17 gal

    First, find the length (in feet) of water in the casing.

    Length of water-filled casing = Depth of well − Depth of water to top of casing

    Length of water-filled casing = 227 ft − 143 ft = 84 ft

    Then, convert the diameter from inches to feet.

    Diameter, ft = 12 in. = 1.0 ft

    Next, determine the volume in gallons of water in the well casing using the following equation:

    Volume, gal = (0.785)(Diameter, ft)^2(Length, ft)(7.48 gal/ft^3)

    Volume, gal = (0.785)(1.0 ft)(1.0 ft)(84 ft)(7.48 gals/ft^3) = 493.23 gal

    Next, determine the number of mil gal.

    mil gal = 493.23 gal ÷ 1,000,000 = 0.000493 mil gal

    Using the "pounds" formula, calculate the number of lb of sodium hypochlorite needed.

    NaOCl, lb = (0.000493 mil gal)(50.0 mg/L)(8.34 lb/gal) = 0.20558 lb NaOCl

    Lastly, calculate the gallons of sodium hypochlorite solution required.

    NaOCl, gal = [(Chlorine, lb)(100%)] ÷ [(NaOCl, lb/gal)(Hypochlorite, %)]

    NaOCl, gal = [(0.20558 lb)(100%)] ÷ [(9.59 lb/gal)(12.5%)]

    = 0.1715 gal, round to **0.17 gal NaOCl**

ANSWERS TO MATH QUESTIONS FOR WATER DISTRIBUTION OPERATORS LEVELS III & IV   389

25. Answer: **c.** 12.6 lb

First, calculate the number of gallons in the pipeline.

Equation: Pipeline, gal = (0.785)(Diameter, ft)^2(Length, ft)(7.48 gal/ft^3)

Pipeline, gal = (0.785)(2.50 ft)(2.50 ft)(1,058 ft)(7.48 gal/ft^3) = 38,827 gal

Convert gallons to mil gal.

Number of mil gal = 38,827 gal ÷ 1,000,000 = 0.038827 mil gal

Next, determine the number of pounds of calcium hypochlorite {$Ca(OCl)_2$} needed using the modified version (accounting for the percent available chlorine) of the "pounds" formula.

$Ca(OCl)_2$, lb = [(mil gal)(Dosage, mg/L)(8.34 lb/gal)(100%)] ÷ 64.5%

$Ca(OCl)_2$, lb = [(0.038827 mil gal)(25.0 mg/L)(8.34 lb/gal)(100%)] ÷ 64.5% = **12.6 lb**

26. Answer: **b.** 350 gal

First, determine the capacity of the tank in gallons.

Equation: Volume, gal = (0.785)(Diameter, ft)^2(Height, ft)(7.48 gal/ft^3)

Volume, gal = (0.785)(84.0 ft)(84.0 ft)(24.25 ft)(7.48 gal/ft^3) = 1,004,712 gal

Next, convert number of gallons to million gallons.

1,004,712 gal ÷ 1,000,000 = 1.0047 mil gal

Using the "pounds" formula, determine the chlorine pounds needed.

Chlorine, lb = (Volume, mil gal)(Dosage, mg/L)(8.34 lb/gal)

Chlorine, lb = (1.0047 mil gal)(50.0 mg/L)(8.34 lb/gal) = 418.96 lb

NaOCl solution, gal = [(Chlorine, lb)(100%)] ÷ [(NaOCl, lb/gal)(Hypochlorite, %)]

= [(418.96 lb)(100%)] ÷ [(9.59 lb/gal)(12.5 %)]

= 349.5 gal, round to **350 gal**

27. Answer: **c.** 14 tablets

First, determine the volume of the pipeline in million gallons (mil gal).

Number of mil gal = [(0.785)(Diameter, ft)^2(Length, ft)(7.48 gal/ft^3)] ÷ 1,000,000

Number of mil gal = [(0.785)(2.50 ft)(2.50 ft)(513 ft)(7.48 gal/ft^3)] ÷ 1,000,000
= 0.0188

Next, determine the number of pounds of calcium hypochlorite needed.

Equation: $Ca(OCl)_2$, lb = [(mil gal)(Dosage, mg/L)(8.34 lb/gal)] ÷ % Purity

$Ca(OCl)_2$, lb = [(0.0188 mil gal)(25.0 mg/L)(8.34 lb/gal)] ÷ [64.0% ÷ 100%]
= 6.12 lb

Lastly, find the number of tablets needed.

Number of Tablets = 6.12 lb ÷ 0.45 lb/tablet = 13.6 tablets, round to **14 tablets**

28. Answer: **a.** 43 oz of NaOCl

First, find the length (in feet) of water in the casing.

Length of water-filled casing = Depth of well − Depth of water to top of casing

Length of water-filled casing = 210 ft − 91 ft = 119 ft

Then convert the diameter from inches to feet.

Diameter, ft = 14.0 in. ÷ 12 in./ft = 1.1667 ft

Next, determine the volume in gallons of water in the well casing using the following formula:

Volume, gal = (0.785)(Diameter, ft)^2(Length, ft)(7.48 gal/ft^3)

Volume, gal = (0.785)(1.1667 ft)(1.1667 ft)(119 ft)(7.48 gal/ft^3) = 951.12 gal

Next, determine the number of mil gal.

mil gal = 951.12 gal ÷ 1,000,000 = 0.000951 mil gal

Lastly, using the "pounds" formula, calculate the number of lb of sodium hypochlorite.

Sodium hypochlorite, lb
$$= \frac{[(0.000951 \text{ mil gal})(50.0 \text{ mg/L})(8.34 \text{ lb/gal})]}{(12.5\% \text{ Available chlorine}/100\%)}$$
$$= 3.173 \text{ lb}$$

Next, find the number of ounces sodium hypochlorite (NaOCl).

Number of ounces = [(3.173 lb)(128 oz/gal)] ÷ [9.50 lb/gal] = 42.75 oz, round to **43 oz**

29. Answer: **c.** 10.69 mg/L

    First, determine the amount of water released from the storage tank in mgd.

    Number of mgd = [(3,075 gpm)(1,440 min/day)] ÷ 1,000,000 = 4.428 mgd

    Next, find the number of pounds of soda ash used for the day.

    Soda ash, lb = (124.5 grams/min)(1,440 min/day)(1 lb/454 grams) = 394.89 lb

    Lastly, calculate the dosage.

    Equation: Number of lb = (mil gal)(Dosage, mg/L)(8.34 lb/gal)

    Solve for dosage.

    Dosage, mg/L = Soda ash, lb ÷ [(mil gal)(8.34 lb/gal)]

    Dosage, mg/L = 394.89 lb, Soda ash ÷ [(4.428 mil gal)(8.34 lb/gal)] = **10.69 mg/L**

30. Answer: **a.** 1,150 gpm

    First, find the water production during the 12-hr interval.

    Gal of water treated in 12-hr interval = [(4.75 mgd)(1,000,000)(12 hr)] ÷ 24 hr

    Gal of water treated in 12-hr interval = 2,375,000 gal

    Next, find the gallons contained in the 7.08 ft drop in water level.

    Volume, tank = (0.785)(Diameter, ft)^2(Height, ft)

    Volume of 7.08 ft in 149.8 ft diameter tank
     = (0.785)(149.8 ft)(149.8 ft)(7.08 ft)(7.48 gal/ft^3)

    Volume of 7.08 ft in 149.8 ft diameter tank = 932,885 gal

    Production plus the loss in level is the amount the discharge pump sent into the distribution system, but first find the number of minutes in 12 hr.

    (12.0 hr)(60 min/hr) = 720 min

    Then determine total gallons discharge pumps moved.

    Total gal discharge pumps moved in 12 hr = 2,375,000 gal + 932,885 gal
    = 3,307,885 gal

    Lastly, divide the number of gallons the discharge pumps moved by the time in minutes and the number of pumps.

Discharge pumps, gpm = 3,307,885 gal ÷ [(720 min)(4 pumps)]
= 1,149 gpm, round to **1,150 gpm**

31. Answer: **c.** 420 hp

    First, determine the pumping head for this pump.

    Pumping head = Elevation of water storage tank − Elevation of pump

    Pumping head = 478.16 ft − 170.84 ft = 307.32 ft

    Next, calculate the friction loss in the pipeline.

    Friction loss = [(2,107 ft)(1.57 ft)] ÷ 1000 ft = 3.31 ft

    Now, calculate the total head.

    Total dynamic head (TDH) = Suction lift + pumping head + friction loss
    + velocity head

    TDH = 2.5 ft + 307.32 ft + 3.31 + 2.38 ft = 315.51 ft

    Lastly, determine the required horsepower of the pump.

    Hp = [(gpm)(TDH)] ÷ [(3960)(Pump efficiency)(Motor efficiency)]
    = [(4,000 gpm)(315.51 ft)] ÷ [(3960)(85%/100%)(89%/100%)]
    = 421.28, round to **420 hp**

32. Answer: **b.** 17 ft, therefore NPSHA < NPSHR so cavitation should occur

    First determine the atmospheric pressure in feet:

    Know: 1.09 ft/in. Hg. Thus:

    Atmospheric pressure = (29.8 in. Hg)(1.11 ft/in. Hg) = 33.078 ft

    Next determine the NPSHA.

    NPSHA = AP, ft − SSL, ft − Hf, ft − VP, ft

    NPSHA = 33.078 ft − 15.1 ft − 0.61 ft − 0.50 ft

    NPSHA = 16.868 ft, round to **17 ft**

    Therefore: **NPSHA 17 ft < NPSHR 18.4 ft, so cavitation should occur.**

33. Answer: **a.** 95

    First, determine the energy loss in feet.

    Energy loss, ft = Upstream gauge read − Downstream gauge read

    Energy loss, ft = 120 ft − 105 ft = 15 ft

    Next, calculate the slope.

    Equation: Slope = (Energy loss, ft) ÷ (Distance, ft)

    Slope = 15 ft ÷ 2,274 ft = 0.0065963 ft

    Lastly, find the C Factor.

    Equation: C Factor = Flow, gpm ÷ (193.75)(Pipe diameter, ft)$^{2.63}$(Slope, ft)$^{0.54}$

    C Factor = 1,225 gpm ÷ [(193.75)(1.0 ft)$^{2.63}$(0.0065963 ft)$^{0.54}$]

    C Factor = 1,225 gpm ÷ [(193.75)(1.0)(0.06644)] = 95.16, round to **95**

34. Answer: **d.** 14.8 mA

Equation: Current process reading

$$= \frac{(\text{Live signal, mA} - 4\text{ mA offset})(\text{Maximum capacity})}{16 \text{ milliamp span}}$$

Substitute known values and solve:

22.89 ft (Storage tank level)

$$= \frac{(\text{Live signal mA} - 4\text{ mA offset})(34.0 \text{ ft Maximum level})}{16 \text{ mA}}$$

Rearrange the equation to solve for live signal in mA.

Live signal mA − 4 mA offset $= \dfrac{(22.89 \text{ ft})(16 \text{ mA})}{34.0 \text{ ft}}$

Live signal mA $= \dfrac{(22.89 \text{ ft})(16 \text{ mA})}{34.0 \text{ ft}} + 4$ mA offset

$= 10.77$ mA $+ 4$ mA offset $= 14.77$ mA, round to **14.8 mA**

Appendix A: Formulas for Water Treatment, Distribution, and Laboratory Exams

ABC standardized exams include the following formula and conversion information as a reference for the examinee. Formulas and conversions in the tables should be used to solve calculations on the exam.

Alkalinity, as mg $CaCO_3$/L = $\dfrac{\text{(Titrant volume, mL)(Acid normality)(50,000)}}{\text{Sample volume, mL}}$

Amps = $\dfrac{\text{Volts}}{\text{Ohms}}$

Area of circle = $(0.785)(\text{Diameter}^2)$ or $(\pi)(\text{Radius}^2)$

Area of cone (lateral area) = $(\pi)(\text{Radius})(\sqrt{\text{Radius}^2 + \text{Height}^2})$

Area of cone (total surface area) = $(\pi)(\text{Radius})(\text{Radius} + \sqrt{\text{Radius}^2 + \text{Height}^2})$

Area of cylinder (total outside surface area) = 2(Surface area of ends)
 + [(π)(Diameter)(Height or Depth)]

Area of rectangle = (Length)(Width)

Area of a right triangle = $\dfrac{\text{(Base)(Height)}}{2}$

Average (arithmetic mean) = $\dfrac{\text{Sum of terms}}{\text{Number of all terms}}$

Average (geometric mean) = $[(X_1)(X_2)(X_3)(X_4)(X_n)]^{1/n}$ The nth root of the product of n numbers

Chemical feed pump setting, % Stroke = $\dfrac{\text{(Desired flow)(100\%)}}{\text{Maximum flow}}$

Chemical feed pump setting, mL/min = $\dfrac{\text{(Flow, mgd)(Dose, mg/L)(3.785 L/gal)(1,000,000 gal/MG)}}{\text{(Liquid, mg/mL)(24 hr/day)(60 min/hr)}}$

Circumference of circle = $(\pi)(\text{Diameter})$

Composite sample single portion = $\dfrac{\text{(Instantaneous flow)(Total sample volume)}}{\text{(Number of portions)(Average flow)}}$

Detention time = $\dfrac{\text{Volume}}{\text{Flow}}$ Note: Units must be compatible.

Electromotive force (EMF), volts = (Current, amps)(Resistance, ohms) or E = IR

Feed rate, lbs/day = $\dfrac{\text{(Dosage, mg/L)(Capacity, mgd)(8.34 lb/gal)}}{\text{(Purity, decimal percentage)}}$

Feed rate, gpm (Fluoride saturator) = $\dfrac{\text{(Plant capacity, gpm)(Dosage, mg/L)}}{(18{,}000 \text{ mg/L})}$

Filter backwash rise rate, in./min = $\dfrac{\text{(Backwash rate, gpm/ft}^2\text{)(12 in/ft)}}{(7.48 \text{ gal/ft}^3)}$

Filter drop test velocity, ft/min = $\dfrac{\text{Water drop, ft}}{\text{Time of drop, minutes}}$

Filter flow rate or backwash rate, gpm/ft^2 = $\dfrac{\text{Flow, gpm}}{\text{Filter area, ft}^2}$

Filter yield, lbs/hr/ft^2 = $\dfrac{\text{(Solids loading, lbs/day)(Recovery, \%/100\%)}}{\text{(Filter operation, hr/day)(Area, ft}^2\text{)}}$

Flow rate, cfs = (Area, ft^2)(Velocity, ft/sec) or Q = AV, where: Q = flow rate, A = area, V = velocity

Force, pounds = (Pressure, psi)(Area, in^2)

Gallons/Capita/Day (gpcd) = $\dfrac{\text{Volume of water produced, gpd}}{\text{Population}}$

Hardness, as mg CaCO$_3$/L = $\dfrac{\text{(Titrant volume, mL)(1,000)}}{\text{Sample volume, mL}}$

Only when the titration factor is 1.00 of EDTA

Horsepower, Brake (bhp) = $\dfrac{\text{(Flow, gpm)(Head, ft)}}{(3{,}960)\text{(Decimal pump efficiency)}}$

Horsepower, Motor (mhp) = $\dfrac{\text{(Flow, gpm)(Head, ft)}}{(3{,}960)\text{(Decimal pump efficiency)(Decimal motor efficiency)}}$

APPENDIX A: FORMULAS FOR WATER TREATMENT, DISTRIBUTION, AND LABORATORY EXAMS 395

$$\text{Horsepower, Water (whp)} = \frac{(\text{Flow, gpm})(\text{Head, ft})}{3{,}960}$$

$$\text{Hydraulic loading rate, gpd/ft}^2 = \frac{\text{Total flow applied, gpd}}{\text{Area, ft}^2}$$

$$\text{Hypochlorite strength, \%} = \frac{(\text{Chlorine required, lb})(100)}{(\text{Hypochlorite solution needed, gal})(8.34 \text{ lb/gal})}$$

$$\text{Leakage, gpd} = \frac{\text{Volume, gal}}{\text{Time, days}}$$

Mass, lb = (Volume, mil gal)(Concentration, mg/L)(8.34 lb/gal)

Mass flux, lb/day = (Flow, mgd)(Concentration, mg/L)(8.34 lb/gal)

Milliequivalent = (mL)(Normality)

$$\text{Molarity} = \frac{\text{Moles of solute}}{\text{Liters of solution}}$$

$$\text{Normality} = \frac{\text{Number of equivalent weights of solute}}{\text{Liters of solution}}$$

$$\text{Number of equivalent weights} = \frac{\text{Total weight}}{\text{Equivalent weight}}$$

$$\text{Number of moles} = \frac{\text{Total weight}}{\text{Molecular weight}}$$

$$\text{Reduction in flow, \%} = \frac{(\text{Original flow} - \text{Reduced flow})(100\%)}{\text{Original flow}}$$

$$\text{Removal, \%} = \frac{(\text{In} - \text{Out})(100)}{\text{In}}$$

$$\text{Slope, \%} = \frac{\text{Drop or rise}}{\text{Distance}} \times 100$$

$$\text{Solids, mg/L} = \frac{(\text{Dry solids, grams})(1{,}000{,}000)}{\text{Sample volume, mL}}$$

$$\text{Solids concentration, mg/L} = \frac{\text{Weight, mg}}{\text{Volume, L}}$$

$$\text{Specific Gravity} = \frac{\text{Specific weight of substance, lb/gal}}{\text{Specific weight of water, lb/gal}}$$

Surface loading rate/Surface overflow rate, gpd/ft² = $\dfrac{\text{Flow, gpd}}{\text{Area, ft}^2}$

Three Normal Equation = $(N_1 \times V_1) + (N_2 \times V_2) = (N_3 \times V_3)$, where $V_1 + V_2 = V_3$

Two Normal Equation = $N_1 \times V_1 = N_2 \times V_2$, where N = normality, V = volume or flow

Velocity, ft/sec = $\dfrac{\text{Flow rate ft}^3/\text{sec}}{\text{Area, ft}^2}$ or $\dfrac{\text{Distance, ft}}{\text{Time, sec}}$

Volume of cone = ⅓ (.785)(Diameter²)(Height)

Volume of cylinder = (.785)(Diameter²)(Height)

Volume of rectangular tank = (Length)(Width)(Height)

Watts (AC circuit) = (Volts)(Amps)(Power Factor)

Watts (DC circuit) = (Volts)(Amps)

Weir overflow rate, gpd/ft = $\dfrac{\text{Flow, gpd}}{\text{Weir length, ft}}$

Wire-to-water efficiency, % = $\dfrac{\text{whp}}{\text{Power input, mph}} \times 100$

Wire-to-water efficiency, % = $\dfrac{(\text{Flow, gpm})(\text{Total dynamic head, ft})(0.746 \text{ kw/HP})(100)}{(3{,}960)(\text{Electrical demand, kW})}$

Alkalinity Relationships

	Alkalinity, mg/L as CaCO₃		
Result of Titration	Hydroxide Alkalinity as CaCO₃	Carbonate Alkalinity as CaCO₃	Bicarbonate Concentration as CaCO₃
P = 0	0	0	P
P = ½T	0	2P	T − 2P
P < ½T	0	2P	0
P > ½T	2P − T	2(T − P)	0
P = T	T	0	0
* Key: P = phenolphthalein alkalinity; T = total alkalinity			

Conversion Factors & Abbreviations

1 acre	=	43,560 square feet (ft^2)
1 acre foot (acre-ft)	=	326,000 gallons (gal)
1 cubic foot (ft^3)	=	7.48 gal
1 ft^3	=	62.4 pounds (lb)
1 cubic foot per second (cfs or ft^3/sec)	=	0.646 mgd
1 foot (ft)	=	0.305 meters
1 ft of water	=	0.433 psi
1 gal	=	3.79 liters (L)
1 gal	=	8.34 lb
1 grain per gallon (gpg)	=	17.1 mg/L
1 horsepower (HP)	=	0.746 kilowatts (kW) or 746 watts
1 HP	=	33,000 ft lb/min
1 mgd	=	1.55 cfs or ft^3/sec
1 mile	=	5,280 ft
1 million gallons per day (mgd)	=	694 gallons per minute (gpm)
1 lb	=	0.454 kilograms
1 lb/in^2	=	2.31 ft of water
1 ton	=	2,000 lb
1%	=	10,000 mg/L
π or pi	=	3.14159
°Celsius	=	(°Fahrenheit − 32) × (5/9)
°Fahrenheit	=	(°Celsius) × (9/5) + 32

Additional Abbreviations

bhp	brake horsepower
DO	dissolved oxygen
EDTA	ethylenediaminetetraacetic acid
g	grams
gpcd	gallons per capita per day
gpd	gallons per day
in.	inch(es)
mg/L	milligrams per liter
mhp	motor horsepower
mil gal	million gallons
min	minute(s)
mL	milliliter
ppb	parts per billion
ppm	parts per million
psi	pounds per square inch
Q	low

SS	settleable solids
TTHM	total trihalomethanes
TOC	total organic carbon
TSS	total suspended solids
VS	volatile solids
whp	water horsepower

Appendix B: Sample CT Tables for *Giardia* Inactivation

CT Values for 1.5 Log Inactivation of *Giardia* Cysts by Free Chlorine

pH = 6.9

TEMP. °C	Chlorine Concentration in mg/L													
	0.4	0.6	0.8	1.0	1.2	1.4	1.6	1.8	2.0	2.2	2.4	2.6	2.8	3.0
0.5	94.80	96.80	99.60	101.60	104.40	107.20	109.40	112.20	114.20	117.00	119.80	121.80	124.60	126.60
1.00	89.40	91.36	93.80	95.80	98.24	100.84	102.80	105.44	107.40	110.00	112.48	114.44	116.88	118.88
2.00	84.00	85.92	88.00	90.00	92.08	94.48	96.20	98.68	100.60	103.00	105.16	107.08	109.16	111.16
3.00	78.60	80.48	82.20	84.20	85.92	88.12	89.60	91.92	93.80	96.00	97.84	99.72	101.44	103.44
4.00	73.20	75.04	76.40	78.40	79.76	81.76	83.00	85.16	87.00	89.00	90.52	92.36	93.72	95.72
5.00	67.80	69.60	70.60	72.60	73.60	75.40	76.40	78.40	80.20	82.00	83.20	85.00	86.00	88.00
6.00	64.32	66.12	67.12	68.92	69.92	71.56	72.72	74.52	76.16	77.96	79.12	80.76	81.76	83.72
7.00	60.84	62.64	63.64	65.24	66.24	67.72	69.04	70.64	72.12	73.92	75.04	76.52	77.52	79.44
8.00	57.36	59.16	60.16	61.56	62.56	63.88	65.36	66.76	68.08	69.88	70.96	72.28	73.28	75.16
9.00	53.88	55.68	56.68	57.88	58.88	60.04	61.68	62.88	64.04	65.84	66.88	68.04	69.04	70.88
10.00	50.40	52.20	53.20	54.20	55.20	56.20	58.00	59.00	60.00	61.80	62.80	63.80	64.80	66.60
11.00	47.12	48.72	49.72	50.72	51.52	52.52	54.12	55.12	56.12	57.72	58.56	59.56	60.52	62.16
12.00	43.84	45.24	46.24	47.24	47.84	48.84	50.24	51.24	52.24	53.64	54.32	55.32	56.24	57.72
13.00	40.56	41.76	42.76	43.76	44.16	45.16	46.36	47.36	48.36	49.56	50.08	51.08	51.96	53.28
14.00	37.28	38.28	39.28	40.28	40.48	41.48	42.48	43.48	44.48	45.48	45.84	46.84	47.68	48.84
15.00	34.00	34.80	35.80	36.80	36.80	37.80	38.60	39.60	40.60	41.40	41.60	42.60	43.40	44.40
16.00	32.24	33.08	34.04	34.88	35.04	35.88	36.68	37.68	38.48	39.32	39.64	40.48	41.28	42.12
17.00	30.48	31.36	32.28	32.96	33.28	33.96	34.76	35.76	36.36	37.24	37.68	38.36	39.16	39.84
18.00	28.72	29.64	30.52	31.04	31.52	32.04	32.84	33.84	34.24	35.16	35.72	36.24	37.04	37.56
19.00	26.96	27.92	28.76	29.12	29.76	30.12	30.92	31.92	32.12	33.08	33.76	34.12	34.92	35.28
20.00	25.20	26.20	27.00	27.20	28.00	28.20	29.00	30.00	30.00	31.00	31.80	32.00	32.80	33.00
21.00	23.64	24.44	25.28	25.44	26.08	26.44	27.08	28.04	28.08	28.88	29.68	29.88	30.68	30.84
22.00	22.08	22.68	23.56	23.68	24.16	24.68	25.16	26.08	26.16	26.76	27.56	27.76	28.56	28.68
23.00	20.52	20.92	21.84	21.92	22.24	22.92	23.24	24.12	24.24	24.64	25.44	25.64	26.44	26.52
24.00	18.96	19.16	20.12	20.16	20.32	21.16	21.32	22.16	22.32	22.52	23.32	23.52	24.32	24.36
25.00	17.40	17.40	18.40	18.40	18.40	19.40	19.40	20.20	20.40	20.40	21.20	21.40	22.20	22.20

Modified from *Guidance Manual for Compliance With the Filtration and Disinfection Requirements for Public Water Systems Using Surface Water Sources*. Copyright 1990, American Water Works Association.

CT Values for 1.5 Log Inactivation of Giardia Cysts by Free Chlorine
pH = 7.6

TEMP. °C	Chlorine Concentration in mg/L															
	0.4	0.6	0.8	1.0	1.2	1.4	1.6	1.8	2.0	2.2	2.4	2.6	2.8	3.0
0.5	122.00	124.60	128.00	132.00	135.40	138.60	142.60	145.80	149.00	154.60	155.40	158.40	161.60	164.60
1.00	115.64	117.52	120.68	124.92	127.48	130.48	134.08	137.08	140.08	144.96	146.20	149.00	152.00	154.80
2.00	108.28	110.44	113.36	116.64	119.56	122.36	125.56	128.36	131.16	135.32	137.00	139.60	142.40	145.00
3.00	100.92	103.36	106.04	108.96	111.64	114.24	117.04	119.64	122.24	125.68	127.80	130.20	132.80	135.20
4.00	93.56	96.28	98.72	101.28	103.72	106.12	108.52	110.92	113.32	116.04	118.60	120.80	123.20	125.40
5.00	86.20	89.20	91.40	93.60	95.80	98.00	100.00	102.20	104.40	106.40	109.40	111.40	113.60	115.60
6.00	82.04	84.68	86.84	88.84	91.00	93.00	95.00	97.20	99.16	101.16	103.96	105.80	107.96	109.80
7.00	77.88	80.16	82.28	84.08	86.20	88.00	90.00	92.20	93.92	95.92	98.52	100.20	102.32	104.00
8.00	73.72	75.64	77.72	79.32	81.40	83.00	85.00	87.20	88.68	90.68	93.08	94.60	96.68	98.20
9.00	69.56	71.12	73.16	74.56	76.60	78.00	80.00	82.20	83.44	85.44	87.64	89.00	91.04	92.40
10.00	65.40	66.60	68.60	69.80	71.80	73.00	75.00	77.20	78.20	80.20	82.20	83.40	85.40	86.60
11.00	61.04	62.20	64.04	65.20	67.04	68.20	70.00	72.00	73.00	74.80	76.80	77.96	79.76	80.92
12.00	56.68	57.80	59.48	60.60	62.28	63.40	65.00	66.80	67.80	69.40	71.40	72.52	74.12	75.24
13.00	52.32	53.40	54.92	56.00	57.52	58.60	60.00	61.60	62.60	64.00	66.00	67.08	68.48	69.56
14.00	47.96	49.00	50.36	51.40	52.76	53.80	55.00	56.40	57.40	58.60	60.60	61.64	62.84	63.88
15.00	43.60	44.60	45.80	46.80	48.00	49.00	50.00	51.20	52.20	53.20	55.20	56.20	57.20	58.20
16.00	41.32	42.36	43.52	44.52	45.68	46.52	47.52	48.68	49.68	50.68	52.32	53.32	54.32	55.92
17.00	39.04	40.12	41.24	42.24	43.36	44.04	45.04	46.16	47.16	48.16	49.44	50.44	51.44	52.44
18.00	36.76	37.88	38.96	39.96	41.04	41.56	42.56	43.64	44.64	45.64	46.56	47.56	48.56	49.56
19.00	34.48	35.64	36.68	37.68	38.72	39.08	40.08	41.12	42.12	43.12	43.68	44.68	45.68	46.68
20.00	32.20	33.40	34.40	35.40	36.40	36.60	37.60	38.60	39.60	40.60	40.80	41.80	42.80	43.80
21.00	30.12	31.28	32.12	33.08	33.92	34.28	35.08	36.08	36.92	37.88	38.08	39.08	39.88	40.88
22.00	28.04	29.16	29.84	30.76	31.44	31.96	32.56	33.56	34.24	35.16	35.36	36.36	36.96	37.96
23.00	25.96	27.04	27.56	28.44	28.96	29.64	30.04	31.04	31.56	32.44	32.64	33.64	34.04	35.04
24.00	23.88	24.92	25.28	26.12	26.48	27.32	27.52	28.52	28.88	29.72	29.92	30.92	31.12	32.12
25.00	21.80	22.80	23.00	23.80	24.00	25.00	25.00	26.00	26.20	27.00	27.20	28.20	28.20	29.20

Modified from Guidance Manual for Compliance With the Filtration and Disinfection Requirements for Public Water Systems Using Surface Water Sources. Copyright 1990, American Water Works Association.

Additional Resources

The following items are all available from the AWWA Bookstore : www.awwa.org/bookstore; 1.800.926.7337; custsvc@awwa.org

Principles and Practices of Water Supply Operations (WSO): *Water Sources*, 4th edition. 2010. 192 pp. Hardback. 978-1-58321-782-5. Catalog No. 1955

WSO: *Water Treatment*, 4th edition. 2010. 512 pp. Hardback. 978-1-58321-777-1. Catalog No. 1956

Water Treatment Student Workbook, 4th edition. 2010. 120 pp. Softbound. 978-1-58321-794-8. Catalog No. 1966

WSO: *Water Transmission and Distribution*, 4th edition. 2010. 546 pp. Hardback. 978-1-58321-781-8. Catalog No. 1957

Water Transmission and Distribution Student Workbook, 4th edition. 2010. 114 pp. Softbound. 978-1-58321-800-6. Catalog No. 1967

WSO: *Water Quality*, 4th edition. 2010. 213 pp. Hardback. 978-1-58321-798-6. Catalog No. 1958

Water Quality Student Workbook, 4th edition. 2010. 63 pp. Softbound. 978-1-58321-794-8. Catalog No. 1968

WSO: *Basic Science Concepts and Applications*, 4th edition. 2010. 626 pp. Hardback. 978-1-58321-778-8. Catalog No. 1959

Basic Science Concepts and Applications Student Workbook, 4th edition. 2010. 200 pp. Softbound. 978-1-58321-799-3. Catalog No. 1954

Basic Chemistry for Water & Wastewater Operators. 2005. Darshan Singh Sarai. 96 pp. Softbound. 978-1-58321-148-9. Catalog No. 20494

Basic Microbiology for Drinking Water Personnel, 2nd edition. 2006. Dennis Hill. 109 pp. Softbound. 978-1-58321-434-3. Catalog No. 20463

Math for Distribution System Operators. 2007. John Giorgi. 250 pp. Softbound. 978-1-58321-455-8. Catalog No. 20628

Math for Water Treatment Operators. 2007. John Giorgi. 250 pp. Softbound. 978-1-58321-454-1. Catalog No. 20618

Water Distribution Operator Training Handbook, 3rd edition. 2005. 286 pp. Softbound. 978-1-58321-372-8. Catalog No. 20428

Water Treatment Made Simple for Operators. 2006. Darshan Singh Sarai. 263 pp. Hardback. 978-0-47174-002-5. Catalog No. 20526

Water Treatment Operator Handbook. 2005. Nicholas G. Pizzi. 251 pp. Softbound. 978-1-58321-371-1. Catalog No. 20481

Water Distribution System Operation and Maintenance: A Field Study Training Program, 5th edition. 2005. 654 pp. Softbound. 978-1-59371-020-0. Catalog No. 20684

Water Treatment Plant Operation: A Field Study Training Program, vol. 1, 6th edition. 2008. Kenneth D. Kerri. 550 pp. Softbound. 978-1-59371-040-8. Catalog No. 20685

Water Treatment Plant Operation: A Field Study Training Program, vol. 2, 5th edition. 2006. Kenneth D. Kerri. 875 pp. Softbound. 978-1-59371-036-1. Catalog No. 20686

Utility Management for Water and Wastewater Operators. 2011. Frederick Bloetscher. 496 pp. Hardback. 978-1-58321-823-5. Catalog No. 20721

WaterMath: Quick Calculator for Water Operators CD. 2011. Kenneth Mercer. 16 spreadsheets. CD-ROM. 978-1-58321-821-1. Catalog No. 60109

Water Distribution Operator Training: Water Mains. 2006. 978-1-58321-625-5. Catalog No. 64323

Operator Chemistry Made Easy. 2009. DVD. 1-58321-638-3. Catalog No. 64128

Operator Math Made Easy. 2009. DVD. 1-58321-637-5. Catalog No. 64127

Secrets to Success: How To Prepare for Operator Certification. 2006. DVD. 978-1-58321-609-5. Catalog No. 64045

Water Distribution Operator Training: Hydrants. 2006. DVD. 978-1-58321-624-8. Catalog No. 64322

Water Distribution Operator Training: Pumps and Motors. 2006. DVD. 978-1-58321-628-6. Catalog No. 64326

Water Distribution Operator Training: Services and Meters. 2006. DVD. 978-1-58321-627-9. Catalog No. 64325

Water Supply Operations: Centrifugal Pumps. 2007. DVD. 978-1-58321-620-0. Catalog No. 64110

Water Supply Operations: Coagulation, Flocculation and Sedimentation. 2005. DVD. 978-1-58321-612-5. Catalog No. 64091

Water Supply Operations: Disinfection Strategies. 2004. DVD. 978-1-58321-614-9. Catalog No. 64093

Water Supply Operations: Distribution Systems. 2007. DVD. 978-1-58321-619-4. Catalog No. 64109

Water Supply Operations: Filtration. 2004. DVD. 978-1-58321-613-2. Catalog No. 64092

Water Supply Operations: Flushing And Cleaning. 2006. DVD. 978-1-58321-621-7. Catalog No. 64111

Water Supply Operations: Jar Testing. 2009. DVD. 978-1-58321-610-1. Catalog No. 64083

Water Supply Operations: Membrane Technology. 2006. DVD. 978-1-58321-616-3. Catalog No. 64096

Water Supply Operations: Ozone and UV. 2007. DVD. 978-1-58321-615-6. Catalog No. 64095

Water Supply Operations: SCADA and Instrumentation. 2005. DVD. 978-1-58321-618-7. Catalog No. 64099

Water Supply Operations: Source Water Protection. 2005. DVD. 978-1-58321-611-8. Catalog No. 64090

Water Supply Operations: Water Loss. 2006. DVD. 978-1-58321-622-4. Catalog No. 64112

Water Supply Operations: Water Sources. 2005. DVD. 978-1-58321-623-1. Catalog No. 64272

AWWA's eLearning Program, an online learning tool for water industry professionals, offers quality training, professional development, and continuing education units (CEUs) via the Internet. Learn on your own time and at your own pace. The many class offerings are listed below.

Conservation
- High-Bill Complaints? Easy Water Conservation Tips for Your Customers   .2 CEUs   **EL83**
- Increasing Need for Water Loss Control   .2 CEUs   **EL106**

Distribution
- Advanced Disinfection of Pipeline and Storage Facilities   .2 CEUs   **EL69**
- Advanced Metering Infrastructure for the Water Industry   .2 CEUs   **EL68**
- Backflow Prevention & Cross-Connection Control   .2 CEUs   **EL89**
- Disinfection of Pipelines & Storage Facilities   .2 CEUs   **EL24**
- Distribution Service to Customers   .3 CEUs   **EL13**
- Distribution System Materials and Equipment   .3 CEUs   **EL14**
- Hot Topics in The Water Industry   .2 CEUs   **EL91**
- Key Elements to Maintaining Distribution System Integrity   .2 CEUs   **EL82**
- Pump Maintenance   .2 CEUs   **EL22**
- Trenchless Technology Applications in the Water Industry   .2 CEUs   **EL74**
- Water Main Installation   .3 CEUs   **EL15**
- Water Storage Tanks   .3 CEUs   **EL80**
- Water System Mechanical Equipment   .2 CEUs   **EL16**

Engineering
- Getting the Most From Your Hydraulic Model   .2 CEUs   **EL104**
- Hydraulics   .8 CEUs   **EL11**
- PE Pipe in the Field   .2 CEUs   **EL26**
- PVC Pipe in the Field   .2 CEUs   **EL19**

Government & Regulations
- How to Prepare Your Utility for USEPA Regulations   .2 CEUs   **EL90**
- The Fundamentals of Drinking Water Regulations: Part 1   .1 CEUs   **EL64**
- The Fundamentals of Drinking Water Regulations: Part 2   .1 CEUs   **EL65**
- Total Coliform Rule   .2 CEUs   **EL18**

Management
- Chlorine Gas: Balancing Public Health   .2 CEUs   **EL21**
- Setting Rates in a Tough Economy   .2 CEUs   **EL75**
- Water Basics for Decision Makers: Course 1   .4 CEUs   **EL95**
- Water Basics for Decision Makers: Course 2   .4 CEUs   **EL96**
- Water Basics for Decision Makers: Course 3   .4 CEUs   **EL97**

Math & Science
- Applied Mathematics .3 CEUs **EL6**
- Basic Mathematics .7 CEUs **EL7**
- Fundamentals of Chemistry for Water Professionals .8 CEUs **EL12**
- Things That Tripped Me Up on the Operator Certification Exam .2 CEUs **EL88**

Quality
- Chlorine Gas: Balancing Public Health .2 CEUs **EL21**
- Disinfection By-Products: Recent Research Raises Concern .2 CEUs **EL105**
- Endocrine Disruptors, Pharmaceuticals, and Personal Care Products Actions and Communication .2 CEUs **EL17**
- Harmful Algal Blooms: Cyanobacteria .2 CEUs **EL20**
- Inorganic Treatment: Avoiding Inadvertent Noncompliance .2 CEUs **EL102**
- New Approaches for Assessing Microbial Threats .2 CEUs **EL103**
- Optimizing Water Distribution System Quality .2 CEUs **EL101**
- Perchlorate and Emerging Contaminants: Where Are We Now? .2 CEUs **EL72**
- Plant to Tap: the Importance of Disinfection .2 CEUs **EL79**
- Quagga/Zebra Mussel Control .2 CEUs **EL70**

Resources & Reuse
- GeoScience in Water Aquifers .2 CEUs **EL78**
- Water Shortages: Finding a Solution .2 CEUs **EL73**

Treatment
- Best of Membrane Conference .2 CEUs **EL81**
- Chemicals: Best Practices for Quality Assurance .2 CEUs **EL76**
- Chlorine Gas an Inherently Safer Technology .2 CEUs **EL27**
- Chlorine Gas: Balancing Public Health .2 CEUs **EL21**
- Coagulation, Flocculation, and Sedimentation Basic .3 CEUs **EL8**
- Disinfection Basics .3 CEUs **EL9**
- Filter Optimization Tips You Can Use .2CEUs **EL107**
- Filtration Basics .3 CEUs **EL10**
- High Tech Operator Course 1: Treatment & Distribution – Process Monitoring and Control 1.2 CEUs **EL84**
- High Tech Operator Course 2: Application & Tools 1.2 CEUs **EL85**
- High Tech Operator Course 3: Data Management 1.2 CEUs **EL86**
- High Tech Operator Certificate Program 3.6 CEUs **EL87**
- Membranes: Emerging Issues & Technologies .2 CEUs **EL23**
- Plant to Tap: The Importance of Disinfection .2 CEUs **EL79**
- Residuals Management and Disposal .2 CEUs **EL77**

For complete course descriptions and outlines, please visit www.awwa.org/elearning. Titles subject to change. For a list of pre-approved courses for continuation education credits, please visit www.awwa.org/elearning/ceus. For questions, please email elearning@awwa.org or call 303.347.6204.